The Handbook of Environmental Chemistry

Founded by Otto Hutzinger

Editors-in-Chief: Damià Barceló · Andrey G. Kostianoy

Volume 26

Advisory Board:
Jacob de Boer, Philippe Garrigues, Ji-Dong Gu,
Kevin C. Jones, Thomas P. Knepper, Alice Newton,
Donald L. Sparks

The Handbook of Environmental Chemistry
Recently Published and Forthcoming Volumes

Urban Air Quality in Europe
Volume Editor: M. Viana
Vol. 26, 2013

Climate Change and Water Resources
Volume Editors: T. Younos and C.A. Grady
Vol. 25, 2013

Emerging Organic Contaminants in Sludges: Analysis, Fate and Biological Treatment
Volume Editors: T. Vicent, G. Caminal, E. Eljarrat, and D. Barceló
Vol. 24, 2013

Global Risk-Based Management of Chemical Additives II: Risk-Based Assessment and Management Strategies
Volume Editors: B. Bilitewski, R.M. Darbra, and D. Barceló
Vol. 23, 2013

Chemical Structure of Pelagic Redox Interfaces: Observation and Modeling
Volume Editor: E.V. Yakushev
Vol. 22, 2013

The Llobregat: The Story of a Polluted Mediterranean River
Volume Editors: S. Sabater, A. Ginebreda, and D. Barceló
Vol. 21, 2012

Emerging Organic Contaminants and Human Health
Volume Editor: D. Barceló
Vol. 20, 2012

Emerging and Priority Pollutants in Rivers: Bringing Science into River Management Plans
Volume Editors: H. Guasch, A. Ginebreda, and A. Geiszinger
Vol. 19, 2012

Global Risk-Based Management of Chemical Additives I: Production, Usage and Environmental Occurrence
Volume Editors: B. Bilitewski, R.M. Darbra, and D. Barceló
Vol. 18, 2012

Polyfluorinated Chemicals and Transformation Products
Volume Editors: T.P. Knepper and F.T. Lange
Vol. 17, 2012

Brominated Flame Retardants
Volume Editors: E. Eljarrat and D. Barceló
Vol. 16, 2011

Effect-Directed Analysis of Complex Environmental Contamination
Volume Editor: W. Brack
Vol. 15, 2011

Waste Water Treatment and Reuse in the Mediterranean Region
Volume Editors: D. Barceló and M. Petrovic
Vol. 14, 2011

The Ebro River Basin
Volume Editors: D. Barceló and M. Petrovic
Vol. 13, 2011

Polymers – Opportunities and Risks II: Sustainability, Product Design and Processing
Volume Editors: P. Eyerer, M. Weller, and C. Hübner
Vol. 12, 2010

Polymers – Opportunities and Risks I: General and Environmental Aspects
Volume Editor: P. Eyerer
Vol. 11, 2010

Chlorinated Paraffins
Volume Editor: J. de Boer
Vol. 10, 2010

Biodegradation of Azo Dyes
Volume Editor: H. Atacag Erkurt
Vol. 9, 2010

Urban Air Quality in Europe

Volume Editor: Mar Viana

With contributions by

F. Amato · C. Ansorge · A. Asmi · J. Borken-Kleefeld ·
C. Borrego · P.G. Boulter · J. Cyrys · R.G. Derwent ·
B.L. van Drooge · J. Ferreira · G. Fuller · J.O. Grimalt ·
R.M. Harrison · H. Herich · O. Hertel · R. Hillamo ·
A.-G. Hjellbrekke · O. Hnninen · G. Hoek · C. Hueglin ·
A.C. John · A. Karanasiou · K. Katsouyanni ·
T.A.J. Kuhlbusch · P. Kumar · N. Mihalopoulos ·
A.I. Miranda · L. Moosmann · L. Morawska · C. Nagl ·
L. Ntziachristos · U. Quass · X. Querol · P. Quincey ·
C. Reche · S. Saarikoski · M. Schaap · C.A. Skjøth ·
W. Spangl · J. Valente · M. Viana · E. Weijers ·
H. Wiesenberger

Springer

Editor
Dr. Mar Viana
Institute for Environmental Assessment
  and Water Research (IDÆA-CSIC)
Spanish Research Council – CSIC
Barcelona, Spain

The Handbook of Environmental Chemistry
ISSN 1867-979X   ISSN 1616-864X (electronic)
ISBN 978-3-642-38450-9   ISBN 978-3-642-38451-6 (eBook)
DOI 10.1007/978-3-642-38451-6
Springer Heidelberg New York Dordrecht London

Library of Congress Control Number: 2013941728

© Springer-Verlag Berlin Heidelberg 2013

This work is subject to copyright. All rights are reserved by the Publisher, whether the whole or part of the material is concerned, specifically the rights of translation, reprinting, reuse of illustrations, recitation, broadcasting, reproduction on microfilms or in any other physical way, and transmission or information storage and retrieval, electronic adaptation, computer software, or by similar or dissimilar methodology now known or hereafter developed. Exempted from this legal reservation are brief excerpts in connection with reviews or scholarly analysis or material supplied specifically for the purpose of being entered and executed on a computer system, for exclusive use by the purchaser of the work. Duplication of this publication or parts thereof is permitted only under the provisions of the Copyright Law of the Publisher's location, in its current version, and permission for use must always be obtained from Springer. Permissions for use may be obtained through RightsLink at the Copyright Clearance Center. Violations are liable to prosecution under the respective Copyright Law.

The use of general descriptive names, registered names, trademarks, service marks, etc. in this publication does not imply, even in the absence of a specific statement, that such names are exempt from the relevant protective laws and regulations and therefore free for general use.

While the advice and information in this book are believed to be true and accurate at the date of publication, neither the authors nor the editors nor the publisher can accept any legal responsibility for any errors or omissions that may be made. The publisher makes no warranty, express or implied, with respect to the material contained herein.

Printed on acid-free paper

Springer is part of Springer Science+Business Media (www.springer.com)

Editors-in-Chief

Prof. Dr. Damià Barceló

Department of Environmental Chemistry
IDAEA-CSIC
C/Jordi Girona 18–26
08034 Barcelona, Spain
and
Catalan Institute for Water Research (ICRA)
H20 Building
Scientific and Technological Park of the
   University of Girona
Emili Grahit, 101
17003 Girona, Spain
dbcqam@cid.csic.es

Prof. Dr. Andrey G. Kostianoy

P.P. Shirshov Institute of Oceanology
Russian Academy of Sciences
36, Nakhimovsky Pr.
117997 Moscow, Russia
kostianoy@gmail.com

Advisory Board

Prof. Dr. Jacob de Boer
IVM, Vrije Universiteit Amsterdam, The Netherlands

Prof. Dr. Philippe Garrigues
University of Bordeaux, France

Prof. Dr. Ji-Dong Gu
The University of Hong Kong, China

Prof. Dr. Kevin C. Jones
University of Lancaster, United Kingdom

Prof. Dr. Thomas Knepper
University of Applied Science, Fresenius, Idstein, Germany

Prof. Dr. Alice Newton
University of Algarve, Faro, Portugal

Prof. Dr. Donald L. Sparks
Plant and Soil Sciences, University of Delaware, USA

The Handbook of Environmental Chemistry Also Available Electronically

The Handbook of Environmental Chemistry is included in Springer's eBook package *Earth and Environmental Science*. If a library does not opt for the whole package, the book series may be bought on a subscription basis.

For all customers who have a standing order to the print version of *The Handbook of Environmental Chemistry,* we offer free access to the electronic volumes of the Series published in the current year via SpringerLink. If you do not have access, you can still view the table of contents of each volume and the abstract of each article on SpringerLink (www.springerlink.com/content/110354/).

You will find information about the

– Editorial Board
– Aims and Scope
– Instructions for Authors
– Sample Contribution

at springer.com (www.springer.com/series/698).

All figures submitted in color are published in full color in the electronic version on SpringerLink.

Aims and Scope

Since 1980, *The Handbook of Environmental Chemistry* has provided sound and solid knowledge about environmental topics from a chemical perspective. Presenting a wide spectrum of viewpoints and approaches, the series now covers topics such as local and global changes of natural environment and climate; anthropogenic impact on the environment; water, air and soil pollution; remediation and waste characterization; environmental contaminants; biogeochemistry; geo-ecology; chemical reactions and processes; chemical and biological transformations as well as physical transport of chemicals in the environment; or environmental modeling. A particular focus of the series lies on methodological advances in environmental analytical chemistry.

Series Preface

With remarkable vision, Prof. Otto Hutzinger initiated *The Handbook of Environmental Chemistry* in 1980 and became the founding Editor-in-Chief. At that time, environmental chemistry was an emerging field, aiming at a complete description of the Earth's environment, encompassing the physical, chemical, biological, and geological transformations of chemical substances occurring on a local as well as a global scale. Environmental chemistry was intended to provide an account of the impact of man's activities on the natural environment by describing observed changes.

While a considerable amount of knowledge has been accumulated over the last three decades, as reflected in the more than 70 volumes of *The Handbook of Environmental Chemistry,* there are still many scientific and policy challenges ahead due to the complexity and interdisciplinary nature of the field. The series will therefore continue to provide compilations of current knowledge. Contributions are written by leading experts with practical experience in their fields. *The Handbook of Environmental Chemistry* grows with the increases in our scientific understanding, and provides a valuable source not only for scientists but also for environmental managers and decision-makers. Today, the series covers a broad range of environmental topics from a chemical perspective, including methodological advances in environmental analytical chemistry.

In recent years, there has been a growing tendency to include subject matter of societal relevance in the broad view of environmental chemistry. Topics include life cycle analysis, environmental management, sustainable development, and socio-economic, legal and even political problems, among others. While these topics are of great importance for the development and acceptance of *The Handbook of Environmental Chemistry,* the publisher and Editors-in-Chief have decided to keep the handbook essentially a source of information on "hard sciences" with a particular emphasis on chemistry, but also covering biology, geology, hydrology and engineering as applied to environmental sciences.

The volumes of the series are written at an advanced level, addressing the needs of both researchers and graduate students, as well as of people outside the field of "pure" chemistry, including those in industry, business, government, research establishments, and public interest groups. It would be very satisfying to see these volumes used as a basis for graduate courses in environmental chemistry. With its high standards of scientific quality and clarity, *The Handbook of*

Environmental Chemistry provides a solid basis from which scientists can share their knowledge on the different aspects of environmental problems, presenting a wide spectrum of viewpoints and approaches.

The Handbook of Environmental Chemistry is available both in print and online via www.springerlink.com/content/110354/. Articles are published online as soon as they have been approved for publication. Authors, Volume Editors and Editors-in-Chief are rewarded by the broad acceptance of *The Handbook of Environmental Chemistry* by the scientific community, from whom suggestions for new topics to the Editors-in-Chief are always very welcome.

Damià Barceló
Andrey G. Kostianoy
Editors-in-Chief

Volume Preface

The link between degrading air quality and adverse health outcomes is widely recognised (Lim et al. 2012). Ambient particulate matter pollution ranks as number 11 to 14 as a risk factor accounting for total burden of disease across Western, Central and Eastern Europe, while ambient ozone appears in positions 36 and 37 in the same ranking. Irrespective of particle size, epidemiological studies have shown a clear association between exposure to airborne pollutants and adverse cardiovascular and respiratory health outcomes. Evidence also shows adverse effects of short- and long-term exposure to gaseous pollutants, e.g. ozone, on all-cause, cardiovascular and respiratory mortality. The urban environment plays a key role in the occurrence of these health effects, given that it is in urban areas where the largest fraction of the population in Europe is exposed to degraded and degrading air quality.

Air quality is, however, an exceptionally broad subject, and it would be pretentious to aim to cover all of its aspects in one single book. Through its 16 chapters, this book aims to provide an overview of air quality in urban environments in Europe mainly from the point of view of the sources of atmospheric pollutants, whether they may be natural or anthropogenic, primary or secondary, and whether pollutants originate from local sources or are transported large distances across the continent. In this context, in Part I authors have contributed with valuable chapters dealing with emission sources such as biomass burning, vehicular traffic, industry and agriculture, but also with African dust and long-range transport of pollutants across the European regions. Assessments are based on measurements and exposure modelling approaches. The impact of these emission sources and processes on atmospheric particulate matter, ozone, NO_x and volatile and semi-volatile organic compounds is discussed. Based on air quality data, criteria for the identification of critical areas for particulate matter and NO_2 in Europe are presented, followed by an analysis of air quality management approaches. In Part II future perspectives are presented, giving insights into potential upcoming air quality monitoring strategies and metrics of interest such as nanoparticles and submicron particle size distribution data. The relevance of indoor and outdoor exposure scenarios is also highlighted.

Certainly a number of aspects are not covered by this overview, such as ultrafine particle or secondary organic aerosol formation processes and their roles on air quality degradation, urban-scale dispersion models for air quality modelling or the impact of air quality in regional background areas on pollutants in the urban environment. In sum, while we hope this book covers major aspects influencing air quality in urban areas, it also evidences the sheer size of this topic and the need to push research forward in this direction.

This book is intended for a broad audience, from environmental specialists working already in this field to newcomers who want to learn more about this issue. I would like to thank all the authors for their time and efforts in preparing their corresponding chapters, as well as my team leaders and I would also like to thank my team leaders and my co-workers for creating a motivating work environment, which allowed this project to come to life.

Barcelona, Spain   Mar Viana

Reference

Lim et al (2012) Lancet 380:2224–2260

Contents

Part I Gaseous and Particulate Pollutants and Their Sources

Critical Areas for Compliance with PM_{10} and NO_2 Limit Values
in Europe 3
C. Nagl, C. Ansorge, L. Moosmann, W. Spangl, and H. Wiesenberger

The Evolution and Control of NO_x Emissions from Road Transport
in Europe 31
P.G. Boulter, J. Borken-Kleefeld, and L. Ntziachristos

Air Pollution by Ozone Across Europe 55
Richard G. Derwent and Anne-Gunn Hjellbrekke

Persistent Organic Pollutants in the European Atmosphere 75
Barend L. van Drooge and Joan O. Grimalt

Wildfires as a Source of Aerosol Particles Transported
to the Northern European Regions 101
Sanna Saarikoski and Risto Hillamo

Residential Wood Burning: A Major Source of Fine Particulate
Matter in Alpine Valleys in Central Europe 123
Hanna Herich and Christoph Hueglin

Ammonia Emissions in Europe 141
Carsten Ambelas Skjøth and Ole Hertel

Road Traffic: A Major Source of Particulate Matter in Europe 165
Fulvio Amato, Martijn Schaap, Cristina Reche, and Xavier Querol

Source Apportionment of Airborne Dust in Germany: Methods
and Results 195
U. Quass, A.C. John, and T.A.J. Kuhlbusch

Air Quality in Urban Environments in the Eastern Mediterranean 219
A. Karanasiou and N. Mihalopoulos

Anthropogenic and Natural Constituents in PM10 at Urban and Rural Sites in North-Western Europe: Concentrations, Chemical Composition and Sources 239
Ernie Weijers and Martijn Schaap

Particulate Matter and Exposure Modelling in Europe 259
A.I. Miranda, J. Valente, J. Ferreira, and C. Borrego

Part II Future Air Quality Monitoring Strategies and Research Directions

Air Pollution Monitoring Strategies and Technologies for Urban Areas 277
Thomas A.J. Kuhlbusch, Ulrich Quass, Gary Fuller, Mar Viana, Xavier Querol, Klea Katsouyanni, and Paul Quincey

Number Size Distributions of Submicron Particles in Europe 297
Ari Asmi

Indoor–Outdoor Relationships of Particle Number and Mass in European Cities 321
Gerard Hoek, Otto Hänninen, and Josef Cyrys

Nanoparticles in European Cities and Associated Health Impacts 339
Prashant Kumar, Lidia Morawska, and Roy M. Harrison

Index 367

Part I
Gaseous and Particulate Pollutants and Their Sources

Critical Areas for Compliance with PM_{10} and NO_2 Limit Values in Europe

C. Nagl, C. Ansorge, L. Moosmann, W. Spangl, and H. Wiesenberger

Abstract The Directive on ambient air quality and cleaner air for Europe (2008/50/EC) requires compliance with limit values inter alia for PM_{10} and nitrogen dioxide (NO_2), to be achieved by Member States by 2005 and 2010, respectively. Member States have been allowed to postpone compliance with these limit values until 2011 and 2015, respectively, under certain conditions.

Within Europe there are several regions where compliance with these limit values is expected to be very difficult even after the postponement period.

In this study criteria for the identification of such critical areas have been developed, and the respective areas are described.

A subset of these areas has been selected and analysed in more detail. Critical factors that hamper compliance in these selected areas have been identified. These include adverse dispersion conditions, sheer size of agglomerations, emission densities from traffic, industry, domestic heating, and deficiencies in air quality assessment and management.

For the selected areas air quality management approaches have been analysed and suggestions for suitable courses of action are made, based on these findings.

Keywords Air quality, Air quality directive, Air quality limit value, NO_2, PM_{10}

Contents

1 Introduction 4
2 Definition of Criticality and Non-criticality 4
3 Datasets and Data Sources Used 5
4 General Methodology 6

C. Nagl (✉), C. Ansorge, L. Moosmann, W. Spangl, and H. Wiesenberger
Umweltbundesamt GmbH, Spittelauer Lände 5, 1090 Vienna, Austria
e-mail: christian.nagl@umweltbundesamt.at

5	Analysis and Results	6
	5.1 Correlation and Threshold Analysis	6
	5.2 Selection of a Subset of Critical Areas for Further Analysis	7
	5.3 Information about Critical and Non-critical Areas	8
6	Overview of Critical Factors	17
	6.1 Natural Factors	17
	6.2 Anthropogenic Factors	18
	6.3 Qualitative Estimate of Critical Factors	19
7	Possible Courses of Action for Critical Areas	20
	7.1 Suggestions for Guidance Documents and Further Support	20
	7.2 Suggestions for Specific Action in Critical Areas and on EU Level	22
	7.3 Action on EU Level	24
8	Conclusions	27
References		28

1 Introduction

According to the Directive on ambient air quality and cleaner air for Europe [1], Member States may postpone attainment of the limit values for PM_{10}, nitrogen dioxide (NO_2) and benzene under certain conditions. Since the Directive came into force in June 2008, 22 Member States have notified the Commission about such a postponement.

Within Europe there are several regions where compliance with these limit values is deemed to be very difficult. In this study the following work has been carried out to identify and analyse such areas:

- Development of a methodology to identify and rank the areas where complying with the limit values might prove to be particularly challenging
- Compilation of information on air quality, emissions and other meteorological, economic and demographic parameters in these areas
- Identification of critical areas and comparison with other comparable zones
- Compilation, evaluation and aggregation of air quality assessment, air quality management and health impact information for critical areas
- Identification of potential actions addressing critical areas

2 Definition of Criticality and Non-criticality

Within this study critical areas are defined as areas within Europe where air quality limit values have been exceeded to a large extent in recent years and are expected to be exceeded in the future (i.e. the second decade of this century) as well.

The reasons for high pollution levels and persistent non-compliance can be divided into natural (external) and anthropogenic reasons.

Natural reasons cover

1. The topographic and climatic situation, which triggers (adverse) dispersion conditions, atmospheric chemistry and the possible contribution of long-range transport
2. Natural sources (deductible in the air quality assessment)

Anthropogenic reasons can be subdivided into

1. Emissions
2. The administrative, legal and political capacity for air quality management

Air quality management covers both national as well as international obligations. From the position of a single Member State, transboundary pollution can be seen as an *"external reason"*. Transboundary pollution is tackled e.g. by the NEC Directive [2] and the UNECE protocols. However, the implementation of these obligations is also a national task.

Some emission categories are interlinked with climatic factors, e.g. emissions from domestic heating, from winter sanding or salting of roads. Therefore, a strict separation of natural and anthropogenic reasons is not possible.

Emissions are discussed with respect to (high) emission densities (related to area or population), but also considered as a result of long-lasting policies and societal developments.

Non-critical areas are those with partly similar conditions – e.g. adverse climatic conditions, population density and emissions in previous time periods – but where air quality problems have been dealt with successfully already or will be addressed in the near future.

3 Datasets and Data Sources Used

In order to identify critical areas according to the methodology described below different sources of information have been used.

These include:

- Datasets describing natural reasons for criticality:
  - Climatic conditions (natural)
  - Topography (natural)
  - Land cover (partly natural, partly anthropogenic)
  - Natural sources
- Datasets describing anthropogenic reasons for criticality:
  - Emission data
  - Population density
  - Resulting pollutant levels (air quality and air quality trends)
  - Transboundary contributions (*"external"* from the point of view of a single MS)

4 General Methodology

The following methodology was applied to identify critical areas (according to the definition of critical areas within this project):

1. *Pollution level*: identification of zones and areas with levels above the limit values for PM_{10} and NO_2 in 2007. Sources of information were the annual reports submitted according to Commission Decision 2004/461/EC [3], AirBase,[1] time extension notifications,[2] model calculations (e.g. EMEP[3]) and reports from ETC/ACC.[4] Further, mostly qualitative information was derived from Europe-wide modelling and earth observation studies.
2. *Trends*: identification of zones and areas with stagnant or increasing PM, NO_x and NO_2 levels. Sources of information were annual reports (2004/461/EC), AirBase and time extension notifications. Separate analysis of the trends of NO_x and NO_2 levels helps to identify areas influenced by increasing primary NO_2 emissions from diesel vehicles.
3. Investigation into the identification of possible *indicators* for "critical areas", relating to the reasons for measured concentrations:

   - Topographic and climatic conditions
   - Transboundary contributions, leading to increased regional background levels
   - Emissions/emission densities
   - Indicators of emission densities, e.g. population density

The indicators can be separated into those that cannot be influenced, such as climatic conditions and topography, and those which can be influenced partly (transboundary contributions, population density) or fully by the Member State (emissions, especially of relevant sources such as traffic, domestic heating, industry). These indicators have been visualised with the help of GIS and correlated with air quality indicators.

5 Analysis and Results

5.1 Correlation and Threshold Analysis

Table 1 gives an overview of the correlation between MACC[5] gridded annual mean PM_{10} levels and the parameters described above. No parameter shows a correlation

---

[1] http://www.eea.europa.eu/data-and-maps/data/airbase-the-european-air-quality-database-3.
[2] http://ec.europa.eu/environment/air/quality/legislation/time_extensions.html.
[3] http://www.emep.int/index.html.
[4] European Topic Centre on Air and Climate Change, now ETC/ACM, European Topic Centre for Air Pollution and Climate Change Mitigation.
[5] Monitoring Atmospheric Composition and Climate, http://www.gmes-atmosphere.eu/.

Table 1 Correlation coefficient between gridded PM_{10} annual mean values and other parameters

	Correlation coefficient
PM_{10} 36th highest daily mean	0.99
Population density	0.25
Wind speed	−0.19
Indicator for valleys[a]	−0.01
Land use[b]: urban fabric	0.14
Land use[b]: traffic	0.33
Land use[b]: industry	0.12
NO_x emissions high level sources	0.07
NO_x emissions low level sources	0.16
NO_x emissions traffic	0.28
NO_x emissions total	0.15
PM_{10} emissions high level sources	0.07
PM_{10} emissions low level sources	0.15
PM_{10} emissions traffic	0.26
PM_{10} emissions total	0.15
Classification[c]: lowlands	−0.18
Classification[c]: mountains	−0.11
Classification[c]: urban	0.13

[a]From the ASSET project (http://www.asset-eu.org/)
[b]Corine land cover (http://www.eea.europa.eu/data-and-maps/data#c12=corine+land+cover+version+13)
[c]Climate classification map by LANMAP (http://www.alterra.wur.nl/UK/research/Specialisation+Geo-information/Projects/LANMAP2/)

coefficient above 0.5. A correlation coefficient of about 0.25 has been calculated for traffic emissions and population density. A relatively high negative correlation has been calculated for wind speed and percentage of lowlands. Nevertheless, these rather low correlation coefficients do not allow a straightforward identification of critical areas or of thresholds for certain parameters to discriminate between critical and non-critical areas.

5.2 Selection of a Subset of Critical Areas for Further Analysis

The analysis of the datasets described above has shown that an unambiguous identification of critical areas allowing for generalisation would require further more sophisticated tools such as extensive air quality modelling with chemical transport models and improved comparable datasets.

Nevertheless, the available data have allowed us to identify several areas where air quality problems are rather severe for specific reasons, and exemplary areas with problems of a more general nature.

A subset of these areas has been selected on the basis of various criteria, one of which was the availability of information about air quality management.[6] The subset included the following cities and regions, for which a more detailed analysis was performed:

- Athens, Greece (pollutants concerned: PM_{10}, NO_2)
- Košice, Slovak Republic (PM_{10})
- Kraków, Southern Poland, Poland (PM_{10}, NO_2)
- Lisbon, Portugal (PM_{10}, NO_2)
- London, United Kingdom (NO_2)
- Milan, Po Valley, Italy (PM_{10}, NO_2)
- Paris, France (PM_{10}, NO_2)
- Sofia, Bulgaria (PM_{10}, NO_2)
- Stuttgart, Germany (PM_{10}, NO_2)

The city of Kraków is considered representative of other cities in east central Europe with high industrial emissions (such as Katowice, Ostrava). Similarly, in the case of Milan, information will be given for the whole Po Valley if available.

To identify comparable non-critical areas, NO_2 and PM levels and the trend of these pollutants were analysed in cities across the EU. Taking the criteria for non-critical areas into account, Berlin was selected as a comparable non-critical area.[7] Other cities or areas which fit the criteria for "non-critical areas" either have a special emission structure (which eases abatement measures), or they are comparably small, or largely comparable to Berlin with respect to their dispersion conditions or modal split of traffic; in several cases the available information about air quality assessment and management is not enough. Therefore, Berlin is the only non-critical area which is discussed here in detail.

5.3 Information about Critical and Non-critical Areas

5.3.1 General Information

The area and population sizes of the selected areas are given for the metropolitan area in most cases, as the city boundaries often denote a smaller area than that of the conurbation (Table 2). The city with the largest population is London, followed by Paris and Milan. About 42 million people are living in these selected critical areas and about 3.4 million people in the non-critical area of Berlin.

---

[6] The latter criteria excluded southern Spain, as no air quality plans and time extension notification were available at the time of writing.

[7] Scandinavian cities have been excluded as the sparse air quality problems are mostly due to winter sanding only. In addition, due the rather low background levels, these cities are not comparable to central European cities.

Table 2 Geographical data of selected areas (*source*: Wikipedia)

MS	Name of city/area	Critical pollutant	Metropolitan area (km^2)	Population (metropolitan area)
BG	Sofia	PM_{10}, NO_2	1,310	1,400,000
DE	Stuttgart	PM_{10}, NO_2	207	600,000
DE	Berlin	PM_{10}	890	3,440,000
FR	Paris	PM_{10}, NO_2	2,700[a]/14,500	11,700,000
GR	Athens	PM_{10}, NO_2	2,900	3,680,000
IT	Milan	PM_{10}, NO_2	2,370	7,500,000
PL	Kraków	PM_{10}	326	750,000
PT	Lisbon	PM_{10}, NO_2	958	2,800,000
SK	Košice	PM_{10}	242	230,000
UK	London	NO_2	1,707	12,300,000–13,945,000

[a]Urban area

Table 3 Wind speed data for critical and non-critical areas from time extension notifications, WMO and MARS datasets in m/s

	Additional information	Notifications	WMO	MARS
Athens	1.4–3.3[a]	1.8		2.6
Berlin	2.6			4.5
Košice		2.9		2.5
Kraków	0.3–0.6[b]	1.2	2.9	2.7
Lisbon	2.3		3.9	3.3
London	4.0–4.4[c]			4.0
Milan	1.3			1.5
Paris				4.5
Sofia		1.2		1.2
Stuttgart	1.2–1.5	1.3		3.2

[a]Highest value in the outskirts of the city (email from 1 July 2010)
[b]http://213.17.128.227/iseo/roczny_main.php
[c]London Heathrow, 2005–2009

5.3.2 Dispersion Conditions – Wind Speed

Average wind speed data is a readily available indicator for dispersion conditions. Data are available from the notifications of time extensions for some areas, from the WMO[8] and MARS datasets[9] of JRC[10] and through additional information from the Member States. From Table 3 it can be seen that the figures obtained with MARS 25 km × 25 km dataset are higher compared to the time extension data (except for Košice).

---

[8] http://www.wmo.int/pages/index_en.html.

[9] http://mars.jrc.ec.europa.eu/mars.

[10] http://ec.europa.eu/dgs/jrc/index.cfm.

Table 4 Annual average PM$_{10}$ values from EMEP for the year 2005 and ETC/ACC [14], change in PM$_{10}$ levels due to emission reductions (*source*: EMEP, ETC/ACC [4])

	EMEP (regional background level) (µg/m^3)	Change in PM$_{10}$ regional background levels (%)	ETC/ACC (urban concentration) (µg/m^3)
Athens	9	7	33–37
Berlin	9–10	7–8	22–23
Košice	11	3	31
Kraków	13	9	49–50
S-Poland	10–17	2–10	27–48
Lisbon	12–13	11	29–35
London	9–11	8	24–27
Milan	20	13	45–46
N-Italy	6–20	1–13	26–47
Paris	11–15	10–11	26–29
Sofia	7–8	5–7	33–59
Stuttgart	11–12	8–9	19–21

Comparably high wind speeds have been measured for Berlin, London and Paris, whereas rather low values have been recorded in Kraków, Milan, Sofia and Stuttgart, indicating adverse dispersion conditions for air pollutants.

5.3.3 National vs. Transboundary Contribution

For this analysis, model calculations provided by EMEP were used. As a rough indicator for transboundary air pollution, Table 4 shows the impact on the concentration within each area (average regional background concentrations calculated on a 50 km × 50 km grid) when reducing emissions of PM and precursors by 15% in the respective Member State. Thus a high impact corresponds to a high national contribution (and thus to a low transboundary contribution).

A large national contribution as indicated by a relatively large change in regional PM$_{10}$ levels due to a national reduction of emissions can be identified for Milan, Paris, Lisbon and Stuttgart, whereas large transboundary contributions have been calculated for Košice.[11]

5.3.4 Natural Sources

A contribution to PM$_{10}$ levels from natural sources in the selected critical areas is claimed for Athens and Lisbon in the questionnaire submitted according to Decision 2004/461/EC. For Athens, 22 up to 56 (out of up to 178) exceedance

---

[11] The range of values in different areas also reflects the size of the regions. Whereas the area of Košice is covered by two EMEP grid cells, northern Italy comprises about 50 grid cells.

Table 5 Population and area affected by exceedances of the limit values (*source*: notifications for time extension)

MS	City	Critical pollutant	Area exceeded (PM_{10}/NO_2) (km^2)	Population affected (PM_{10}/NO_2)
BG	Sofia	PM_{10}	Unavailable	Unavailable
DE	Stuttgart	PM_{10}, NO_2	Unavailable/unavailable[d]	2,000/unavailable[d]
EL	Athens	PM_{10}, NO_2	216/unavailable[d]	500,000/unavailable[d]
FR	Paris[a]	PM_{10}, NO_2	417/260	3,900,000/3,200,000
UK	London	NO_2	Unavailable	Unavailable[d]
IT	Milan	PM_{10}, NO_2	1,954/unavailable[d]	4,437,716/unavailable[d]
IT	Northern Italy[c]	PM_{10}, NO_2	91,000/unavailable[d]	25,900,000/unavailable[d]
PL	Kraków	PM_{10}	Unavailable	100,000
PT	Lisbon	PM_{10}, NO_2	Unavailable	~ 36,000/unavailable[d]
SK	Košice	PM_{10}	295	~ 240,000
DE	Berlin[b]	PM_{10}	450	190,000

[a]DRIRE [15], notification FR (Réf.04A00001-fiche récapitulative.doc)
[b]Data for 2002, Berlin [16]
[c]Data for northern Italy based on ETC/ACC maps
[d]No notification of time extension available at the time of writing

days could be attributed to natural sources in the year 2007, and in 2008 15 up to 67. 122 exceedance days remained after subtracting the natural contribution. For Lisbon, the reason for 52 out of 149 exceedances in the year 2007 is claimed to be natural sources, as well as for 22 out of 88 exceedance days in 2008.

For all other critical and comparable non-critical areas[12] no contribution is claimed to have come from natural sources.

5.3.5 Population and Area Affected in Selected Areas

Data on the population and area affected in selected areas are given in Table 5 as stated in the notifications of time extensions.

While Table 5 shows officially reported data for both hot spots and urban background areas, Table 6 shows the population affected by exceedances of the daily mean PM_{10} limit value in the urban background of the critical areas. These data were derived from the gridded ETC/ACC PM_{10} maps. Berlin, Stuttgart, London and Paris do not show exceedances on a 10 km × 10 km scale. The largest number of people affected live in northern Italy and southern Poland.

---

[12] For Košice one exceedance was claimed to have been caused by the Ukraine in 2007.

Table 6 Population affected by exceedances of the PM_{10} daily mean limit values (*source*: ETC/ACC [4])

MS	City	Population affected
BG	Sofia	1,261,000
EL	Athens	3,605,000
IT	Milan/northern Italy	2,730,000/25,943,000
PL	Kraków/S-Poland	808,000/8,410,000
PT	Lisbon	351,000
SK	Košice	65,000

Table 7 $PM_{2.5}$ urban increment (*source*: IIASA [17])

	$PM_{2.5}$ urban increment
Athens	14.6
Berlin	4.2
Košice	11.5
Krakow	9.8
S-Poland	14.2
Lisboa	11.5
London	4.8
Milan	17.5
N-IT	17.5
Paris	11.9
Sofia	19
Stuttgart	1.7

5.3.6 Urban Increment

Table 7 shows the urban increment (difference between regional and urban background level for a given city) of $PM_{2.5}$ for the critical and non-critical areas. Large urban increments have been calculated for Sofia, Milan and northern Italy, Athens and southern Poland. A large urban increment may indicate adverse dispersion conditions and/or high local emission densities.

5.3.7 Emissions

Data relating to emissions are available from the MACC project[13] and, for on some areas, from the notifications of time extension. Table 8 shows PM_{10} emissions from MACC categorised into low level sources,[14] high level sources[15] and traffic. Among urban agglomerations, the highest total emissions have been calculated

---

[13] http://www.gmes-atmosphere.eu/.

[14] Source categories residential, commercial and other combustion, industrial processes, extraction distribution of fossil fuels, solvent use, road transport, other mobile sources, waste treatment and disposal, agriculture (SNAP 2, 4–10).

[15] Source categories power generation and industrial combustion (SNAP 1 and 3).

Table 8 PM_{10} emissions in critical and non-critical areas as a sum for the whole area (in t. *Source*: MACC)

City	High level	Low level	Traffic	Total
Athens	359	8,140	1,496	9,996
Berlin	320	3,003	575	3,898
Košice	1,602	1,439	56	3,097
Kraków/S-PL, NE-CZ	1,277/13,725	2,527/32,916	249/2,921	4,054/49,562
Lisbon	960	4,889	1,141	6,990
London	911	7,484	3,144	11,539
Milan/N-IT	192/11,479	2,257/33,359	1,717/22,352	4,167/67,160
Paris	1,385	22,257	3,477	27,118
Sofia	1,793	6,898	236	8,927
Stuttgart	37	745	272	1,054

Table 9 NO_x emissions in critical and non-critical areas on 10 km × 10 km (in t. *Source*: MACC)

City	High level	Low level	Traffic	Total
Athens	5,165	8,855	22,707	36,726
Berlin	6,118	6,039	6,620	18,777
Košice	9,497	378	793	10,668
Kraków/S-PL, NE-CZ	10,599/91,468	1,297/21,322	2,315/33,954	14,210/146,744
Lisbon	17,552	6,679	14,121	38,352
London	20,369	34,031	40,048	94,448
Milan/N-IT	2,375/118,988	6,490/96,236	16,920/243,429	25,785/458,654
Paris	12,906	18,389	25,850	57,145
Sofia	4,855	1,423	5,719	11,996
Stuttgart	1,334	1,957	3,976	7,267

for Paris, followed by London and Athens. On a regional level, the highest emissions, for areas examined as a whole, can be found in southern Poland. Rather high emissions from high level sources occur in Košice, Kraków and Sofia.

NO_x emissions from MACC are shown in Table 9. The highest emissions in agglomerations can be found in London, Paris and Athens.

5.3.8 Air Quality in Selected Areas

PM_{10}, NO_2 and NO_x levels for the year 2009 are given in Table 10 and Fig. 1. For PM_{10} the highest levels have been observed in Kraków, and for NO_2 in Athens, London, Paris and Stuttgart.

5.3.9 Trends in PM_{10}, NO_2 and NO_x in Selected Areas

Table 11 shows the trend in PM_{10} annual mean levels for the years 2000–2009 at the stations in each of the critical and non-critical areas with the highest levels and

Table 10 Pollution levels in selected areas in 2009 (*source*: AirBase)

MS	City/area	NO$_2$ Annual mean (µg/m^3)	NO$_x$ Annual mean (µg/m^3)	PM$_{10}$ Annual mean (µg/m^3)	# of exceedances
BG	Sofia	5–58	7–171	20–65	1–161
FR	Paris	27–96	Unavailable	25–59	16–222
EL	Athens	11–91	19–279	26–49	17–122
DE	Stuttgart	34–112	60–174[a]	20–45	15–112
IT	Milan	37–80	100–171	39–47	83–116
PL	Kraków	26–70	Unavailable	54–61[a]	137–168
SK	Košice	Unavailable	Unavailable	51	168
PT	Lisbon	25–70	32–153	26–40	15–94
UK	London	22–107	34–303	19–35	5–36
DE	Berlin	13–63	16–149	21–39	7–73

[a]No data for station showing the highest values in previous years (highest PM$_{10}$ annual mean level in Kraków in 2008: 81 µg/m^3, highest NO$_x$ level in Stuttgart in 2008: 343 µg/m^3)

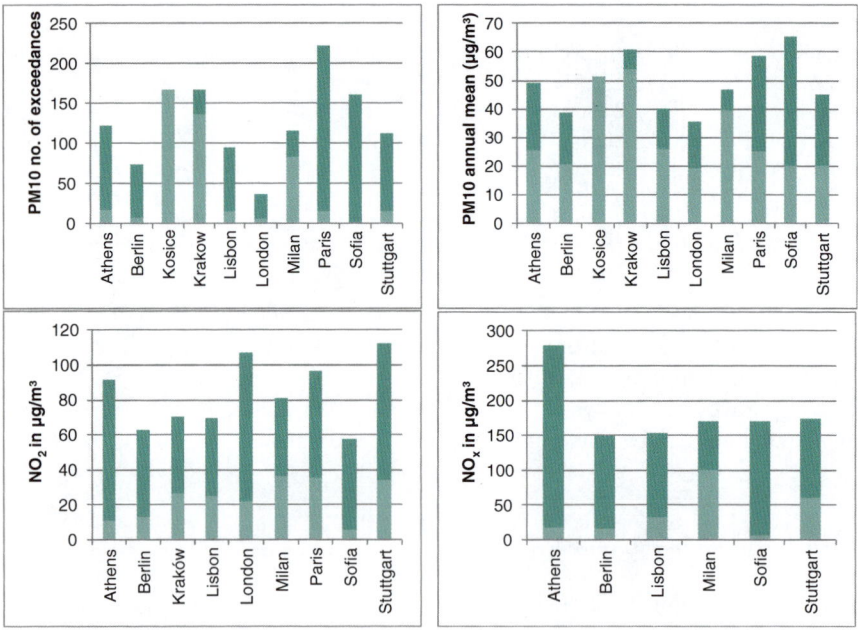

Fig. 1 Minimum and maximum: number of exceedances of the PM$_{10}$ daily mean limit value, PM$_{10}$ NO$_2$ and NO$_x$ annual mean values for 2009 (*source*: AirBase)

where data for at least 4 years are available. Figure 2 shows the mean annual change for these cities. In Sofia and Kraków a non-significant increase of more than 3 µg/m^3 per year has been observed. For the station in Sofia data between 2003 and 2008 are available, and for the station in Kraków between 1998 and 2008. In

Table 11 PM$_{10}$ annual mean trend statistics for the station with the highest level within the agglomeration; PM$_{10}$ concentration in 2009 in µg/m^3 (Q: slope in µg/m^3 per year. Sign.: level of significance. *Source*: AirBase, FMI [18])

MS	City	Station	Station name	PM$_{10}$	Sign.	Q
EL	Athens	GR0035A	Lykovrisi	43	0.05	−1.5
DE	Berlin	DEBE065	Friedrichshain–Frankfurter Allee	34	<0.1	−0.7
SK	Košice	SK0018A	Velká Ida–Letná	51	<0.1	−2.4
PL	Kraków	PL0012A	MpKrakówWIOSAKra6117	81[a]	<0.1	3.5
PT	Lisbon	PT03075	Avenida da Liberdade	39	0.001	−2.5
UK	London	GB0682A	London Marylebone Road	45[a]	<0.1	0.2
IT	Milan	IT0770A	Arese	41	<0.1	−0.7
FR	Paris[b]	FR04053	Bd Periph Auteuil	50	<0.1	−0.5
BG	Sofia	BG0052A	Drujba	65	<0.1	3.8
DE	Stuttgart	DEBW099	Stuttgart-Mitte	26	0.1	−1.0

[a]2007 value
[b]PM$_{10}$ levels used only until 2006 as correction model was changed in 2007 and 2008

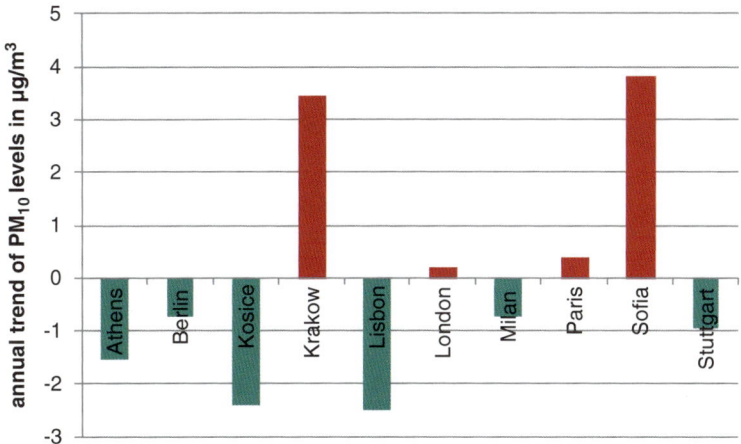

Fig. 2 Trend in annual mean PM$_{10}$ levels at the station with the highest level within the city and agglomeration

London, the station with the maximum levels shows a slight, non-significant increase as well. For the other stations a decrease has been observed.

In Table 12 the same analysis for the station with the highest level is shown for NO$_2$; in Table 13 for NO$_x$. For NO$_2$ an increase has been observed in Sofia, Stuttgart, Paris, Lisbon, London and Berlin, whereas a decrease has been recorded in Athens, Milan, Kraków and Košice.

Figure 3 shows a comparison of the range of trends at urban traffic sites in the selected areas. Whereas for most areas a decrease in NO$_x$ has been observed at all sites, NO$_2$ values have increased in Berlin, Lisbon, London and Stuttgart. This is an

Table 12 NO_2 annual mean trend statistics for the station with the highest level within the city and agglomeration; NO_2 annual mean concentration in 2009 in µg/m³ (Q: slope in µg/m³ per year. Sign.: level of significance. *Source*: AirBase, FMI [18])

MS	City	station	Station name	NO_2	Sign.	Q
EL	Athens	GR0032A	Patision	91	<0.1	−0.6
DE	Berlin	DEBE064	Neukölln-Karl-Marx-Str. 76	55	<0.1	0.8
SK	Košice	SK0014A	Košice–Štúrova	36[a]	<0.1	−0.8
PL	Kraków	PL0012A	MpKrakówWIOSAKra6117	70	<0.1	−0.2
PT	Lisbon	PT03075	Avenida da Liberdade	70	0.05	1.5
UK	London	GB0682A	London Marylebone Road	107	0.05	2.4
IT	Milan	IT0477A	Milano–V. Le Marche	80	<0.1	−0.9
FR	Paris	FR04053	Bd Periph Auteuil	113[a]	0.01	1.7
BG	Sofia	BG0054A	AMS Orlov most	58	<0.1	1.7
DE	Stuttgart	DEBW099	Stuttgart-Mitte	76	0.05	0.5

[a]Data for 2009 from www.airparif.fr

Table 13 NO_x annual mean trend statistics for the station with the highest level within the city and agglomeration; NO_x annual mean level in 2009 in µg/m³ (Q: slope in µg/m³ per year. Sign.: level of significance. *Source*: AirBase, FMI [18])

MS	City	Station	Station name	NO_x	Sign.	Q
GR	Athens	GR0032A	Patision	279	>0.1	−0.5
DE	Berlin	DEBE064	Neukölln-Karl-Marx-Str. 76	135	0.01	−2.5
PL	Kraków	PL0012A	MpKrakowWIOSAKra6117	203[a]	0.01[a]	−7.2[a]
PT	Lisbon	PT03075	Avenida da Liberdade	153	>0.1	−1.0
UK	London	GB0682A	London Marylebone Road	303	0.01[b]	−6.5[b]
IT	Milan	IT0477A	Milano–V. Le Marche	171	0.01	−10.4
DE	Stuttgart	DEBW099	Stuttgart-Mitte	174	0.001	−8.1

[a]2007 value
[b]Data from 1998 until 2011 from http://uk-air.defra.gov.uk

indication of the influence of primary NO_2 emissions from diesel passenger cars and buses. For the other areas no data are available.

5.3.10 Population Weighted Exposure

As detailed and comparable information about health impacts in the critical areas is not available for all areas, the population weighted exposure[16] has been calculated on the basis of gridded air quality and population data [4, 5]. The values obtained in this way give a rough estimate of potential health effects (Table 14). The highest levels have been calculated for Sofia, Kraków and Milan, where the levels for the annual mean are more than twice as high as in Berlin.

---

[16] Average concentration per inhabitant.

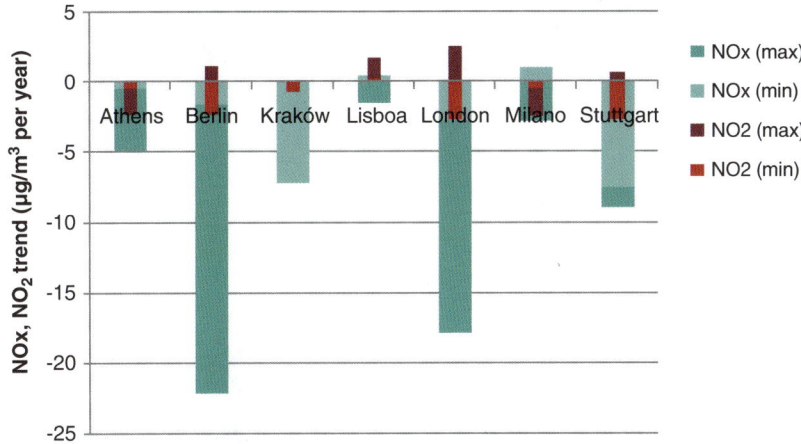

Fig. 3 Comparison of the slope of the trend in NO_x and NO_2 annual mean concentrations at urban traffic sites (*green*: NO_x, *red*: NO_2. *Source*: AirBase, FMI [18])

Table 14 Population weighted exposure to annual mean and 36th highest daily mean levels in urban and regional background areas in $\mu g/m^3$ (*source*: ETC/ACC [4])

City/area	PM_{10} annual mean ($\mu g/m^3$)	PM_{10} 36th highest daily mean[a] ($\mu g/m^3$)
Athens	36	55
Berlin	22	37
Košice	31	51
Lisbon	31	48
London[b]	26	39
Milan/N-IT	45/39	83/69
Paris	28	47
Kraków/S-PL, NE-CZ	49/39	95/71
Sofia	56	96
Stuttgart	20	36

[a]If above 50 $\mu g/m^3$, more than 35 exceedances of the daily mean PM_{10} limit value have occurred
[b]According to re-notification of 7 May 2010 PM_{10} levels at this site are considerably lower compared to those reported to AirBase, due to changes in the correction model [19]

6 Overview of Critical Factors

In the following the critical factors discussed in the chapters above are summarised.

6.1 Natural Factors

The natural factors cover dispersion conditions, which are influenced by climate and (regional and local) topography, and natural PM_{10} sources.

Of the selected cities, the following cities have a continental or Mediterranean climate, characterised by low wind speeds and frequent high-pressure situations:

- Athens
- Kraków
- Košice
- Milan
- Sofia

The following cities have a more or less oceanic climate:

- Berlin
- Lisbon
- London
- Paris
- Stuttgart (transitional climate)

The following cities are severely affected by adverse dispersion conditions, due to the local or regional topography of the surrounding area:

- Athens
- Košice
- Sofia (basin between high mountains)
- Stuttgart (narrow basin)
- Milan (Po Valley surrounded by Alps and Apennines in N, W, S)

Natural sources (Sahara dust) contribute significantly to PM_{10} levels in Lisbon, a factor which can be considered (deducted) in the air quality assessment.

6.2 Anthropogenic Factors

Source apportionment data are provided in the notifications of time extensions.[17] The most important contribution to PM_{10} levels on both the urban and local scale in most cities originates from road traffic, and for NO_x road traffic is the by far predominant source. The highest contributions of road traffic to urban background PM_{10} levels have been reported for Kraków (62%), Paris, Stuttgart, Berlin and Lisbon (about 40–45%), although the figures do not seem to be completely comparable (Table 15).

Significantly high industrial contributions (25–30% to urban background) affect Kraków and Milan, and Košice (for which no appropriate source apportionment has been provided).

Forest fires have been reported as a relevant PM_{10} source for Lisbon.

---

[17] http://ec.europa.eu/environment/air/quality/legislation/time_extensions.html.

Table 15 Summary table for critical factors

	Dispersion conditions climate	Topography	Share of private cars (modal split)	AQ Plan
Athens	Adverse (Mediterranean)	Adverse	65%	2001, 2005
Berlin			29%	2005
Košice	Adverse (continental)		No data	2004, 2007, 2009
Kraków	Adverse (continental)		27%	2006, 2009
Lisbon			40%	2006
London			41%	2003, 2010
Milan	Adverse (Mediterranean)	Very adverse	47%	2001, 2007
Paris			44%	2000, 2009
Sofia	Adverse (continental)	Very adverse	No plausible data	2004/2005, incomplete
Stuttgart		Adverse	45%	2005, 2010

The data on the modal split of urban traffic do not seem to be completely comparable (e.g. not all the data cover pedestrian traffic). However, significantly high shares of private cars have been reported for Athens, Paris, Stuttgart, Milan and London (40–50%). Distinctly low shares of private cars have been reported for Kraków and Berlin (about 30%).

A semi-external critical factor is the total size of an agglomeration, which causes high urban background concentrations and is related to a high amount of urban traffic. This is a critical factor which applies especially to Paris and London.

The reasons for the non-attainment of the PM_{10} limit values were also stated in the notifications of time extensions. In most cases, it was stated that non-attainment by 2005 could be attributed inter alia to a delay in the implementation of measures, or certain measures leading to a lower level of emission reductions than expected. An increase in road traffic is stated as a reason for non-attainment especially in eastern European countries. An increase of primary NO_2 emissions due to an increasing share of diesel vehicles has been reported e.g. by France, Germany, Italy and UK.

In all cities, most measures are targeted at road traffic. In Kraków the most effective measures concern domestic heating, and in Košice industrial facilities.

6.3 Qualitative Estimate of Critical Factors

A "qualitative" summary of the critical factors for the cities discussed above is given in Table 16.

The comparably low pollution levels in Berlin – the "non-critical area" in this comparison – is due to favourable dispersion conditions (flat terrain and moderate oceanic climate), comparably low emissions from industry and domestic heating, low emission densities and effective air quality management.

Table 16 Summary of critical factors

	Dispersion conditions	Size of agglomeration	Traffic emissions	Industrial and/or domestic emissions	Difficulties in AQ management/missing data[a]
Athens	+	+	++	+	
Berlin		+	+		
Košice	+			++	+
Kraków	+		+	++	+
Lisbon			+		
London		++	++		
Milan	++	+	++	+	
Paris		++	++	+	
Sofia	++		+		++
Stuttgart	+		++		

[a]For example emission data, traffic data, modelling

7 Possible Courses of Action for Critical Areas

In the analysis presented above, critical factors have been identified. This knowledge about critical factors and the experience gained from air quality management in the selected areas is used to propose possible courses of action for dealing with air quality issues in critical areas.

7.1 Suggestions for Guidance Documents and Further Support

7.1.1 Air Quality Assessment

Air quality assessment is undertaken with different levels of detail in the Member States. There are also indications that assessment is incomplete in some zones. The analysis of non-critical areas has shown that for some cities only information about urban background stations is available and that the data in AirBase are different from other datasets. Whereas compliance is achieved in urban background areas, exceedances might occur at hotspot sites, which may remain undetected if no further assessments are undertaken. Therefore guidance documents and further support, e.g. via dedicated workshops, might be helpful.

7.1.2 Emission Data, Traffic Analysis, Modal Split, etc.

Data necessary to analyse the contributions from different sources, and to define indicators to prove the effects of measures, are missing or incomplete in some cities and areas. In order to support the relevant authorities, a (guidance) document should be developed which should cover the following aspects:

- Best practice examples
- List of contact points for inquiries
- Minimum requirements for data quality and availability

7.1.3 Modelling for Source Apportionment

For some cities and areas no detailed source apportionment was available. As with the datasets described above, which are important input parameters for source apportionment, a guidance document or further support might be helpful.

In order to support the relevant authorities, a (guidance) document should be developed which should cover the following aspects:

- Best practice examples
- List of contact points for inquiries
- Minimum requirements for data quality and availability
- Requirements for the outputs of the analysis

7.1.4 Implementation of Measures

Usually, numerous measures are – in principle – available to reduce air pollution and to achieve compliance with limit values. However, there are often barriers, even to implementing cost-effective measures.

In order to facilitate effective communication between the Commission and the Member States authorities, standardised materials and formats should be developed. These should include:

- Checklists for possible measures
- Checklists or implementation status of measures

7.1.5 Administrative and Financial Deficiencies

Deficiencies in the development of the above-mentioned datasets and in the implementation of measures might be due to administrative and/or financial deficiencies, especially in the course of the economic crisis in recent years. Possible ways to overcome these hurdles depend on the specific situation in the concerned Member States. Suitable approaches might be to encourage transnational projects so as to exchange best practice examples and to support institution building such as projects under the South East Europe Transnational Cooperation Programme,[18] the Central

---

[18] http://www.southeast-europe.net/.

Europe Operational Programme,[19] the MED programme for Mediterranean countries[20] or the Operational Programme for Southwest Europe.[21]

7.2 Suggestions for Specific Action in Critical Areas and on EU Level

7.2.1 Athens, Greece

Compared to other cities, the modal split of Athens shows a rather high share of private vehicles. Freight transport is also mostly done by road traffic. Even though a ban on private diesel vehicles has been in place since 1992, emissions from traffic are of major importance. The measures proposed in the notification of time extensions and the air quality plan for Athens address the main sources and are aimed at improving public transport. Hence, if these measures are far-reaching enough, if they cover a substantial part of the relevant sources and are implemented in an effective way, a substantial decrease of the pollutant levels can be expected.

7.2.2 Košice (Industrial Region), Slovakia

The main source of PM_{10} in Košice is the U.S. Steel company. Within this facility the blast furnace and coke production are most important.

An assessment of the reduction potential is not possible on a detailed level, since only aggregated emission data are available. Further measures should be planned – but more detailed information about specific processes, fugitive emissions and existing de-dusting facilities would be necessary.

7.2.3 Kraków, Poland

The main sources of PM_{10} in Kraków (and also in southern Poland) are domestic heating, industry and road traffic; for NO_2 it is road traffic.

The replacement of solid fuels for domestic heating by gas or district heating is considered the most effective measure to reduce residential emissions.

Traffic emissions in the city centre can be effectively dealt with by an "environmental zone" or low emission zone, including improvements in public transport.

---

[19] http://www.central2013.eu/.
[20] http://www.programmemed.eu/index.php?id=5175&L=1.
[21] http://www.interreg-sudoe.org/castellano/index.asp.

Compliance of all industrial facilities with IPPC and LCP requirements should be fostered and scrutinised. Measures going beyond "best available technology" should also be encouraged.

7.2.4 Lisbon, Portugal

The main sources are road traffic and natural sources. Various high impact measures such as a low emission zone and the renewal of the public transport fleet are foreseen in Lisbon [6]. Further measures might be implemented to reduce the share of diesel vehicles, and to increase the share of bicycles.

7.2.5 London, United Kingdom

Traffic is the dominant source on the urban and local level. For PM_{10}, small contributions from industry, as well as from the off-road and commercial and residential sectors, were identified. Several high impact measures such as a congestion charge, a low emission zone and the renewal of the public transport fleet have already been implemented.

The updated Mayor's air quality strategy addresses various measures which are proposed in the strategy and should be implemented in the near future [7]. This includes inter alia "no-idling zones", more efficient freight movements, promotion of smarter travel, lane rental schemes for construction work in congested roads, supporting the uptake of low emission and electric vehicles, hybrid buses for public transport, renewal of the taxi fleet and enforced implementation of best practice guidance for construction sites.

7.2.6 Milan Agglomeration/Northern Italy, Italy

The main sources in Milan, of both PM_{10} and NO_2, are traffic (amounting to about half of the contribution), other mobile sources and non-industrial combustion. A large number of measures have been implemented according to the air quality plan [8]. Due to the high share of private car traffic for trips between Milan and the Metropolitan area, improvements of public transport and further restrictions for vehicles entering the city may prove to be effective.

7.2.7 Paris, France

The main sources in Paris, based on the emission inventory, are traffic as well as the commercial and residential sectors and industry. At a large number of monitoring sites, traffic is the dominant local source. Various high impact measures have already been implemented. Examples are traffic restrictions during pollution episodes, improvements of public transport and reductions of industrial emissions.

The measures already implemented cover the most important source categories. With the experience gained from other critical areas, it may be beneficial to expand the existing measures for reducing the share of diesel vehicles and to implement further traffic restrictions and pedestrian zones, to stimulate the early uptake of Euro 6/VI vehicles and to apply measures for construction sites and off-road machinery.

7.2.8 Sofia, Bulgaria

The main sources of PM_{10} in Sofia, according to the air quality plan and the notifications of time extensions, are traffic and industry.

Suggestions for concrete measures are hampered by the lack of precise data on the share of different sources. Nevertheless it can be stated that the use of private cars has strongly increased in recent years, the share of bicycles is rather low and the vehicle fleet is aged. Therefore, as a first step, datasets should be improved with respect to emissions, the share of different sources and activity data (especially private cars and buses as well as industry and domestic heating). Based on these data, actions should be implemented to address the main source categories.

7.2.9 Stuttgart, Germany

According to the source apportionment, the main source for PM_{10} and NO_2 on a local level is traffic. A low emission zone has been introduced, and heavy duty vehicles have been banned from the city. Further action might be taken to tighten the requirements for the low emission zone, stimulating the early uptake of Euro 6/VI vehicles and introducing measures to increase the share of non-motorised traffic.

7.3 Action on EU Level

7.3.1 Traffic Measures

For compliance with NO_2 limit values, low real-world emissions in urban areas (of Euro 5/V and especially Euro 6/VI vehicles) will be very important. Recent findings have shown that emissions of HDVs equipped with SCR might be higher than expected under urban driving conditions [9]. Therefore it has to be ensured that emission levels are low under real-world conditions. As regulations concerning general vehicle standards require extensive negotiations, voluntary agreements and guidance for the public procurement of these vehicles might be a more feasible option.

Another EU strategy, also targeted at lowering traffic emissions, is to accelerate the introduction of alternative transmission concepts by establishing homologation standards, e.g. for electric vehicles, research and development funding, as well as setting up public procurement guidance (see below).

The EU green car initiative is targeted at the encouragement of all road transport stakeholders to move towards more sustainable transport, mainly by funding industry research and development.

7.3.2 Public Procurement

Green Public Procurement (GPP) is defined in the Communication (COM [10], 400) "Public procurement for a better environment" as "[...] a process whereby public authorities seek to procure goods, services and works with a reduced environmental impact throughout their life cycle [...]". Since public authorities are major consumers in Europe and green public procurement is also targeted at vehicle fleets, public authorities should also provide industry with real incentives for developing green technologies and products that offer environmental benefits. GPP is a voluntary instrument, which means that individual Member States and public authorities can determine the extent to which they implement it themselves.

7.3.3 Counterproductive Community Measures for Traffic

Besides the critical measures mentioned above, it has to be kept in mind that other community activities have counterproductive effects on air quality. The European Single Market has fostered growth of mobility, with side effects. The objective of the Trans-European Transport Networks (TEN-T[22]) is to make traffic efficient and safe, but it has been shown that they bring about additional traffic with negative ecological consequences.

Liberalisation of rail networks may lead to a decrease of public transport offers in more remote areas and to an increase in car traffic.

The liberalisation of air traffic leads to increased air traffic volumes, with effects on the air quality in areas around airports.

7.3.4 Measures for Off-Road Machinery incl. Ships

Europe-wide emission standards for off-road machinery are laid down in Directive 97/68/EC and Directive 2004/26/EC. Due to the introduction of more challenging

---

[22] Planned set of transport networks designed to serve entire Europe, see http://tentea.ec.europa.eu/.

emission standards for vehicles, emissions from off-road machinery will become more important. This is especially true for construction machinery, and also for other means of off-road transport. In ports ships can contribute to PM_{10} and NO_x levels considerably. As this machinery often has a long lifetime, stringent emission limit values are rather important.

In the foreseen revision of Directive 2004/26/EC stringent emission limit values should be foreseen that are more in line with those for heavy duty vehicles than the previous ones.

In addition, regular inspections should be foreseen.

For ships, the foreseen revision of Directive 2005/33/EC amending Directive 1999/32/EC should include stringent regulations for the sulphur content in marine fuels for certain areas. Furthermore, limits for NO_x and PM emissions should be imposed, which are possible due to recent IMO amendments of MARPOL Annex VI. Also, Emission Control Areas should be enlarged to the extent possible. PM and NO_x emissions of existing ships should be addressed e.g. via an emission trading scheme.

7.3.5 Measures for Domestic Heating

For PM_{10} emissions from domestic heating the ecodesign requirements for heating and water heating equipment are of great importance. Stringent emission limit values for PM_{10} from heating appliances should be foreseen in these ecodesign requirements for all domestic heating facilities.

In addition, the establishment of a suitable type of framework directive should be considered, regulating the periodical inspection of small combustion facilities EU-wide and, in case of emission limit value exceedances or for old facilities (e.g. PM_{10}, NO_x, CO, etc.), stipulating the obligation to maintain, renovate or replace an insufficient heating system.

The Energy Performance for Buildings Directive [11] foresees inspections of boilers with respect to boiler efficiency and boiler sizing. These inspection requirements could be extended to include air pollutants.

Binding targets for energy efficiency and sanctioning mechanisms should also be foreseen.

7.3.6 Measures for Industrial Sources

In some areas industrial sources are of importance as well. A rigid adherence to Directives regulating industrial facilities such as the IPPC and LCP Directives is therefore of great importance.

In some countries a national emission reduction plan according to Article 4 (4) of the LCP Directive [12] has been implemented instead of applying emission limit values for individual installations. High local concentrations at specific sites might therefore occur. As laid down in Article 10 of the IPPC Directive [13], stricter conditions beyond BAT may be introduced to comply with environmental quality standards. The implementation of this Article could be scrutinised on a European level.

7.3.7 Revision of the NEC Directive

A revised NEC Directive, which includes emission ceilings for $PM_{2.5}$, is of utmost importance in order to prepare the path for emission reductions up to 2020 and to achieve various environmental objectives. It will also contribute to attaining the PM_{10} and NO_2 limit values. It will furthermore be an important input for the foreseen revision of the Gothenburg Protocol, designed to also reduce transboundary pollution from countries outside the EU. Therefore a proposal for a revised NEC Directive should be published as soon as possible. The emission ceilings included therein should be ambitious so as to facilitate the attainment of air quality limit values.

8 Conclusions

Critical areas are defined as areas where compliance with limit values, in particular for PM_{10} and NO_2, is difficult for various reasons.

This study has shown that basic available data such as levels of air pollution (including background levels), trends of air pollution and population (size of cities, population density) are suitable indicators to characterise most critical areas. It has been investigated if certain additional factors known to influence air quality show a correlation with air pollution levels and are therefore appropriate proxy data to identify such areas.

However, no clear pattern has emerged that allows a clear discrimination between critical and non-critical areas.

Besides high background concentrations – partly originating from transboundary pollution – a number of different sources exist for PM_{10} which require specific action. Nevertheless, traffic is a dominant local source in most areas. For NO_2 traffic is the only predominant source. NO_2 levels have often remained on a high level in recent years, which points to the non-delivery of Euro standards for diesel vehicles and increasing primary NO_2 emissions.

This study has shown that various critical areas across Europe share similar problems and that it is therefore helpful to look at factors influencing air quality on a European scale, and to share knowledge on specific factors across different areas. However, some gaps have been identified, especially with respect to the

- Completeness of air quality assessments throughout the territory of all Member States
- Completeness and quality of datasets on activities and emissions
- Documentation of the status and success of measures

Even though a number of high impact measures have been implemented successfully, some critical areas are far from complying with the limit values. Generally speaking, there is no single measure with which compliance can be achieved. For the critical areas which have been analysed in this study further measures have been suggested. On a European level, special attention should be paid to a reduction of real-world driving cycle and vehicle emissions.

Acknowledgments This study has been financed by the European Commission under framework contract ENV.C.3/FRA/2009/0008. The opinions expressed are those of the authors only and do not represent the Commission's official position

References

1. Air Quality Directive: Directive 2008/50/EC of the European Parliament and of the Council of 21 May 2008 on ambient air quality and cleaner air for Europe
2. NEC Directive: Directive 2001/81/EC of the European Parliament and of the Council of 23 October 2001 on national emission ceilings for certain atmospheric pollutants
3. Commission Decision 2004/461/EC: Commission Decision of 29 April 2004 laying down a questionnaire to be used for annual reporting on ambient air quality assessment under Council Directives 96/62/EC and 1999/30/EC and under Directives 2000/69/EC and 2002/3/EC of the European Parliament and of the Council. C(2004) 1714
4. ETC/ACC – European Topic Centre on Air and Climate Change, de Smet P, Horálek J, Coňková M, Kurfürst P, de Leeuw F, Denby B (2009a) European air quality maps of ozone and PM10 for 2007 and their uncertainty analysis, ETC/ACC Technical Paper 2009/9. Bilthoven
5. ETC/ACC – European Topic Centre on Air and Climate Change, de Leeuw F, Horálek J (2009b) Assessment of the health impacts of exposure to PM2.5 at a European level. ETC/ACC Technical Paper 2009/1. Bilthoven
6. DCEA – Departamento de Ciências e Engenharia do Ambiente, Universidade Nova de Lisboa/ Faculdade de Ciências e Tecnologia (2006) Planos e Programas para a melhoria da qualidade do ar na Região de Lisboa e Vale do Tejo. Edição Revista, December 2006. Lisboa
7. GLA – Greater London Authority (2010) Clearing the air. The Mayor's Air Quality Strategy. London
8. Regiona Lombardia (2007) Aggiornamento del Piano Regionale di Risanamento della Qualità dell'Aria (P.R.Q.A.) anche ai fini dell'accesso al Programma di finanziamenti di cui al Decreto del Ministro dell'Ambiente e della Tutela del Territorio e del Mare del 16 ottobre 2006
9. TNO (2010) Real world NO_x emissions of Euro V vehicles. MON-RPT-2010-02777. Den Haag
10. COM (2008) 400 final: Communication from the Commission to the European Parliament, the Council, the European economic and social committee and the committee of the regions. Public procurement for a better environment
11. EPB Directive: Directive 2010/31/EU of the European Parliament and of the Council of 19 May 2010 on the energy performance of buildings
12. LCP Directive: Directive 2001/80/EC of the European Parliament and of the Council of 23 October 2001 on the limitation of emissions of certain pollutants into the air from large combustion plants
13. IPPC Directive: Council Directive 1996/61/EC of 24 September 1996 concerning integrated pollution prevention and control

14. ETC/ACC – European Topic Centre on Air and Climate Change, Horálek J, de Smet P, de Leeuw F, Denby B, Swart R, Kurfürst P (2007) European air quality mapping 2005 including uncertainty analysis, ETC/ACC Technical Paper 2007/7. Bilthoven
15. DRIRE – Direction régionale de l'industrie, de la recherche et de l'environnement (2009) L'environnement industriel en Ile de France – Édition 2009. La prevention de la pollution atmosphérique. Chapitre 6
16. Berlin – Senatsverwaltung für Stadtentwicklung (2005) Luftreinhalteplan und Aktionsplan für Berlin 2005–2010. Berlin
17. IIASA – International Institute for Applied Systems Analysis (2007) Estimating concentrations of fine particulate matter in urban background air of European cities. IIASA Interim Report IR-07-001. Laxenburg
18. FMI – Finnish Meteorological Institute (2002) MAKESENS 1.0. Mann–Kendall Test and Sen's Slope Estimates for the Trend of Annual Data. Version 1.0 Freeware
19. UK Representation to the EU Brussels (2010) Letter from 3 May 2010 concerning Air Quality Directive 2008/50/EC. Greater London agglomeration zone, UK: exemption until 2011 from the obligation to apply the daily limit values for PM_{10}. Ref. Ares(2010)246219 – 07/05/2010

The Evolution and Control of NO_x Emissions from Road Transport in Europe

P.G. Boulter, J. Borken-Kleefeld, and L. Ntziachristos

Abstract Road transport is the largest contributor to NO_x emissions in the EU. This chapter discusses NO_x formation mechanisms, control strategies, trends in emissions and possible future developments. Control strategies include vehicle emission legislation, engine design, exhaust after-treatment, modification of fuel properties, alternative fuels and new powertrain technologies. Calculations show that NO_x emissions from the sector decreased substantially between 1990 and 2010. Such calculations are based on the assumption that the systematic tightening of emission limits has been effective. However, there is evidence that modern diesel vehicles are not delivering the expected reductions in emissions during real-world driving. Moreover, diesel vehicles emit more NO_x than petrol vehicles (with a larger proportion of "primary" NO_2), and their market share has increased in many countries. These factors partly explain the observation that ambient NO_2 concentrations continue to exceed health-based limits in urban areas. Up to 2020 there is a need for a more effective regulation of emissions, and the chapter proposes several measures that can be taken. Beyond 2020 emissions of NO_x from the sector will depend on the market penetration of low-carbon technologies.

Keywords Air quality, Alternative fuels, Emission-control technologies, Transport, Vehicle emissions

---

P.G. Boulter
PAEHolmes, Sydney, Australia

J. Borken-Kleefeld
International Institute for Applied Systems Analysis, Laxenburg, Austria

L. Ntziachristos (✉)
Laboratory of Applied Thermodynamics, Aristotle University Thessaloniki, PO Box 458, 54124 Thessaloniki, Greece
e-mail: leon@auth.gr

Contents

1 Introduction 32
2 Evolution of NO_x Emissions to date 33
3 NO_x Emissions from Road Vehicles 35
   3.1 Formation Mechanisms 35
   3.2 Emission-Control Technologies 36
   3.3 Factors Affecting On-Road NO_x Emissions 40
4 Reducing NO_x Emissions from Road Transport 41
   4.1 Vehicle and Engine Type Approval 41
   4.2 Emission-Control Technology 44
   4.3 Fuel Quality and Alternative Fuels 45
   4.4 New Powertrain Technologies 47
   4.5 Eco-Driving 47
   4.6 Traffic Management 48
5 Future Evolution of Emissions and Remaining Challenges 48
References 51

1 Introduction

For many years the presence of nitric oxide (NO) and nitrogen dioxide (NO_2) in the atmosphere has been a cause for concern on account of both the scale of anthropogenic emissions of these compounds and their impacts on health and the environment. There are various atmospheric reactions which cycle NO and NO_2, and it is therefore convenient to think of the two compounds as a group. By convention the sum total of oxides of nitrogen (i.e. NO + NO_2) is termed NO_x and is expressed as NO_2 mass equivalents.

The compound of more interest in relation to local air quality and human health is NO_2. It is an irritant and oxidant gas which has been linked to a range of adverse health effects, including cancer, although it is possible that NO_2 may be acting as an indicator of other traffic-related carcinogens. The most consistent association has been found with respiratory outcomes [1, 2]. NO_2 is also a precursor for a number of harmful secondary air pollutants, such as nitric acid, the nitrate part of secondary inorganic aerosols and photo-oxidants including ozone. In addition, NO_2 absorbs visible solar radiation, thus contributing to impaired atmospheric visibility.

NO_x emissions are implicated in phenomena such as acidification and eutrophication, with their subsequent impacts on the biodiversity of habitats, as well as radiative forcing of climate through the formation of nitrate aerosol and tropospheric ozone, and impacts on the carbon cycle [3].

Because of these adverse effects of NO_x compounds on both human health and the natural environment, several regulatory steps have been introduced to control NO_x emissions from different sources and to limit ambient NO_2 concentrations.

Nitrogen oxides are primarily produced when nitrogen and oxygen combine at high temperatures and pressures. Such conditions are reached in internal combustion engines (ICEs). Consequently, motorised transport modes – and in particular road vehicles – are major contributors to overall NO_x emissions. This chapter deals with

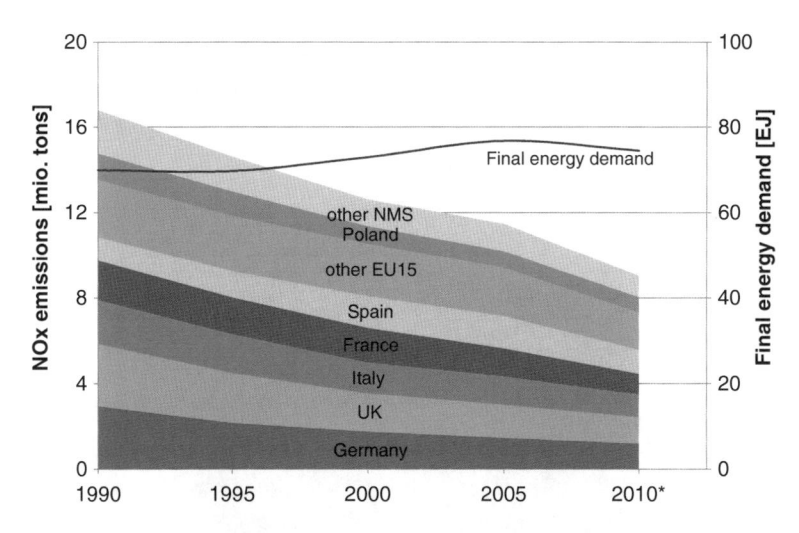

Fig. 1 NO_x emissions in Europe (EU-27) by country versus total final energy demand during the period 1990–2010. The data for 2010 are still provisional [4]

the formation and control of NO_x emissions from road transport, explains historical trends in emissions from the sector and discusses possible future developments.

2 Evolution of NO_x Emissions to date

It can be seen from Fig. 1 that NO_x emissions in the European Union (EU-27) decreased by more than 40% between 1990 and 2010, although total final energy demand grew by almost 10% during the same period [4]. Emissions were reduced in almost all countries, and notably in the larger countries (Germany, UK, Italy, France and Poland) with the exception of Spain, where total NO_x emissions increased between 1990 and 2005. The general decreases resulted from progressively stringent emission controls across all sectors, a restructuring of the power supply, and an overall increase in energy efficiency.

Figure 2 shows sectoral NO_x emissions in the EU-27 countries in 2008 based on submissions to the UNECE Convention on Long-Range Transboundary Air Pollution (CLRTAP). Almost all of the oxides of nitrogen emitted to air were from combustion sources, and road transport was the single largest contributor (41%) [5]. In urban areas the emissions from road transport are proportionally higher, and the local impacts are exacerbated, due to the density of the road network, the volume of traffic, the close proximity of the population to the emission source, and the larger distances to other relevant sources.

Legislation and strategies to reduce exhaust emissions from road vehicles have been in place since the early 1970s. Calculations have established that emissions of

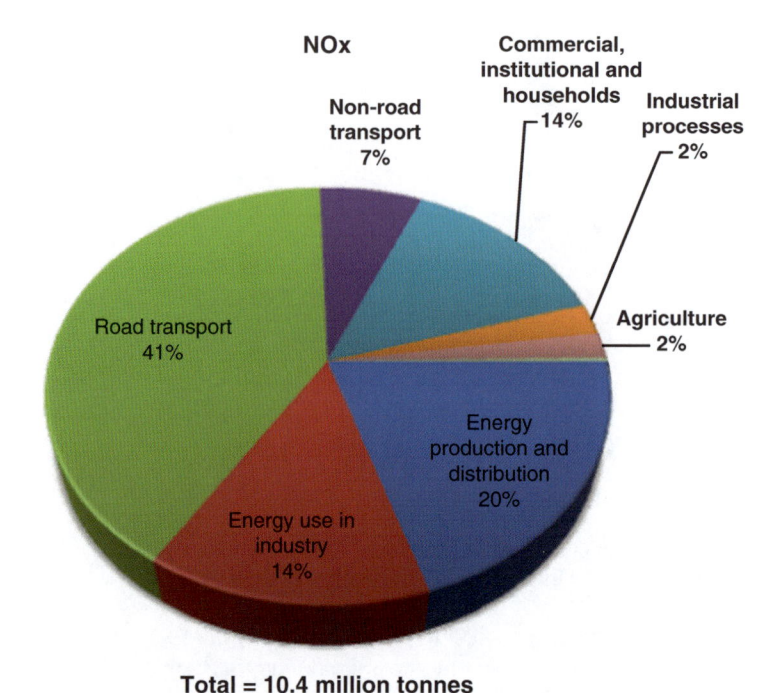

Fig. 2 Sectoral emissions in the EU-27 in 2008 [5]

Pie chart: NOx, Total = 10.4 million tonnes
- Road transport 41%
- Energy use in industry 14%
- Energy production and distribution 20%
- Agriculture 2%
- Industrial processes 2%
- Commercial, institutional and households 14%
- Non-road transport 7%

most regulated pollutants (including NO_x) from road transport in the EU peaked in the early 1990s [6]. NO_x emissions from road vehicles decreased by 45% between 1990 and 2010, while the volumes of passenger transport (in passenger-km) and freight transport (in tonne-km) increased by more than 40% and 80%, respectively (see Fig. 3). This effective decoupling of transport activity from NO_x emissions was achieved through the interplay between legislation and technology. Notably, the mandatory use of the three-way catalytic converter (often referred to as a "three-way catalyst" – TWC) in 1992 reduced emissions from petrol passenger cars rapidly and substantially. With some delay, emission-reduction measures for heavy-duty vehicles also became effective. These two vehicle categories together contribute around 80–90% of all road transport NO_x emissions. However, the reduction in NO_x emissions in Europe has slowed down in recent years due to increasing numbers of diesel cars in the fleet. The emission characteristics of diesel vehicles pose a challenge to any emission-reduction target, as will be explained below.

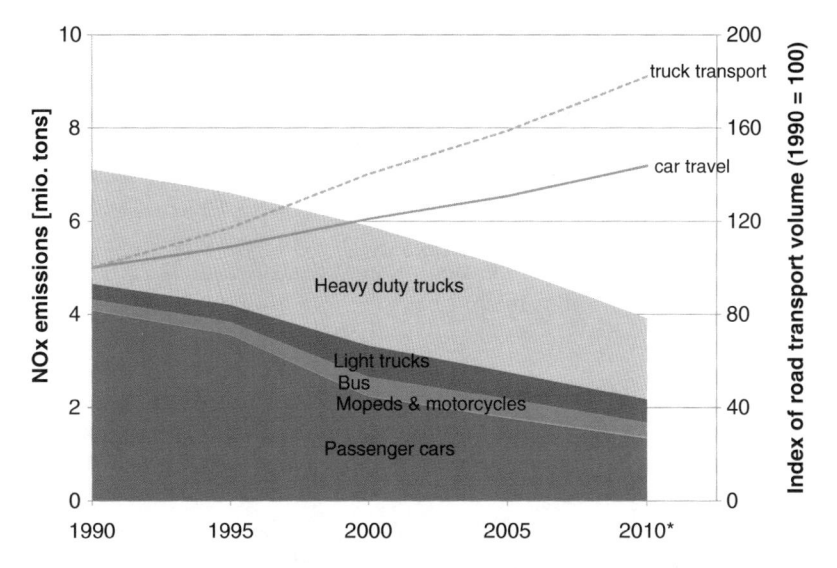

Fig. 3 NO$_x$ emissions by road vehicles in Europe (EU-27) versus transport demand during the period 1990–2010. The 2010 data are still provisional [4, 7]. The calculations have been made at 5-year intervals. The changes in 1995 and 2000 are mainly due to the competing effects of decreasing emission limits and increasing dieselisation

3 NO$_x$ Emissions from Road Vehicles

3.1 Formation Mechanisms

Almost all road vehicles are powered by ICEs. The only notable exceptions include vehicles powered by electricity drawn from an external grid (e.g. trolley buses), from on-board batteries (e.g. electric cars) or, possibly in the future, from fuel cells.

In an ICE energy is derived from the burning of fuel in air, with the main oxidation products being CO_2 and water vapour. However, some of the nitrogen in the combustion air is also oxidised, leading mainly to the formation of NO. NO formation is favoured by high temperatures and pressures found in the combustion chamber, as well as lean (i.e. oxygen rich) fuelling conditions.

NO formation proceeds via two main mechanisms known as "thermal" (or Zel'dovich) and "prompt" [8], with the former being responsible for more than 90% of emissions [9]. "Fuel NO" may also be formed from nitrogen chemically bound in fuels. However, for road vehicles this is responsible for only a small proportion of total NO due to the negligible nitrogen content of fuels.

The thermal mechanism is shown in reactions (1), (2) and (3) [10]:

$$N_2 + O \rightarrow NO + N \tag{1}$$

$$N + O_2 \rightarrow NO + O \tag{2}$$

$$N + OH \rightarrow NO + H \tag{3}$$

Reaction (1) is the rate-determining step, influencing the amount of NO which is formed, and is highly dependent on combustion temperature due to the high activation energy of the reaction (320 kJ/mol). Increasing the temperature from 1,200°C to 2,000°C increases the rate of this reaction by a factor of 10,000 [11].

The prompt-NO mechanism forms NO earlier in the flame than the thermal mechanism and is initiated by reaction (4) [12]. Both N and HCN react rapidly with oxidant to form NO in the flame:

$$CH + N_2 \rightarrow HCN + N \tag{4}$$

Whilst NO is the dominant NO_x species formed during engine combustion, significant amounts of NO_2 can also be produced under certain conditions. The NO_2 which is emitted directly from vehicle exhaust is commonly referred to as "primary NO_2". As will be seen later, the amount of NO_2 emitted from the tailpipe is dependent upon the type of exhaust after-treatment used.

3.2 Emission-Control Technologies

Various engine and after-treatment technologies have been developed for controlling emissions. These technologies, which are summarised below, are often used in combination to ensure compliance with increasingly stringent legislation. For this reason emission-control systems have become increasingly complex and expensive. Whilst after-treatment technologies are generally fitted during manufacture, retrofitting to older vehicles is also a common pollution-reduction strategy [e.g. in low-emission zones (LEZs)].

3.2.1 Control of Combustion

The quantity of NO formed in a petrol engine or diesel engine depends on the combustion parameters and, to a certain extent, can be controlled by adjustment of the engine operation. For example, NO formation may be limited by retarding the spark timing in petrol engines and the fuel injection in diesel engines [8].

For diesel vehicles reductions in "engine-out" NO_x emissions are harder to achieve, mainly due to the high combustion temperatures and oxygen-rich operational regime of the engine. Moreover, the effects of engine adjustments are limited due to a trade-off between NOx and particulate matter (PM); combustion-related measures which aim to reduce emissions of one pollutant (e.g. NO_x) lead to an increase in emissions of the other (PM), and vice versa. Engine calibration is usually determined by the need to balance emissions of the two pollutants and to maximise fuel efficiency. Alternatively, emissions of either NO_x or PM can be reduced via engine calibration, with the other pollutant being controlled using an after-treatment device.

A relatively recent development has been the gasoline direct injection (GDI) engine, in which fuel injection takes place in the cylinder. This allows better control of the combustion process. Early (1990s) GDI engines operated using the lean-burn principle over their complete range, in which proportionally more air is fed in the cylinder [13]. Such engines benefited from better fuel consumption but the oxygen abundance led to higher NO_x emissions. Exhaust aftertreatment was therefore needed to bring NO_x emissions within the legislative limits [14]. Modern GDI engines can switch between lean-burn (stratified) and stoichiometric combustion depending on the operational mode to achieve low emissions and high fuel efficiency.

3.2.2 Exhaust Gas Recirculation

Exhaust gas recirculation (EGR) is a NO_x-reduction technology which has been used in both petrol and diesel engines for some years (since the mid-1990s in the case of the latter). It works by redirecting a portion of the engine exhaust gas back into the combustion chamber where it is mixed with the fresh fuel-air mixture. The EGR gases act as a diluent, thereby lowering the peak flame temperature and hence the rate of NO formation. Increasing the amount of gas recirculated reduces the rate of NO formation. However, it also reduces the combustion rate, making stable combustion more difficult to achieve [15].

In petrol engines the "internal" EGR concept is often implemented. This involves adjusting the exhaust valve timing so that some of the combusted gas is trapped in the cylinder [16]. This residual gas acts as a diluent for the next combustion cycle, hence lowering the combustion temperature. Internal EGR can only reduce NO_x slightly, because no more than 5–10% of the exhaust gas can be trapped without significantly affecting combustion.

In modern diesel engines the reduction in NO_x emissions required by legislation cannot normally be achieved using internal EGR alone. In this case an "external" EGR loop is the preferred option, whereby some of the exhaust gas is fed back into the cylinder by means of a pump, again reducing the combustion temperature. This EGR configuration has the advantage of higher EGR rates (up to 40–50%) and the possibility of adjusting the quantity of exhaust gas recirculated independently of the valve timing. It also permits the introduction of a heat exchanger between the outlet and inlet pipes (the so-called "intercooler"). This decreases the temperature of the recirculated gas – and hence the combustion temperature – even further. On the other hand, external EGR increases the complexity, size and cost of the system.

3.2.3 Three-Way Catalytic Converter

The TWC has been mandatory on all new petrol cars and vans sold in Europe since the introduction of the Euro 1 emission standard in 1992 and has proved to be a very robust technology for reducing emissions of NO_x, carbon monoxide (CO) and hydrocarbons (HC) (hence the name "three-way").

The TWC is a flow-through device consisting of a ceramic or metal substrate which is coated with an active catalytic layer of precious metals, such as platinum (Pt), palladium (Pd) and rhodium (Rh). The first two oxidise CO and HC, and Rh is used to reduce NO_x to nitrogen. High conversion efficiencies for both reduction and oxidation of pollutants in a TWC can be only achieved through stoichiometric combustion (i.e. maintaining the air-to-fuel ratio at the minimum necessary for complete combustion). A shortage of air (rich fuelling conditions) would make oxidation impossible, whilst air excess (lean fuelling conditions) would inhibit the reduction mechanism. Stoichiometry is maintained by means of closed-loop control, in which the oxygen concentration in the exhaust gas is measured using a so-called "lambda" sensor. The information from the sensor is fed to the engine control unit, and the fuel injection system is adjusted to correct the air-to-fuel ratio. The closed-loop system operates on the basis of fast correction algorithms which achieve oxygen adjustment in real time. Conversion efficiencies in excess of 90% can be achieved with a properly functioning closed-loop TWC.

3.2.4 Diesel Oxidation Catalyst

Whilst NO_x emissions from petrol vehicles can be controlled by catalytic reduction, this is not very effective under the oxygen-rich conditions of diesel combustion. A diesel oxidation catalyst (DOC) is similar to a TWC in terms of structure and configuration but is only capable of oxidation. As the exhaust gases pass through the catalyst CO, unburnt HC and volatile PM are oxidised. The conversion efficiency is a function of cell size, reactive surface, catalyst load and catalyst temperature, although emissions of CO and HC are typically reduced with an efficiency of more that 95%.

DOCs offer no NO_x-reduction capability but can lead to a conversion of NO to NO_2 in the tailpipe, thereby resulting in an increase in primary NO_2 emissions. The extent of the conversion depends on the catalyst specification and the exhaust gas temperature. Typical NO_2/NO_x ratios range from ~10% for diesel vehicles without oxidation after-treatment to more than 50% for DOC- or DPF-equipped vehicles [17].

3.2.5 Selective Catalytic Reduction

Selective catalytic reduction (SCR) is currently the main technology for enabling diesel vehicles to comply with the latest NO_x emission standards. SCR systems became standard in Euro V heavy-duty vehicles (launched in Oct. 2010), and their use is gradually being extended to light-duty vehicles. The method involves the introduction of ammonia (NH_3) into the exhaust stream to chemically reduce NO_x to nitrogen. Typically, an aqueous solution of urea ($CO(NH_2)_2$) is used as the reagent, with the ammonia being generated via thermolysis. The urea is fed into the system in defined doses upstream of the SCR catalyst. An oxidation catalyst downstream of the SCR catalyst may be used to eliminate the possibility of "ammonia slip". The following equations describe the different reactions in these systems:

$$CO(NH_2)_2 \rightarrow NH_3 + HNCO \text{ (thermolysis of urea)} \quad (5)$$

$$HNCO + H_2O \rightarrow NH_3 + CO_2 \text{ (hydrolysis)} \quad (6)$$

$$NO + NO_2 + 2NH_3 \rightarrow 2N_2 + 3H_2O \text{ (reduction)} \quad (7)$$

Typical SCR systems may achieve on-road NO_x conversion efficiencies of 60–70% [18]. However, the thermolysis of urea is an endothermic reaction that is favoured at high temperatures; SCR is inefficient at temperatures below around 200°C [19]. Hence, NO_x emissions from SCR-equipped vehicles can often increase during urban driving where traffic conditions result in low exhaust temperatures [20]. Supplementary systems, such as EGR, are required to maintain acceptable emission-control performance under such conditions.

3.2.6 Lean NO_x Trap

A lean NO_x trap (LNT) (or NO_x adsorber) is similar to a three-way catalyst. However, part of the catalyst contains some sorbent components which can store NO_x. Unlike catalysts, which involve continuous conversion, a trap stores NO and (primarily) NO_2 under lean exhaust conditions and releases and catalytically reduces them to nitrogen under rich conditions. The shift from lean to rich combustion, and vice versa, is achieved by a dedicated fuel control strategy. Typical sorbents include barium and rare earth metals (e.g. yttrium). An LNT does not require a separate reagent (urea) for NO_x reduction and hence has an advantage over SCR. However, the urea infrastructure has now developed in Europe and USA, and SCR has become the system of choice for diesel vehicles because of its easier control and better long-term performance compared with LNT. NO_x adsorbers have, however, found application in GDI engines where lower NO_x-reduction efficiencies are required, and the switch between the lean and rich modes for regeneration is easier to achieve.

3.2.7 Diesel Particulate Filter

Diesel particle filters are a very efficient means of reducing PM mass emissions from diesel vehicles, but do not directly target NO_x. However, most diesel particulate filter (DPF) systems contain catalytic materials which assist in the DPF "regeneration" (the combustion of PM accumulated on the filter to clean the DPF and prevent blockage). Such catalytic materials can have a similar impact to DOCs

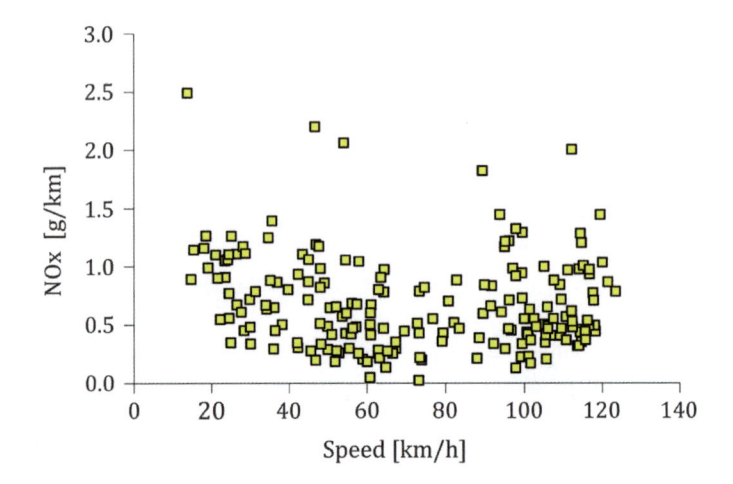

Fig. 4 NO$_x$ emission factors for a Euro 5 diesel passenger car [23]

(i.e. the conversion of NO to NO$_2$ as exhaust gas flows through the filter). In some systems the NO oxidation is intentional, as NO$_2$ has been shown to be an effective agent for oxidising PM [21]. Again, the conversion results in higher primary NO$_2$ emissions and may be one of the reasons for the observed increase in the NO$_2$/NO$_x$ ratio in ambient air at roadside locations in cities (e.g. [22]).

3.3 Factors Affecting On-Road NO$_x$ Emissions

The main technological factors which influence on-road exhaust emissions are the vehicle type (e.g. passenger car, heavy goods vehicle), the fuel type (e.g. petrol, diesel) and the vehicle technology. The latter usually refers to either a specific type of engine or exhaust after-treatment or, more generally, compliance with a particular emission standard. Other considerations include the age and condition of a vehicle's engine and exhaust after-treatment system. High emission rates are often a result of component ageing, component failure, or generally poor maintenance.

Important operational factors include vehicle weight, road gradient, vehicle load and the use of auxiliary equipment such as air conditioning, the thermal state of the engine and exhaust emission-control system, and the way in which a vehicle is driven (e.g. speed, or the so-called "dynamics" of driving).

Much attention has focussed on driving behaviour, as this has a large impact on the emission level. Whilst "calm" and steady-speed driving reduces pollutant emissions, transient operation and frequent speed changes have the opposite effect. Such differences in vehicle operation account for part of the (high) variability of real-world vehicle emissions. For example, NO$_x$ emissions from a single vehicle can vary by more than an order of magnitude, depending on the driver behaviour. Figure 4 shows a typical graph of emission factors (in g/km) for a Euro 5 diesel passenger car, plotted as a function of mean travelling speed. Road vehicle emission models in Europe, such as COPERT, HBEFA and VERSIT+, take driving dynamics into account to predict emission factors [24].

4 Reducing NO_x Emissions from Road Transport

Various technical and non-technical measures have been introduced by different authorities to reduce air pollution from road transport, and these can be grouped according to two general philosophies: "prevention" and "mitigation". Prevention measures are designed to reduce or eliminate emissions at the source, whereas mitigation measures are designed to remove pollutants which have already been emitted, to convert them to more benign compounds, or to modify their dispersion. As this chapter is concerned with emissions, the emphasis here is on prevention rather than mitigation. Mitigation measures typically involve the use of physical barriers, vegetation or some form of air treatment, and further information is available from the literature (e.g. [25–28]).

Some examples of prevention (i.e. emission-reduction) measures are described in the following paragraphs. It is worth noting that an important tool is air quality legislation, but this does not fall conveniently into the prevention or mitigation categories. Although air quality legislation acts as a driver for the development of pollution-reduction policies and technologies, and can thus be considered as "prevention", compliance with the legislation (and the need for further action) can only be determined from historical data. Moreover, air quality legislation does not specifically target road transport.

4.1 Vehicle and Engine Type Approval

The primary tool for combating air pollution from road transport is vehicle emission legislation. In the EU emission tests are required for the type approval of all new passenger car and light-duty vehicle models, and for the engines used in heavy-duty vehicles. Emission limits have been applied to vehicles and engines at the type approval stage since the early 1970s. The exhaust pollutants which are regulated are CO, HC, NO_x, PM, and, recently, particle number (PN). Whilst emissions of NO_x from vehicle exhaust are regulated at type approval, NO_2 emissions per se are not. The limits have been reduced in stages since they were first introduced (through progressive "Euro" standards), and changes have been made to the test methods to make them more realistic and effective. Emission-control technologies have developed accordingly.

For cars and light-duty vehicles the current and future test procedures and limit values have been consolidated in the Euro 5 and Euro 6 legislation [Regulation (EC) No. 692/2008]. In the exhaust emission test a production vehicle is placed on a power-absorbing chassis dynamometer. The driver must follow a driving cycle and the vehicle's emissions are collected and analysed. Emissions are measured over the New European Driving Cycle (NEDC), which is composed of low-speed "urban" segments and one high-speed "extra-urban" segment. The vehicle exhaust gases are diluted with filtered air to prevent condensation or reactions between the exhaust gas components. In addition to the regulated pollutants, carbon dioxide is measured to allow fuel consumption to be calculated using the carbon balance method. For diesel vehicles up to and including Euro 4, PM was collected separately from the other pollutants on a filter. For Euro 5 and Euro 6 vehicles, PM mass and PN are measured using a new procedure. The PN limit is designed to prevent the possibility of the PM mass limit being met using technologies that would enable a high number of ultrafine particles (<0.1 μm diameter) to be emitted. The emission limits are stated in grammes of pollutant (or number of particles) per kilometre. The NO_x and PM limits for cars are shown in Table 1, with the dates shown corresponding to new models. The EC Directives also specify a second date – usually 1 year later – which applies to first registration of existing, previously type-approved vehicle models.

The emission standards for heavy-duty vehicles apply to all vehicles with a maximum laden mass of more than 3,500 kg. The responsibility for compliance is borne by the engine manufacturer, and it is therefore the engine that is subject to type approval. The engine is operated on a test bed, with the exhaust emission limits being expressed in g/kWh (Table 2). The legislation is consolidated in the Euro V/VI standards [Regulation (EC) No. 595/2009]. In addition to introducing more stringent emission limits, the Euro V/VI regulation includes a concentration limit of 10 ppm for ammonia (NH_3), which can be emitted due to the use of additive-based control systems. A particle number limit is also planned in addition to the mass-based limit, and a maximum limit for the NO_2 component of NO_x emissions may also be defined.

The type approval of the engine rather than the complete vehicle introduces difficulties for current and future heavy-duty vehicles, as they are equipped with advanced after-treatment systems. In such cases the complete engine, after-treatment system and electronic control system have to be set-up in the laboratory for type approval to take place. To avoid this complication, type approval of the complete vehicle at the Euro VI level can be performed on the road using portable emission measurement systems (PEMS). Regulation 582/2011 specifies the technical details of the measurement.

The type approval regulations also lay down rules for in-service conformity, durability of pollution-control devices, on-board diagnostic (OBD) systems, measurement of fuel consumption, and accessibility of vehicle repair and maintenance information.

Emissions from in-use vehicles are controlled by legislation relating to periodic technical inspection (PTI) (Directives 2009/40/EC and 2010/48/EC). Here, compliance is the responsibility of the vehicle owner and, given the number of vehicles on the road, PTI tests are by necessity much simpler, shorter and cheaper than type approval tests. However, at present the PTI legislation does not cover NO_x or PM

Table 1 Type approval limits for NO$_x$ and PM from cars

		Limit value							
		Diesel				Petrol			
Stage	Date	HC + NO$_x$ (g/km)	NO$_x$ (g/km)	PM (g/km)	PN (#/km)	HC + NO$_x$ (g/km)	NO$_x$ (g/km)	PM (g/km)	PN (#/km)
Euro 1	1992.07	0.97	–	0.14	–	0.97	–	–	–
Euro 2 IDI	1996.01	0.7	–	0.08	–	0.50	–	–	–
Euro 2 DI	1996.01	0.9	–	0.10	–	–	–	–	–
Euro 3	2000.01	0.56	0.50	0.050	–	–	0.15	–	–
Euro 4	2005.01	0.30	0.25	0.025	–	–	0.08	–	–
Euro 5	2009.09	0.23	0.18	0.005^a	6 × 10^{11b}	–	0.06	0.005^{a,c}	–
Euro 6	2014.09	0.17	0.08	0.005^a	6 × 10^{11}	–	0.06	0.005^{a,c}	6 × 10^{11d}

^{a}0.0045 g/km using PMP measurement procedure
^{b}Added on 2011.09
^{c}Direct injection only
^{d}6 × 10^{12} per km within first 3 years from Euro 6 effective date

Table 2 Type approval limits for NO_x, PM and smoke from heavy-duty engines

Stage	Date	Limit value[a]		
		NO_x (g/kWh)	PM (g/kWh)	Smoke (m^{-1})
Euro I	1992	8.0	0.612[b]	–
Euro II	1996.10	7.0	0.25/0.15[c]	–
Euro III	1999.10[d]	2.0	0.02	0.15
	2000.10	5.0	0.10[e]	0.8
Euro IV	2005.10	3.5	0.02	0.5
Euro V	2008.10	2.0	0.02	0.5
Euro VI	2013.01	0.4	0.01	–

[a]European Stationary Cycle. Smoke is measured over European Load Response test
[b]0.36 g/km for engines >85 kW
[c]New limit introduced in October 1998
[d]For "enhanced environmentally friendly vehicles" (EEVs) only
[e]0.13 g/km for smaller low-speed engines

mass emissions, largely because of the difficulties associated with measuring these pollutants in a simple, low-cost test.

4.2 Emission-Control Technology

As a result of the type approval legislation manufacturers have been required to develop increasingly effective emission-control technologies, and these were described earlier in this chapter. Table 3 provides an overview of the different devices which are typically required for light-duty diesel vehicles in each Euro category. In modern vehicles various elements are used in combination, and these have different effects on the properties and composition of the exhaust gas.

The two major steps for diesel cars were the introduction of oxidation catalysts at the Euro 2 level and the effective mandatory introduction of DPFs at the Euro 5 level. For NO_x, EGR has been the main tool to control emissions up to Euro 5. SCR systems for light-duty vehicles are starting to appear for passenger cars at the Euro 6 level.

In the case of petrol light-duty vehicles the strict control of combustion, together with improvements in TWC efficiency, has proven sufficient for compliance with the emission limits up to Euro 6.

The technology for controlling emissions from heavy-duty vehicles has, until recently, focussed on in-cylinder measures such as direct injection (DI) and high-pressure injection (HPI, >150 MPa). However, at current and future emission levels SCR has become the NO_x emission-control technology of choice, whilst DPFs will effectively become mandatory at the Euro VI level (Table 4).

In addition to engine optimisation measures, Euro V NO_x control has been achieved by implementation of two alternative configurations – either the use of EGR or the use of SCR. Ligterink et al. [20] described the different performance of

The Evolution and Control of NO$_x$ Emissions from Road Transport in Europe

Table 3 Typical exhaust after-treatment for diesel light-duty vehicles

Emission standard		Emission control			
Stage	Date	EGR	DOC	DPF	SCR[a]
Euro 1 and earlier		–	–	–	–
Euro 2	1996	(✓)	✓	–	–
Euro 3	2000	✓	✓	(✓)	–
Euro 4	2005	✓	✓	(✓)	–
Euro 5	2009	✓	✓	✓	–
Euro 6	2014	✓	✓	✓	(✓)

Brackets correspond to application to some vehicles of the particular emission standard only
[a]SCR or NO$_x$ trap

Table 4 Typical exhaust after-treatment for diesel heavy-duty trucks

Emission standard		Emission control					
Stage	Date	DI[a]	HPI[b]	EGR[c]	DOC[c]	DPF	SCR
Pre-Euro I		–	–	–	–	–	
Euro I	1993	✓	–	–	–	–	
Euro II	1996	✓	✓	–	–	–	
Euro III	2000	✓	✓	–	(✓)	–	
Euro IV	2005	✓	✓	(✓)	(✓)	–	(✓)
Euro V	2009	✓	✓	(✓)	✓	–	(✓)
Euro VI	2014	✓	✓	✓	✓	✓	✓

Brackets correspond to application to some vehicles of the particular emission standard only
[a]Direct injection
[b]High-pressure injection
[c]EGR and DOCs were used in buses earlier than stated in the table, and in some cases as early as the Euro I stage

the two systems in terms of NO$_x$ control efficiency as a function of exhaust temperature. Figure 5 shows an example of the emission behaviour of vehicles equipped with the two systems. EGR performs much better at low speeds, whereas SCR becomes more efficient at high speeds as exhaust gas temperature increases. The actual performance of the two systems in real-world terms will depend on the operational patterns of vehicles, such as the frequency of use in urban or highway situations.

4.3 Fuel Quality and Alternative Fuels

Engine and vehicle technologies normally achieve their best emissions performance with high quality fuels. One property on which a great deal of attention has focussed is the sulphur content, partly because of the need to reduce PM and SO$_2$ emissions and partly because fuel sulphur has an adverse effect on certain types of engine and exhaust after-treatment technology. In Europe the controls on fuel

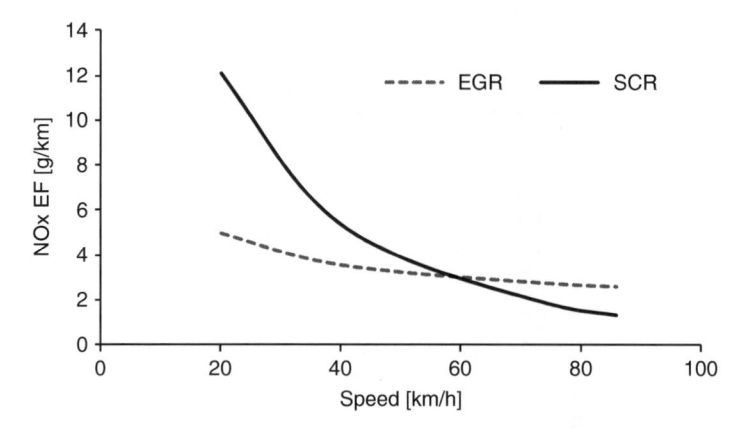

Fig. 5 NO_x emission factors for Euro V trucks equipped with SCR and EGR systems. The example refers to articulated trucks in the 34–40 t gross vehicle weight range. Source: COPERT [29], original data derived from Hausberger et al. [30]

sulphur content have been tightened in a stepwise fashion, and in the latest step Directive 2003/17/EC required a full transition to "sulphur-free" petrol and diesel (having less than 10 ppm of sulphur) by 2009. This should enable advanced technologies – such as lean-burn engines, particle traps and regenerative NO_x storage systems – to meet stringent exhaust emission limits.

In an effort to tackle climate change, biofuels such as fatty-acid methyl esters (FAME or biodiesel), paraffinic fuels from biomass-to-liquid procedures (BTL) or hydrotreated vegetable oil (HVO) and bioethanol have also been introduced into the energy mix, mostly in blends with conventional fuels. Current European regulations (Directive 2009/30/EC) permit up to 10% vol. and 7% vol. bioethanol and biodiesel mixing in fossil fuels, respectively. Higher bioethanol blends (up to 85%) require dedicated vehicle technology (so-called "flexi-fuel" vehicles) due to the need to adjust combustion parameters to the new oxygenated fuel and to use materials that can withstand the corrosive character of ethanol.

The impacts of biofuels on NO_x emissions are variable, and depend heavily on vehicle technology, operational conditions, blending ratio and biodiesel feedstock. On average, low biodiesel blends (up to 10%) do not affect NO_x emissions, or lead to a slight increase (less than 5%), but they do reduce PM emissions [31–33]. Effects at the individual vehicle level may be higher [34]. All these studies propose several possible explanations for the biodiesel impact on emissions, including the effects of higher density and viscosity, and lower compressibility and heating value, on fuel metering. Biodiesel also contains more oxygenated molecules than fossil diesel, which may also affect the combustion chemistry. It has been shown that HVO fuels can lead to reductions in both NO_x (~10%) and PM (~30%) due to their 100% paraffinic composition, and the absence of aromatics and other trace elements [35]. However, due to the unique character of these fuels (high cetane number, low density) engine recalibration may be needed to maximise their benefits [36].

The effects of bioethanol use on NO_x emissions are not consistent between different vehicles and studies. Larsen et al. [37] provided an overview of available studies, and concluded that bioethanol use can lead to either increases or decreases in NO_x emissions during tests, depending on the experimental conditions. This is consistent with the non-linear behaviour of the emission-control system in petrol vehicles, where the smallest deviations from stoichiometry greatly affect NO_x emissions. Leaner mixtures lead to an increase in NO_x emissions, and richer mixtures lead to a decrease in NO_x emissions. This erratic behaviour suggests that any direct fuel effects are masked by the ability of the fuelling system to maintain stoichiometry when changing from petrol to bioethanol blends in the different vehicles tested.

4.4 New Powertrain Technologies

Several new powertrain technologies are currently reaching niche markets in Europe. These include vehicles that use electricity for propulsion, either neat (such as battery-electric vehicles) or in combination with ICEs (such as hybrid or plug-in hybrid systems). These technologies are designed to reduce fuel consumption and greenhouse gas emissions, although they also offer significant benefits in terms of air pollutant reduction. For example, Fontaras et al. [38] showed that emissions of NO_x (and other pollutants) from two hybrid cars were distinctly lower than emissions from conventional vehicles complying with the same emission standards. The benefit comes as a result of the much more efficient use of the ICE. In the case of a battery-electric vehicle air pollutant emissions at the point of use are zero; pollutants are only emitted during electricity production, and depend on the energy mix used in each country. Assuming that power generation occurs away from urban areas, electric vehicles offer the potential for significant air quality improvements in city centres, but may also increase pollution over wider areas [39].

4.5 Eco-Driving

"Eco-driving" has been widely publicised as a means of reducing the fuel consumption and emissions of road vehicles. It is aimed at both private motorists and fleet operators, and typically involves either a simple set of rules to be followed or a programme of training. The advice or training varies considerably in terms of the level of detail, but it generally features a number of common actions, including keeping the tyres at the correct pressure, reducing the vehicle weight, avoiding sharp acceleration and heavy braking, driving in the highest gear, and avoiding unnecessary engine idling. Average overall reductions in fuel consumption of around 5–10% are typically reported for eco-driving. However, it should be noted that some adverse effects of eco-driving have been observed, such as an increase in NO_x emissions from diesel cars during urban driving [40].

4.6 Traffic Management

Numerous forms of traffic management offer the possibility of reducing emissions from road vehicles. Local restrictions to the access and/or circulation of traffic have been introduced in many cities. Such restrictions can take a range of forms, including tolls, congestion charges, alternate number plate schemes and weight restrictions. Cities which have implemented these types of scheme include London, Stockholm, Athens, Budapest and Prague. LEZs are one of the more effective types of restriction. Entry to a LEZ is usually conditional on a vehicle meeting specified standards. These standards can be set in various different ways, but are typically based on emission legislation, the use of specific exhaust after-treatment (e.g. DPF), or vehicle age.

The potential benefits of such measures can be illustrated by reference to a trial road charging scheme introduced in Stockholm city centre in 2006. It was estimated that the scheme resulted in a 15% reduction in total road use within the charging zone. Emissions of NO_x and PM_{10} from road traffic in the zone fell by 8.5% and 13%, respectively [41].

Various technologies and concepts involving the use of information and communication technology (ICT) and intelligent transport systems (ITS) are currently under development. Such systems include dynamic on-trip routing, "green" routing, adaptive traffic control with vehicle-to-infrastructure communication, and others. The main aims are to optimise traffic conditions and maximise mean travel speed, which are both likely to reduce fuel consumption and CO_2 emissions. However, emissions of regulated pollutants, including NO_x, may also be affected (e.g. [42]). The actual impacts of these technologies will depend on their real-world implementation and penetration.

5 Future Evolution of Emissions and Remaining Challenges

In order to protect health and the environment, vehicle exhaust emission standards will continue to be tightened in the EU, and increasing numbers of vehicles will be fitted with the latest exhaust emission-control technologies. By 2020 about one quarter of the total mileage in EU-27 is likely to be covered by cars and trucks certified to Euro 5/V, and more than half by vehicles certified to Euro 6/VI, according to scenarios examined in the LIFE + EC4MACS project (www.ec4macs.eu).

If the latest technologies are effective under real-world driving conditions, then a reduction in NO_x emissions of almost 60% over the period 2010–2020 is projected. However, there is evidence that Euro 5 diesel cars are not delivering the expected reductions in NO_x emissions during real-world driving (see discussion below). The future emission reduction therefore crucially depends on the performance and rate of introduction of Euro 6/VI technologies. So far, only a few tests on prototype Euro

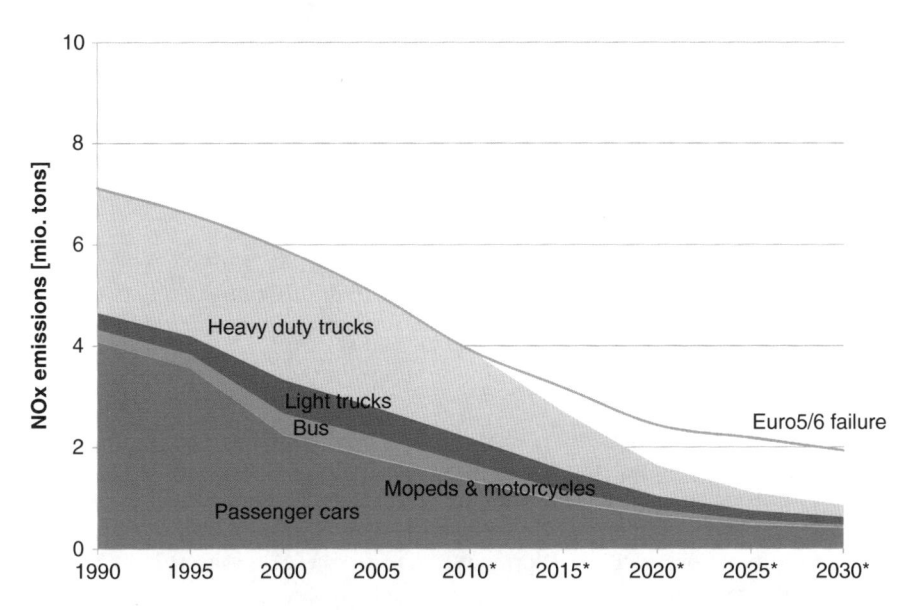

Fig. 6 Projection of NO_x emissions from road vehicles in Europe (EU-27). Together with a business-as-usual scenario, a trend scenario up to 2030 – in which Euro 5 and Euro 6 emission limits do not reduce car NO_x emissions during real-world driving below the Euro 4 level – has been included (Euro 5/6 failure line). Source: Amann et al. [44]

6 technologies equipped with SCR have been conducted [43]. The vehicles achieved high overall NO_x-reduction efficiencies, even over real-world driving cycles. However, these results were based on a non-representative sample of pre-production vehicles. If the Euro 6 emission controls for diesel passenger cars also prove to be ineffective (for example, due to poor SCR performance at low-speed conditions – as noted earlier for heavy-duty vehicles), then the actual overall reductions may not even reach 40%. Such a failure scenario is shown in Fig. 6.

The emphasis on real-world driving follows on from evidence suggesting that diesel vehicles, and in particular Euro 5 cars, have failed to deliver the expected NO_x reductions. Hausberger [45] showed that Euro 5 diesel cars may have similar emissions to Euro 3 cars. The same study showed that none of the Euro 2 to Euro 4 emission steps actually delivered the expected emission reductions. Hence, although the NO_x emission limit for a Euro 5 car is around 5 times lower than that for a Euro 1 car, Hausberger [45] showed that over real-world conditions the former may actually emit more than the latter. This unexpected behaviour is confirmed by PEMS tests performed by Kousoulidou et al. [23]. The discrepancy between type approval and real-world emissions arises from the selective optimisation of vehicle emissions over the type approval test. When vehicles operate outside the rather limited range of conditions at type approval, emissions can be largely uncontrolled. This has led to significant implications for the total NO_x emissions reported by European countries [46].

For these reasons, during the last 20 years NO_x emissions have decreased less than would have been expected given the systematic tightening of the emission limits. A direct implication of this is that NO_2 concentrations still frequently exceed health-based limits in many urban areas. According to the European Environment Agency, in 2004 more than 20% of the European urban population were exposed to ambient NO_2 concentrations above the annual mean limit value of 40 μg/m^3 [47]. Furthermore, NO_2 concentrations at many monitoring sites are not decreasing [15, 47, 48]. Although NO_2 is only a fraction of the total NO_x emitted from vehicles, analyses have indicated that a significant proportion of ambient NO_2 is actually primary exhaust from vehicles, and that the road traffic contribution to ambient NO_2 has increased in recent years [22, 49–52]. Two contributing factors have been cited:

1. Diesel vehicles emit more NO_x than petrol vehicles, and with a larger proportion of primary NO_2. In parallel, the market share of diesel vehicles has increased in many European countries. For example, the share of first registrations of diesel passenger cars in Finland increased from 17% in 2005 to 52% in 2008 [53].
2. The average proportion of primary NO_2 in diesel exhaust is increasing with changes in technology. This appears to be linked to the growth in the use of specific after-treatment technologies in modern diesel vehicles which involve in situ generation of NO_2, such as catalytically regenerative DPFs and DOCs [22, 45].

This increase in primary NO_2 emissions is compounded by atmospheric processes; background concentrations of ozone are also increasing [54], which increases the amount of atmospheric NO converted to NO_2.

It is clear that effectively addressing the NO_x emission problem involves not only more stringent emission limits over the current type approval procedure, but also a more effective regulatory policy as a whole. Measures that can be taken include:

- A revised type approval procedure that introduces a more realistic driving cycle covering a wider area of engine operation.
- The selection of random engine modes during type approval emission testing, and ensuring that emission limits are not exceeded (the "not-to-exceed" limit approach).
- The direct regulation of NO_2 emissions, independently of NO_x.
- Revised and advanced in-use compliance testing/inspection and maintenance schemes, involving OBD checks to control emissions over the lifetime of the vehicle.
- Extension of PEMS-based regulation in all vehicle categories, so as to effectively measure emissions in the field.

Several of these items are already being discussed within the UNECE working group on a Worldwide Harmonized Light vehicle Test Procedure.

Beyond the 2020–2030 horizon, reductions in NO_x will depend heavily on more effective regulations and actions, as well as the rate of introduction of low-carbon technologies such as electric and hydrogen vehicles. Vehicles operating on concepts other than ICEs have the potential for zero emissions at the point of use. However, the wide penetration of such technologies is expected to continue to be hindered by performance issues, such as limited operational range and drivability concerns. For these reasons, but also due to the relatively high cost, hybrid and electric vehicles are still considered a niche market despite the relatively wide range of models offered. Overcoming technological and infrastructural barriers remains a key future challenge for the wider penetration of such vehicles into the fleet.

References

1. US Environmental Protection Agency (2008) Integrated Science Assessment for Oxides of Nitrogen – Health Criteria (Final Report). EPA/600/R-08/071. http://cfpub.epa.gov/ncea/cfm/recordisplay.cfm?deid=194645
2. COMEAP (2009) Long-term exposure to air pollution: effect on mortality. A report by the Committee on the Medical Effects of Air Pollutants
3. Uherek E, Halenka T, Borken-Kleefeld J, Balkanski Y, Berntsen T, Borrego C, Gauss M, Hoor P, Juda-Rezler K, Lelieveld J, Melas D, Rypdal K, Schmid S (2010) Transport impacts on atmosphere and climate: land transport. Atmos Environ 44:4772–4816
4. Amann M et al (2007) Updated baseline projections for the revision of the Emission Ceilings Directive of the European Union. IIASA Report 2007
5. EEA (2010) European Union emission inventory report 1990–2008 under the UNECE Convention on Long-range Transboundary Air Pollution (LRTAP). EEA Technical Report No 7/2010. European Environment Agency, Copenhagen
6. Keuken M, Sanderson E, van Aalst R, Borken J, Schneider J (2005) Contribution of traffic to levels of ambient air pollution in Europe. In: Krzyzanowski M et al (eds) Health effects of transport-related air pollution. Regional Office for Europe of the World Health Organization, Copenhagen. ISBN 92 890 1373 7
7. Mantzos L, Capros P (2006) European Energy and Transport – Trends to 2030 - update 2005. Luxembourg: Office for Official Publications of the European Communities, 2006. ISBN 92-79-02305-5
8. Heywood JB (1988) Internal combustion engine fundamentals. McGraw-Hill, London
9. Vestreng V, Ntziachristos L, Semb A, Reis S, Isaksen ISA, Tarrasón L (2009) Evolution of NO_x emissions in Europe with focus on road transport control measures. Atmos Chem Phys 9:1503–1520
10. Zel'dovich YB (1946) The oxidation of nitrogen in combustion and explosions. Acta PhysioChemica URSS 21:577–628, Cited in AQEG (2004)
11. Baulch DL, Cobos CJ, Cox RA, Frank P, Hayman GD, Just T, Kerr JA, Murrells T, Pilling MJ, Troe J, Walker RW, Warnatz J (1994) Evaluated kinetic data for combustion modelling supplement 1. J Phys Chem Ref Data 23(6):847–1033, Cited in AQEG (2004)
12. Fenimore CP (1971) Formation of nitric oxide in premixed hydrocarbon flames. In: 13th symposium (international) in Combustion. The Combustion Institute, Pittsburgh, pp 373–380. Cited in AQEG (2004)
13. Iwamoto Y, Noma K, Nakayama O, Yamauchi T, Ando H (1997) Development of gasoline direction injection engine. SAE Paper 970541, doi:10.4271/970541
14. Fekete N, Kemmler R, Voigtlaender D, Krutzsch B, Zimmer E et al (1997) Evaluation of NO_x storage catalysts for lean burn gasoline fueled passenger cars. SAE Technical Paper 970746, doi:10.4271/970746
15. AQEG (2004) Nitrogen dioxide in the United Kingdom. Air Quality Expert Group. Published by the Department for Environment, Food and Rural Affairs. DEFRA publications, London

16. Hong H, Parvate-Patil GB, Gordon B (2004) Review and analysis of variable valve timing strategies—eight ways to approach. Proc Inst Mech Eng D J Automobile Eng 218(10):1179–1200
17. Kousoulidou M, Ntziachristos L, Mellios G, Samaras Z (2008) Road-transport emission projections to 2020 in European urban environments. Atmos Environ 42:7465–7475
18. Gabrielson P (2004) Urea-SCR in automotive applications. Top Catal 28(1–4):177–184
19. Sluder CS, Storey JME, Lewis SA, Lewis LA (2005) Low temperature urea decomposition and SCR performance. SAE Paper 2005-04-11
20. Ligterink N, de Lange, R, Vermeulen R, Dekker H (2009) On-road NO_x emissions of Euro-V trucks. TNO Report MON-RPT-033-DRS-2009-03840, Delft, Netherlands, p 19
21. Cooper BJ, Thoss JE (1989) Role of NO in diesel particulate emission control, SAE Technical Paper 890404, doi:10.4271/890404
22. Carslaw DC (2005) Evidence of an increasing NO_2/NO_x emissions ratio from road traffic emissions. Atmos Environ 39:4793–4802
23. Kousoulidou M, Ferrara F, Ntziachristos L, Gkeivanidis S, Franco V, Dilara P, Samaras Z (2010) Use of portable emissions measurement systems (PEMS) for validation and development of passenger car emission factors. 18th International Symposium Transport and Air Pollution. May 18–19, 2010, Duebendorf, Switzerland
24. Smit R, Ntziachristos L, Boulter P (2010) Validation of road vehicle and traffic emission model - a review and meta-analysis. Atmos Environ 44:2943–2953
25. Baldauf R, Thoma E, Khlystov A, Isakov V, Bowker G, Long SR (2008) Impacts of noise barriers on near-road air quality. Atmos Environ 42:7502–7507
26. Yang J, Yu Q, Gong P (2008) Quantifying air pollution removal by green roofs in Chicago. Atmos Environ 42:7266–7273
27. Litschke T, Kuttler W (2008) On the reduction of urban particle concentration by vegetation – a review. Meteorol Z 17(3):229–240
28. Laufs S, Burgeth G, Duttlinger W, Kurtenbach R, Maban M, Thomas C, Wiesen P, Kleffmann J (2010) Conversion of nitrogen oxides on commercial photocatalytic dispersion paints. Atmos Environ 44:2341–2349
29. COPERT (2012) Computer model to calculate emissions from road traffic. Available to download at www.emisia.com
30. Hausberger S, Rexeis M, Zallinger M, Luz R (2009) Emission Factors from the Model PHEM for the HBEFA Version 3. TUG Report Nr. I-20/2009 Haus-Em 33/08/679, Graz, Austria, p 76
31. Lapuerta M, Armas O, Rodríguez-Fernández J (2008) Effect of biodiesel fuels on diesel engine emissions. Prog Energy Combust Sci 34:198–223
32. Kousoulidou M, Fontaras G, Mellios G, Ntziachristos L (2008) Effect of biodiesel and bioethanol on exhaust emissions. ETC/ACC Technical Paper 2008/5, Thessaloniki. http://acm.eionet.europa.eu/docs/ETCACC_TP_2008_5_biofuels_emissions.pdf
33. Xue J, Grift TE, Hansen AC (2011) Effect of biodiesel on engine performances and emissions. Renewable Sustainable Energy Rev 15:1098–1116
34. Fontaras G, Kousoulidou M, Karavalakis G, Tzamkiozis Th, Pistikopoulos P, Ntziachristos L, Stournas S, Samaras Z (2010) Effects of low concentration biodiesel blend application on modern passenger cars. Part 1: feedstock impact on regulated pollutants, fuel consumption and particle emissions. Environ Pollut 158:1451–1460
35. Nylund N-O, Erkkila K, Ahtiainen M, Murtonen T, Saikkonen P, Amberla A, Aaatola H (2011) Optimized usage of NexBTL renewable diesel fuel. OPTIBIO. Espoo 2011. VTT Tiedotteita – Research Notes 2604. 167 p. + app. 5 p
36. Murtonen T, Aakko-Saksa P, Kuronen M, Mikkonen S, Lehtoranta K (2010) Emissions with heavy-duty diesel engines and vehicles using FAME, HVO and GTL fuels with and without DOC + POC aftertreatment. SAE Int J Fuels Lubr 2(2):147–166
37. Larsen U, Johansen T, Schramm J (2009) Ethanol as a fuel for road transportation. Main Report. IEA-AMF report, p 100, Copenhagen, Denmark. Available online at http://www.iea-amf.vtt.fi/pdf/annex35report_final.pdf

38. Fontaras G, Pistikopoulos P, Samaras Z (2008) Experimental evaluation of hybrid vehicle fuel economy and pollutant emissions over real-world simulation driving cycles. Atmos Environ 42:4023–4035
39. Ji S, Cherry CR, Mechle BJ, Wu Y, Marshall JD (2011) Electric vehicles in China: emissions and health impacts. Environ Sci Technol. doi:10.1021/es202347q
40. Vermeulen RJ (2006) The effects of a range of measures to reduce the tail pipe emissions and/or the fuel consumption of modern passenger cars on petrol and diesel. TNO report IS-RPT-033-DTS-2006-01695. TNO Science and Industry, Delft
41. Johansson C, Burman B, Forsberg B (2009) The effects of congestions tax on air quality and health. Atmos Environ 43:4843–4854
42. Klunder GA, Malone K, Mak J, Wilmink IR, Schirokoff A, Sihvola N, Holmén C, Berger A, de Lange R, Roeterdink W, Kosmatopoulos E (2009) Impact of information and communication technologies on energy efficiency in road transport – Final Report. TNO report for the European Commission, Delft, Netherlands, p 126
43. Hausberger S, Rexeis M, Luz R, Sturm, PJ (2011) Local and European scopes to reduce emissions from traffic. The challenge of air quality: a regional perspective conference. Brussels, Nov 10, 2011
44. Amann M et al (2011) An Updated Set of Scenarios of Cost-effective Emission Reductions to Improve Air Quality in Europe in 2020 - Background paper for the 49th Session of the Working Group on Strategies and Review Geneva, September 12–15, 2011
45. Hausberger S (2010) Fuel consumption and emissions of modern passenger cars. TUG Report I-25/10 Haus-Em 07/10/676, Graz, Austria, p 24
46. EEA (2011) NEC Directive status report 2010. Technical Report No 3/2011, European Environment Agency, Copenhagen
47. EEA (2007) Air pollution in Europe 1990–2004. EEA Report No 2/2007. European Environment Agency, Copenhagen
48. Harrison RM, Stedman J, Derwent D (2008) Why are PM_{10} concentrations in Europe not falling? New directions, atmospheric science perspectives special series. Atmos Environ 42:603–606
49. Jenkin ME (2004) Analysis of sources and partitioning of oxidant in the UK—Part 2: contributions of nitrogen dioxide emissions and background ozone at a kerbside location in London. Atmos Environ 38(30):5131–5138
50. Carslaw DC, Beevers SD (2004) Investigating the potential importance of primary NO_2 emissions in a street canyon. Atmos Environ 38:3585–3594
51. Hueglin C, Buchmann B, Weber RO (2006) Long-term observation of real-world road traffic emission factors on a motorway in Switzerland. Atmos Environ 40(20):3696–3709
52. Grice S, Stedman J, Kent A, Hobson M, Norris J, Abbott J, Cooke S (2009) Recent trends and projections of primary NO_2 emissions in Europe. Atmos Environ 43:2154–2167
53. Lappi M, Laurikko J, Erkkilä K, Pellikka A-P, Koskentalo T (2008). Specific NO and NO_2 emissions from a wide range of current and future LD and HD vehicles in urban driving conditions. In: Proceedings of 16th International Conference 'Transport and Air Pollution' 2008, Graz
54. Keuken M, Roemer M, van den Elshout S (2009) Trend analysis of urban NO_2 concentrations and the importance of direct NO_2 emissions versus ozone/NO_x equilibrium. Atmos Environ 43:4780–4783

Air Pollution by Ozone Across Europe

Richard G. Derwent and Anne-Gunn Hjellbrekke

Abstract Episodic peak ozone levels in rural areas have been declining during the last three decades due to regional pollution emission controls applied to the VOC and NO_x emissions from petrol-engined motor vehicles. Long-term downwards trends have been observed at many long-running rural monitoring stations in the EMEP ozone monitoring network. Downwards trends appear to be more pronounced at those stations where initial episodic peak levels were highest and insignificant at those stations where initial levels were lowest. This behaviour has been interpreted as resulting from the combined effect of regional pollution controls and increasing hemispheric ozone levels. Hemispheric ozone levels have been steadily rising in the northern hemisphere because of growing man-made emissions of tropospheric ozone precursors. Episodic ozone levels in major European towns and cities are rising towards the levels found in the rural areas surrounding them, as exhaust gas catalysts fitted to petrol-engined motor vehicles reduce the scavenging of ozone by chemical reaction with emitted nitric oxide.

Keywords Hemispheric background, Long-term trends, NO_x and VOCs, Ozone, Photochemical ozone formation, Regional pollution controls

Contents

1   Introduction to Ozone Air Quality Across Europe 56
2   Spatial Distribution of Episodic Peak Ozone Levels Across Europe 58
3   Trends in Episodic Peak Ozone Levels Across Europe over the 1980–2009 Period 63

R.G. Derwent (✉)
rdscientific, Newbury, Berkshire, UK
e-mail: r.derwent@btopenworld.com

A.-G. Hjellbrekke
Norwegian Institute for Air Research, Kjeller, Norway
e-mail: anne-Gunn.hjellbrekke@nilu.no

4 Why Are the Downwards Trends Largest for the Stations with the Highest Episodic Peak Levels? 67
5 Discussion and Conclusions 70
References 72

1 Introduction to Ozone Air Quality Across Europe

The first reported observations of elevated ozone levels in a European population centre date back over 40 years to July 1972 in London when hourly ozone levels in excess of 110 ppb were recorded [1]. Shortly after this, elevated ozone levels were widely reported throughout northwest Europe during summertime anticyclonic conditions [2]. It soon emerged that a characteristic of these episodes was a rather uniform spatial distribution of elevated ozone levels over a large area of northwest Europe. Observations of elevated ozone levels at Adrigole, Ireland on the western seaboard of Europe confirmed the importance of the long-range transboundary formation and transport of elevated ozone levels [3].

Over the several decades since these early observations, understanding has grown enormously as a result of the establishment of coordinated ozone monitoring networks, field campaigns and atmospheric chemistry studies. The elevated ozone levels have been shown to be caused by sunlight-driven photochemical ozone formation reactions involving oxides of nitrogen (NO_x) and organic compounds (VOCs) emitted largely from man-made sources. Observations have demonstrated that long-range transboundary formation and transport is an important aspect of these regional-scale episodes. The expansion of air quality monitoring networks into central and Eastern Europe and into the Mediterranean basin has shown that elevated ozone levels are a feature of much of the European continent in most summers [4].

In addition to the occurrence of elevated ozone levels during the summer months, there are a number of common features that almost all ozone monitoring datasets exhibit. Ozone is invariably present on every day of the year at each monitoring station. That is to say, there is a clearly observable "background" level and levels below this "background" level are relatively rarely observed. This "background" level is generally found to lie between 20 and 40 ppb across much of Europe. It arises from transport from the stratosphere and photochemical ozone formation throughout the troposphere [5]. Figure 1 illustrates the continental source attribution of the ozone observed during each day of 2008 at the rural European Monitoring and Evaluation Programme (EMEP) monitoring station GB0049R at Weybourne on the north Norfolk coast of eastern England. There are substantial contributions to the ozone from intercontinental transport from North America and Asia and from transport from the stratosphere. These contributions demonstrate the importance of hemispheric scale ozone formation and transport. The evidence is that "background" levels of ozone at the surface have been rising in air masses advected across the North Atlantic Ocean and arriving at the remote EMEP monitoring station IE 0031R at Mace Head on the west coast of Ireland [6] over the last few decades.

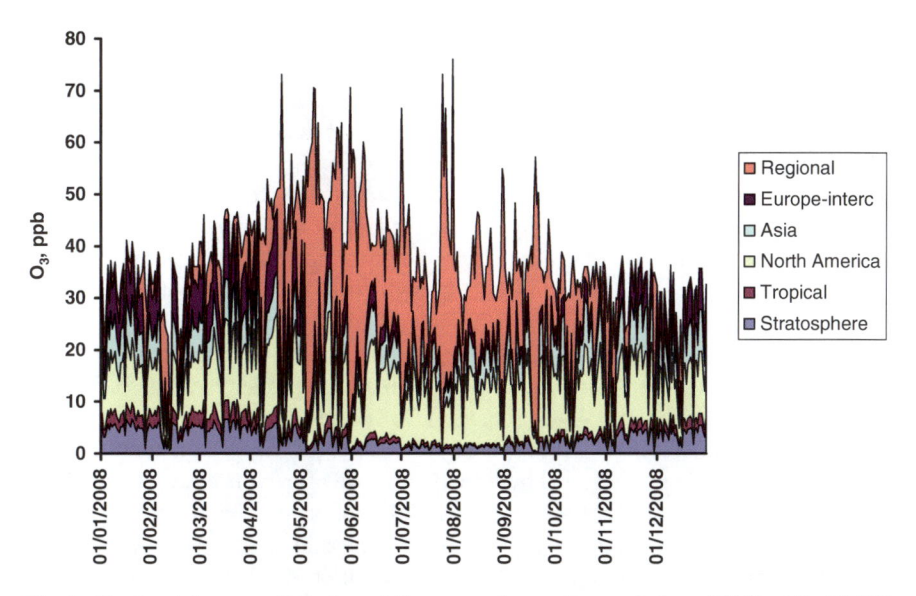

Fig. 1 Continental source attribution of the ozone observed on each day of 2008 at the EMEP rural monitoring station GB0049R at Weybourne on the North Norfolk coast of eastern England using ozone labelling techniques in a global and a regional photochemical model. Key: regional refers to the ozone advected directly over the local- and regional-scales to the location; North America to that formed over that continent and over the North Atlantic and east Pacific; Asia to that formed over that continent and over the western Pacific; Europe-interc to that advected intercontinentally around latitude circles and back into Europe; Tropical to that from the southern hemisphere and tropics

A further feature of the ozone monitoring network observations has been that ozone levels are always higher outside of urban areas than inside [7]. This occurrence has been explained by the presence of motor vehicular sources of $NO_x = NO + NO_2$ in towns and cities. Nitric oxide NO reacts rapidly with ozone to produce NO_2:

$$NO + O_3 = NO_2 + O_2,$$

leaving ozone levels in urban areas severely depleted relative to those in the rural areas surrounding them.

Policy-makers are concerned about the public health impacts associated with elevated ozone levels because ozone has long been recognised as an aggressive and irritating air pollutant [8]. The World Health Organization (WHO) has set a guideline level of 50 ppb as a daily maximum 8-h average mean as a target for air quality policy actions [9]. Policy-makers have recognised that such a guideline is regularly exceeded across much of Europe in most years, and they have taken steps with the goal of ultimately bringing ozone level below internationally accepted guidelines. In this chapter, the annual maximum recorded 8-h average ozone level is the chosen metric as recommended by WHO for the assessment and evaluation of ozone monitoring data.

It was accepted at an early stage that the problems associated with the occurrence of elevated ozone levels require concerted international action [10]. This is because elevated levels are widespread across Europe and impact on almost all European countries. Furthermore, efficient long-range transboundary formation and transport means that countries benefit not only from the results of their own policy actions but also those of their neighbouring countries. Policy actions to reduce episodic peak ozone levels were first coordinated through the United Nations Economic Commission for Europe (UNECE) and its international Convention on Long-Range Transboundary Air Pollution (CLRTAP). Initially, the CLRTAP focused on man-made NO_x emissions through its NO_x Protocol [11]. Then the focus shifted to VOC emissions through the VOC Protocol [12]. The agreed reductions in man-made NO_x and VOC emissions were delivered by a series of mandatory motor vehicle emissions Directives promulgated by the Commission of the European Communities [13]. Subsequently, the CLRTAP agreed its multi-pollutant multi-effect Gothenburg Protocol with the aim of reducing simultaneously acid rain, ground-level ozone formation and eutrophication [14]. The Commission of the European Communities formulated its National Emissions Ceilings [15] and Solvent Emissions [16] Directives to control emissions of NO_x and VOCs from stationary sources. Subsequently, the European Union (EU) has brought all of its air quality policy formulation activities together under its Clean Air for Europe Thematic Strategy [17].

In this chapter, the impact of these policy actions on elevated episodic peak ozone levels is assessed over the 1980–2009 period. The annual maximum 8-h average ozone level is the chosen air quality metric, and the focus is mainly on rural and remote monitoring stations as recorded in the EMEP ozone database.

2 Spatial Distribution of Episodic Peak Ozone Levels Across Europe

Over recent years, several annual reviews of European ozone data have been compiled, and these give an authoritative assessment of the state of European ozone air quality. The Chemical Coordination Centre of EMEP publishes an annual digest of the rural ozone monitoring data [18], based on the EMEP ozone monitoring network. The European Topic Centre on Air and Climate Change of the European Environment Agency also publishes summaries [19] of the AIRBASE database which holds ozone monitoring data for a wide range of urban and rural ozone monitoring stations. There is a degree of overlap between the EMEP and AIRBASE ozone databases in that they both contain largely the same reported rural ozone monitoring data.

In this chapter, focus is given to the EMEP ozone monitoring network which began collecting rural ozone monitoring data in the 1980s as a continuation of the OXIDATE project [20] of the Organization for Economic Cooperation and Development (OECD). The vast majority of the rural ozone observations have been made using the UV absorption method. Details of the calibration and maintenance procedures, siting criteria and site locations are given elsewhere [18, 21]. Data capture is generally good, above 90%, but missing data can be critical in air quality assessment studies. Station data have therefore not been analysed for those years where data capture was below 75%. Ozone levels have been characterised here using, as a metric, the maximum 8-h average ozone mean concentration recorded during each year. Non-overlapping 8-h running mean concentrations have been calculated for each hour of the year for each monitoring station. If there are more than two missing hours during any 8-h period, then that 8-h mean was rejected. On this basis, the EMEP annual 8-h ozone maximum database contains 2,291 station-year entries covering 178 separate stations over the 1980–2009 period. All of the stations are classified as rural or remote stations and as such have minimal influence from local sources of pollution. This is a particularly important characteristic of a station's location because of the dramatic influence that local motor vehicular traffic sources of NO_x can exert on ozone levels. All observations reported and assessed here are as mixing ratios (or mole fractions) and are described in units of ppb where 1 ppb represents 1 part of ozone in 10^9 parts dry air.

The EMEP annual 8-h ozone maximum database contains 133 entries covering the years 1980–1989, 904 over 1990–1999 and 1,254 for the 2000s. The highest recorded maximum 8-h average ozone mixing ratio was 178 ppb during 1981 at the GB0036R station Sibton on the North Sea coast of Eastern England. The lowest was 41 ppb during 1993 at the ES0003R station Tortosa on the Mediterranean coast of north East Spain and during 1998 at the PT0004R station Monte Velho on the North Atlantic coast of Portugal. Out of all the 2,291 entries in the EMEP annual 8-h ozone maximum database, all but 9, that is 99.6% exceeded the WHO air quality guideline for ozone which has been set at 50 ppb. 1,328 entries, 58% were in excess of 75 ppb, 0.5% were in excess of 100 ppb and 0.4% in excess of 125 ppb. The highest entry post-2000 was 130 ppb recorded at the GB0045R station Wicken Fen in eastern England during 2006.

A feature of the EMEP annual 8-h maximum ozone database is pronounced year-by-year variability. Some years were characterised by relatively high maximum 8-h ozone levels at a large number of monitoring stations. In particular the years 1990, 2003 and 2006 fit into this category. Equally well, the years 1988, 2008 and 2009 showed relatively low levels at an equally large number of stations. This year-by-year variability may well have many explanations because the number and spatial distribution of available stations is changing every year but meteorological variability is clearly an important factor. In any review and assessment of the maximum 8-h average ozone data, year-by-year variability must be carefully taken into account. To plot and understand the spatial distribution of episodic peak ozone levels across Europe, we have therefore constructed the 2000–2009 decadal average of the annual maximum 8-h average ozone levels for only those monitoring stations with complete records for the 10-year period. Table 1 presents these decadal averages for the available 78 monitoring stations across Europe.

Table 1 Decadal 2000–2009 average annual 8-h maximum ozone levels for 78 EMEP monitoring stations, together with their station codes, latitudes and longitudes

Station code	Latitude (deg min sec)	Longitude (deg min sec)	Decadal average 8-h ozone maximum, ppb
DE0003R	47 54 53 N	07 54 31 E	95.3
IT0001R	42 06 00 N	12 38 00 E	92.0
CH0005R	47 04 03 N	08 27 50 E	91.9
BE0032R	51 27 27 N	06 00 10 E	90.9
CH0004R	47 02 59 N	06 58 46 E	89.0
SI0032R	46 17 58 N	14 32 19 E	88.9
DE0008R	50 39 00 N	10 46 00 E	88.8
BE0035R	50 30 12 N	04 59 22 E	88.2
AT0032R	47 31 45 N	09 55 36 E	86.2
CH0003R	47 28 47 N	08 54 17 E	85.9
AT0041R	47 58 23 N	13 00 58 E	85.7
AT0005R	46 40 40 N	12 58 20 E	85.1
FR0009R	49 54 00 N	04 38 00 E	85.0
AT0046R	48 20 05 N	16 43 50 E	84.7
DE0002R	52 48 08 N	10 45 34 E	84.5
AT0038R	46 41 37 N	13 54 54 E	84.3
AT0002R	47 46 00 N	16 46 00 E	84.2
AT0042R	48 52 43 N	15 02 48 E	84.0
AT0047R	48 03 03 N	16 40 36 E	83.9
BE0001R	49 52 40 N	05 12 13 E	83.8
AT0030R	48 43 16 N	15 56 32 E	83.7
SI0033R	46 07 43 N	15 06 50 E	83.2
PL0003R	50 44 00 N	15 44 00 E	83.1
GB0038R	50 47 34 N	00 10 46 E	83.1
SI0008R	45 34 00 N	14 52 00 E	83.1
CH0002R	46 48 47 N	06 56 41 E	82.7
AT0043R	48 06 22 N	15 55 10 E	82.1
AT0044R	47 06 47 N	15 28 14 E	82.1
DE0007R	53 10 00 N	13 02 00 E	82.0
NL0010R	51 32 28 N	05 51 13 E	81.9
ES0010R	42 19 10 N	03 19 01 E	81.6
AT0040R	47 20 53 N	15 52 56 E	81.2
CZ0003R	49 35 00 N	15 05 00 E	80.3
AT0034G	47 03 16 N	12 57 30 E	79.8
AT0037R	47 08 13 N	11 52 12 E	79.5
DE0001R	54 55 32 N	08 18 35 E	78.1
GB0036R	51 34 23 N	01 19 00 W	77.7
SK0007R	47 57 36 N	17 51 38 E	77.7
ES0009R	41 16 52 N	03 08 34 W	77.0
CZ0001R	49 44 00 N	16 03 00 E	76.5
GB0013R	50 35 47 N	03 42 47 W	76.1
GB0031R	52 30 14 N	03 01 59 W	75.4
SI0031R	46 25 43 N	15 00 12 E	75.4
FR0010R	47 16 00 N	04 05 00 E	75.3

(continued)

Table 1 (continued)

Station code	Latitude (deg min sec)	Longitude (deg min sec)	Decadal average 8-h ozone maximum, ppb
FR0014R	47 18 00 N	06 50 00 E	75.3
SK0006R	49 03 00 N	22 16 00 E	75.0
SE0011R	56 01 00 N	13 09 00 E	74.5
PL0002R	51 49 00 N	21 59 00 E	74.5
ES0012R	39 05 10 N	01 06 07 W	74.4
ES0007R	37 14 00 N	03 32 00 W	74.3
DE0009R	54 26 00 N	12 44 00 E	74.2
ES0011R	38 28 33 N	06 55 22 W	74.1
SK0004R	49 09 00 N	20 17 00 E	73.1
EE0009R	59 30 00 N	25 54 00 E	73.0
PL0004R	54 45 00 N	17 32 00 E	73.0
SE0032R	57 49 00 N	15 34 00 E	72.8
PL0005R	54 09 00 N	22 04 00 E	72.3
NL0009R	53 20 02 N	06 16 38 E	71.1
DK0041R	55 41 13 N	12 07 34 E	71.1
FR0013R	43 37 00 N	00 11 00 E	71.0
SE0012R	58 48 00 N	17 23 00 E	70.6
NO0043R	59 00 00 N	11 32 00 E	68.8
FI0017R	60 31 36 N	27 41 10 E	68.2
NO0052R	59 12 00 N	05 12 00 E	67.7
FI0037R	62 35 00 N	24 11 00 E	66.9
NO0039R	62 47 00 N	08 53 00 E	66.8
NO0015R	65 50 00 N	13 55 00 E	66.6
FI0022R	66 19 13 N	29 24 06 E	66.6
ES0008R	43 26 32 N	04 51 01 W	66.0
SE0013R	67 53 00 N	21 04 00 E	65.9
NO0055R	69 28 00 N	25 13 00 E	64.2
NO0001R	58 23 00 N	08 15 00 E	63.9
IE0031R	53 10 00 N	09 30 00 W	63.9
SE0035R	64 15 00 N	19 46 00 E	63.8
GB0002R	55 18 47 N	03 12 15 W	63.1
CH0001G	46 32 51 N	07 59 06 E	62.9
GB0033R	55 51 31 N	03 12 18 W	61.8
NO0042G	78 54 00 N	11 53 00 E	55.0

Decadal 2000–2009 average annual maximum 8-h average levels peak at 95.3 ppb for the DE0003R station Schauinsland which is an elevated station in the Black Forest region of Germany. In addition to DE0003R, there are three further stations with decadal averages over 90 ppb. These stations are IT0001R Montelibretti in central Italy, CH0005R Rigi, an elevated station in Switzerland and BE0032R Eupen in Belgium. There are an additional four stations with decadal averages just below 90 ppb and these comprise CH0004R Chaumont, an elevated station in Switzerland, SI0032R Kovk in Slovenia, DE0008R Schmucke, an elevated station in central

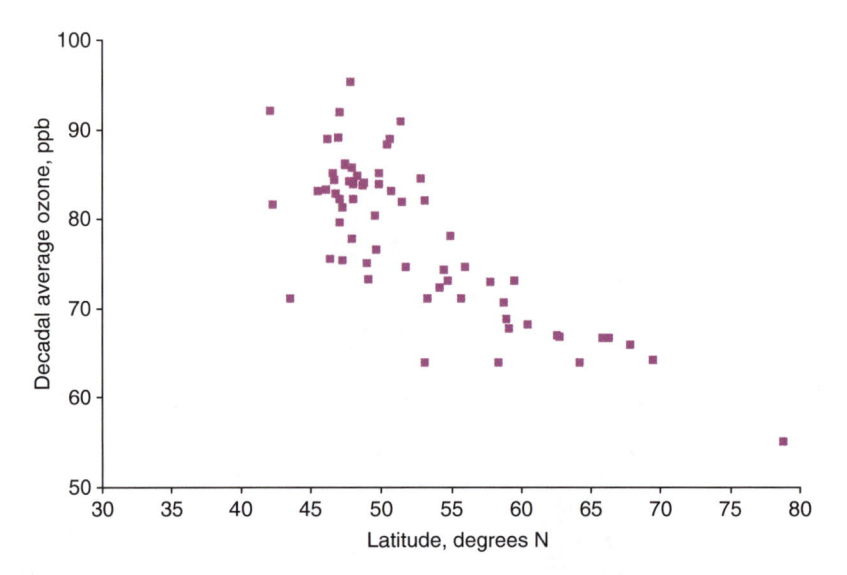

Fig. 2 Decadal average maximum 8-h mean ozone levels along a north–south transect through central Europe

Germany and BE0035R Vezin in Belgium. Together these eight stations provide a picture of an "ozone maximum region" encompassing Belgium, Germany, Switzerland, Italy and Slovenia in the heart of Europe. Surrounding this "ozone maximum region", there are further 25 stations with decadal averages in the range from 80.3 to 86.2 ppb. These stations are located additionally in Austria, France, Poland, United Kingdom, Netherlands, Spain and the Czech Republic. They, together, provide a ring of countries surrounding the "ozone maximum region".

Decadal averages are lowest, 55 ppb, at the NO0042G station Zeppelinfjell on Spitzbergen in the Arctic. Even at this most remote European station, decadal average maximum 8-h average ozone levels exceed the 50 ppb WHO air quality guideline for ozone. There are 16 stations with decadal averages in the range 61.8–68.8 ppb. These include two stations in the United Kingdom: GB0033R Bush and GB0002R Eskdalemuir, both in Scotland; the station CH0001G Jungfraujoch in Switzerland which at 3,578 m is the highest altitude station in the EMEP monitoring network; two Swedish stations SE0035R Vindeln and ES0013R Esrange, both in northern Sweden; the Mace Head, Ireland station IE0031R on the North Atlantic Ocean coastline; six stations in Norway, one station in Spain and three stations in Finland.

Figure 2 constructs a north–south transect through the decadal averages in Table 1 by plotting them out against station latitude. The plot uses only those stations that are east of the Greenwich meridian and also excludes the high-altitude alpine station CH0001G. The points appear to fall within a curved area with a tendency towards higher decadal averages towards the south of Europe and lower

Air Pollution by Ozone Across Europe 63

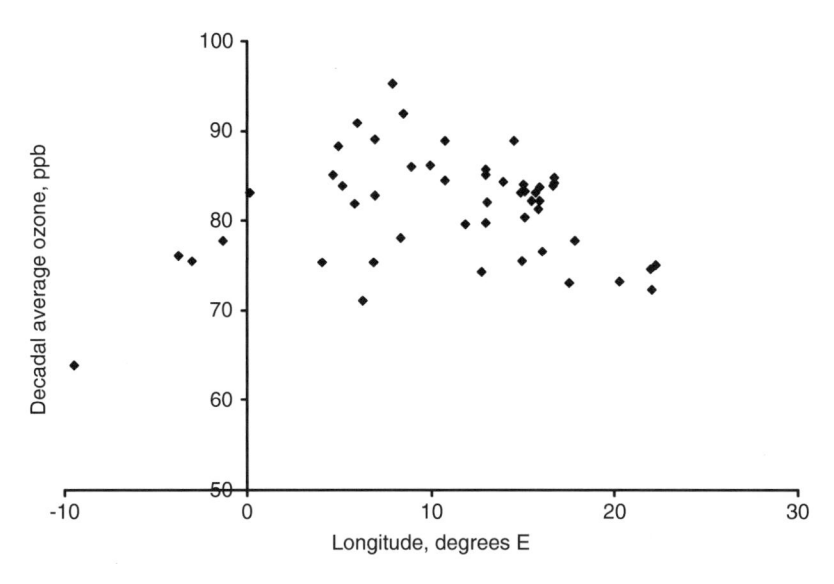

Fig. 3 Decadal average maximum 8-h mean ozone levels along a west–east transect through central Europe

values to the north in Scandinavia. These tendencies are consistent with the concept of an "ozone maximum region" developed above.

Figure 3 constructs a west–east transect through Europe using the stations with latitudes in the range 45°–55° N and also excluding the CH0001G station. The points appear to fall within a curved area with a maximum in the 5°–10° E range, falling off to lower values on either side of the maximum towards the fringes of Europe. Again, this is consistent with the concept of an "ozone maximum region".

3 Trends in Episodic Peak Ozone Levels Across Europe over the 1980–2009 Period

A clear-cut feature of the EMEP 8-h maximum ozone database is the presence of apparent trends. There is a tendency for the highest recorded values to be found in the early years and for the lowest values in the latest years. The aim of this section is to quantify any trends and to offer explanations as to their origins. Trend analysis is not straightforward in the case of the EMEP 8-h maximum ozone database because of a number of difficulties. The first difficulty is the marked year-on-year variability found in the time series data for the individual stations. The second difficulty arises from the presence of gaps in the datasets which result from the strict criteria that have been applied in this analysis based on data capture. Finally, the number of stations operating in a given year has changed markedly through the 1980–2009 period as interest in ground-level ozone has developed. The approach adopted has

Table 2 Trend analysis for the five long-running EMEP monitoring stations over the 1980–2009 period, showing the Sen's slope estimates, initial 1980 levels and significance levels

Station	Slope, ppb per year	Initial 1980 level, ppb	Significance level
GB0039R	-1.4 ± 0.9	118	$\alpha = 0.01$
GB0036R	-1.2 ± 0.7	109	$\alpha = 0.01$
DE0003R	-0.0	95	$\alpha > 0.1$
DE0002R	-0.2	89	$\alpha > 0.1$
DE0001R	-0.1	81	$\alpha > 0.1$

Notes: the confidence limits on the Sen's slope estimates are shown as 2-σ intervals

therefore been to split the 1980–2009 period into sub-periods with consistent and homogeneous sets of stations in each.

Table 2 summarises the available trend data for the complete 1980–2009 period at the five longest-running stations in the EMEP 8-h maximum ozone database. Application of the Mann–Kendall test [22] confirmed that all five stations exhibited downwards trends. The two United Kingdom stations: GB0039R Sibton and GB0036R Harwell, the latter in the south east of England, showed highly statistically significant downwards trends at the $\alpha = 0.01$ level of significance. The Sen's slope estimates [22] were -1.4 ± 0.9 ppb per year and -1.2 ± 0.7 ppb per year, respectively. The downwards trends at the three German stations: DE0003R, DE0002R and DE0001R were not highly significant and showed a significance level which was greater than 0.1. It appears that the downwards trends were greater for the stations with the higher initial annual 8-h maximum ozone levels extrapolated back to 1980. For the stations with the lower annual 8-h maximum ozone levels, no highly statistically significant trends were apparent.

Table 3 summarises the trends found for the 1990–2009 period ranked in descending order of their annual values extrapolated back to 1990 using the trend analysis. Of the 40 stations analysed using the Mann–Kendall test [22], 16 stations showed highly statistically significant ($\alpha \leq 0.1$) downwards trends, 19 stations showed downwards trends that were not high statistically significant ($\alpha > 0.1$) and 5 showed no or slightly positive trends that, again, were not highly statistically significant ($\alpha > 0.1$). The Sen's slope estimates [22] revealed that the highest downwards trend of -2.3 ± 1.1 ppb per year was exhibited by the IT0004R Ispra station, the station with the highest initial 1990 value. Eleven stations showed highly statistically significant ($\alpha \leq 0.1$) downwards trends that exceeded -1 ppb per year, in addition to the Montelibretti station: GB0013R Yarner Wood, GB0038R Lullington Heath, NL0010R Vredepeel, GB0036R Harwell, GB0031R Aston Hill, AT0002R Vorhegg, SE0011R Vavihill, GB0002R Eskdalemuir, NO0001R Birkenes and GB0006R Lough Navar. There was a clear tendency for the stations with the largest annual values extrapolated back to 1990 to show the largest downwards trends.

Table 4 summarises the trends found for the somewhat shorter period from 1993 to 2009, ranked in order of their annual values extrapolated back to 1990 using the trend analyses. Of the 29 stations analysed, the Mann–Kendall test [22] showed that 7 stations exhibited highly statistically significant ($\alpha \leq 0.1$) downwards trends,

Table 3 Trend analysis for 40 EMEP monitoring stations over the 1990–2009 period ranked in order of their 1990 extrapolated level

Station	Slope, ppb per year	1990 level, ppb	Significance level
IT0004R	-2.3 ± 1.1	118 ± 8	$\alpha = 0.001$
GB0013R	-2.0 ± 1.4	104 ± 14	$\alpha = 0.01$
GB0038R	-1.7 ± 1.1	104 ± 13	$\alpha = 0.01$
NL0010R	-1.6 ± 1.1	104 ± 11	$\alpha = 0.05$
AT0032R	-0.7 ± 0.6	103 ± 9	$\alpha = 0.05$
GB0036R	-1.6 ± 1.2	100 ± 10	$\alpha = 0.05$
AT0045R	-0.9 ± 0.8	100 ± 7	$\alpha = 0.1$
DE0003R	-0.5	98	$\alpha > 0.1$
DE0002R	-0.7	94	$\alpha > 0.1$
AT0041R	-0.5	93	$\alpha > 0.1$
GB0031R	-1.3 ± 1.0	93 ± 10	$\alpha = 0.1$
AT0002R	-1.0 ± 0.7	93 ± 5	$\alpha = 0.05$
AT0043R	-1.0	92	$\alpha > 0.1$
AT0046R	-0.5	92	$\alpha > 0.1$
AT0047R	-0.8 ± 0.7	91 ± 7	$\alpha = 0.1$
SE0011R	-1.0 ± 0.8	86 ± 9	$\alpha = 0.05$
CH0003R	-0.1	86	$\alpha > 0.1$
AT0042R	-0.3	86	$\alpha > 0.1$
CH0002R	-0.3	85	$\alpha > 0.1$
GB0002R	-1.4 ± 0.9	85 ± 12	$\alpha = 0.05$
DE0001R	-0.4	84	$\alpha > 0.1$
NL0009R	-0.9 ± 0.7	83 ± 6	$\alpha = 0.01$
AT0034G	-0.2	81	$\alpha > 0.1$
SE0032R	-0.6	79	$\alpha > 0.1$
AT0037R	-0.0	78	$\alpha > 0.1$
NO0001R	-1.0 ± 0.6	77 ± 7	$\alpha = 0.05$
GB0006R	-1.2 ± 0.8	75 ± 8	$\alpha = 0.05$
GB0033R	-0.9 ± 0.8	73 ± 9	$\alpha = 0.1$
PT0004R	0.5	72	$\alpha > 0.1$
IE0031R	-0.5	70	$\alpha > 0.1$
GB0015R	-0.4	70	$\alpha > 0.1$
NO0043R	-0.2	69	$\alpha > 0.1$
SE0012R	-0.1	69	$\alpha > 0.1$
FI0022R	-0.1	68	$\alpha > 0.1$
FI0009R	-0.2	68	$\alpha > 0.1$
FI0017R	0.3	65	$\alpha > 0.1$
NO0039R	0.0	65	$\alpha > 0.1$
SE0035R	-0.0	62	$\alpha > 0.1$
NO0015R	0.5	56	$\alpha > 0.1$
NO0042G	0.1	51	$\alpha > 0.1$

Table 4 Trend analysis for 29 EMEP monitoring stations over the 1993–2009 period ranked in order of their extrapolated 1990 levels

Station	Slope, ppb per year	1990 level, ppb	Significance level
GB0039R	−1.6 ± 1.4	101 ± 15	$\alpha = 0.1$
SI0032R	−0.7	96	$\alpha > 0.1$
DE0008R	−0.7 ± 0.7	95 ± 8	$\alpha = 0.05$
CH0005R	−0.2	94	$\alpha > 0.1$
CZ0001R	−1.3 ± 1.3	92 ± 11	$\alpha = 0.05$
BE0001R	−0.9	93	$\alpha > 0.1$
BE0032R	−0.5	91	$\alpha > 0.1$
GB0014R	−1.7 ± 1.5	91 ± 7	$\alpha = 0.01$
DE0007R	−0.5	89	$\alpha > 0.1$
CH0004R	−0.3	88	$\alpha > 0.1$
CZ0003R	−0.6	88	$\alpha > 0.1$
AT0042R	−0.6 ± 0.6	87 ± 6	$\alpha = 0.1$
AT0038R	−0.4	87	$\alpha > 0.1$
AT0044R	−0.5	86	$\alpha > 0.1$
AT0040R	−0.4	85	$\alpha > 0.1$
GB0037R	−1.0	85	$\alpha > 0.1$
SI0031R	−0.7	84	$\alpha > 0.1$
AT0030R	0.0	84	$\alpha > 0.1$
SI0033R	−0.5	84	$\alpha > 0.1$
DK0041R	−1.0 ± 1.0	82 ± 12	$\alpha = 0.05$
DE0009R	−1.0 ± 0.7	81 ± 6	$\alpha = 0.05$
DK0031R	−0.4	79	$\alpha > 0.1$
SK0004R	−0.2	78	$\alpha > 0.1$
SK0007R	0.1	73	$\alpha > 0.1$
SK006R	0.3	70	$\alpha > 0.1$
LT0015R	−0.3	69	$\alpha > 0.1$
LV0010R	−0.4	67	$\alpha > 0.1$
SE0013R	0.2	61	$\alpha > 0.1$
CH0001G	0.2	59	$\alpha > 0.1$

Notes: the confidence limits on the Sen's slope estimates are shown as 2-σ intervals

17 stations showed downwards trends that were not highly statistically significant ($\alpha > 0.1$) and 5 stations showed slightly positive trends that were not highly statistically significant ($\alpha > 0.1$). Five stations showed highly statistically significant ($\alpha \leq 0.1$) downwards trends that exceeded −1 ppb per year and these were GB0039R Sibton, CZ0001R Svratouch, GB0014R High Muffles, DK0041R Lille Valby and DE0009R Zingst. Again, there was a clear tendency for the stations with the largest annual values extrapolated back to 1990 to show the largest downwards trends. Those stations with the smallest 1990 values tended to show slightly positive trends that were not highly statistically significant ($\alpha > 0.1$).

Table 5 summarises the trends over the 2000–2009 period for those 13 stations that showed highly statistically significant ($\alpha \leq 0.1$) downwards trends. There were an additional 65 stations that showed no highly statistically significant ($\alpha > 0.1$) trends at all. The station in Table 5 which exhibited the largest downwards trend of

Table 5 Trend analysis for the nine EMEP monitoring stations over the 2000–2009 period that exhibited highly statistically significant downwards trends

Station	Slope, ppb per year	Initial 2000 level, ppb	Significance level
CH0003R	−1.6 ± 1.4	101 ± 15	$\alpha = 0.1$
GB0002R	−2.4 ± 2.4	100 ± 11	$\alpha = 0.05$
FR0014R	−1.9 ± 1.9	97 ± 14	$\alpha = 0.1$
AT0047R	−0.7 ± 0.9	95 ± 8	$\alpha = 0.1$
DE0002R	−2.0 ± 1.7	95 ± 10	$\alpha = 0.1$
AT0043R	−1.3 ± 1.3	92 ± 11	$\alpha = 0.05$
CH0001G	−1.7 ± 1.3	91 ± 8	$\alpha = 0.01$
ES0010R	−2.1 ± 1.6	90 ± 6	$\alpha = 0.1$
DK0041R	−1.3 ± 1.4	90 ± 7	$\alpha = 0.1$
AT0032R	−0.6 ± 0.8	87 ± 6	$\alpha = 0.1$
BE0035R	−1.0 ± 1.0	82 ± 13	$\alpha = 0.05$
BE0001R	−1.0 ± 0.7	81 ± 5	$\alpha = 0.05$
NL0009R	−0.9 ± 1.0	75 ± 6	$\alpha = 0.1$

Notes: the confidence limits on the Sen's slope estimates are shown as 2-σ intervals

−2.4 ± 2.4 ppb per year was GB0002R Eskdalemuir. Altogether, all but 3 of the 13 stations in Table 5 showed downwards trends that were greater than 1 ppb per year. The stations that showed the largest downwards trends were AT0043R Forsthof, BE0001R Offagne, BE0035R Vezin, CH0001G Jungfraujoch, CH0003R Tanikon, DE0002R Langenbrugge, DK0041R Lille Valby, ES0010R Cabo de Creus and FR0014R Montandon, in addition to GB0002R Eskdalemuir.

4 Why Are the Downwards Trends Largest for the Stations with the Highest Episodic Peak Levels?

In summarising the observed trends in episodic peak ozone levels over the 1980–2009 period, a clear tendency has emerged for those stations with the highest episodic peak levels to show the largest downwards trends. To attempt to illustrate this apparent tendency, a scatter plot has been constructed, see Fig. 4, with the trend data over the 1990–2009 and 1993–2009 sub-periods, taken from Tables 3 and 4. In this scatter plot, a point is plotted for each station, showing along the x-axis, the annual maximum 8-h ozone maximum at the start of the trend analysis in 1990, and along the y-axis, the Sen's slope estimates [22] from the trend analysis. The plotted points are shown larger for those stations where the Mann–Kendall tests [22] were highly statistically significant ($\alpha \leq 0.1$) and smaller where the trends were not highly statistically significant ($\alpha > 0.1$). The scatter plot shows that, indeed, the downwards trends were larger for the stations with higher episodic peak ozone levels and smaller for the least exposed stations.

These episodic peak ozone levels have unquestionably responded to European air quality policy initiatives that have secured reductions in man-made NO_x and VOC

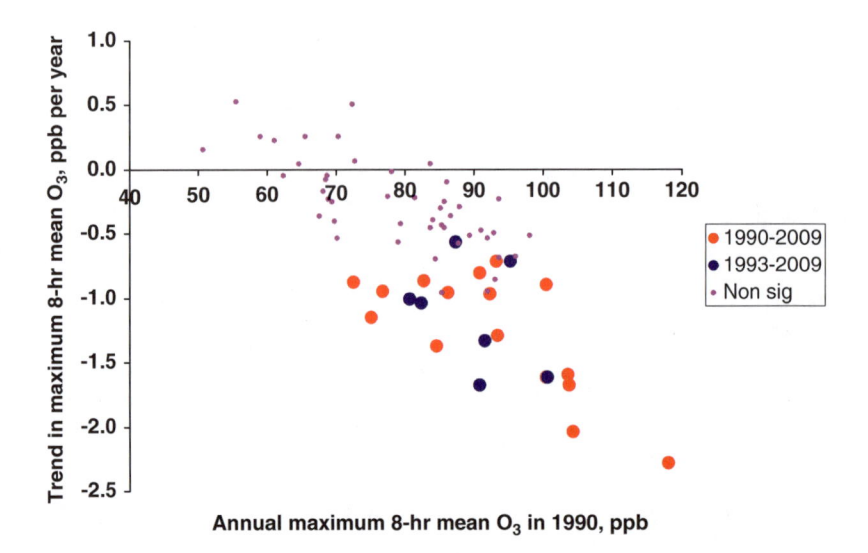

Fig. 4 Scatter plot of the trend in the maximum 8-h mean ozone in ppb per year against the extrapolated annual maximum at the start of the 1990 or 1993–2009 period for 73 EMEP monitoring stations. The *large filled symbols* are for the stations whose downwards trends are highly statistically significant ($\alpha \leq 0.1$) and the *small filled symbols* (Non sig) are for those that are less than highly statistically significant ($\alpha > 0.1$)

precursor emissions. There are two monitoring stations with long-running time series covering the three decades from the 1980s to the 2000s. These stations GB0036R Harwell and GB0039R Sibton showed highly statistically significant downwards trends of -1.2 ± 0.7 ppb per year and -1.4 ± 0.9 ppb per year, respectively. Examination of the residuals showed no evidence of any let up in these downwards trends during the 2000s compared with the 1980s or 1990s. These downwards trends are fully consistent with expectations based on computer modelling studies utilising published European emission inventories for man-made NO_x and VOCs over the 1990–2009 period. These modelling studies focused on the GB0036 Harwell monitoring station in the southern United Kingdom [23]. However, the results should be similar for other monitoring stations in northwest Europe, particularly those in Belgium, the Netherlands, Denmark, northern Germany and northern France.

A simple explanation of why the trends are more pronounced for the stations with the highest initial episodic peak levels would be that these stations show the highest ozone responses to VOC and NO_x emission reductions. Whilst this is certainly feasible, it is not a common finding in photochemical ozone modelling. Ozone responses to changing VOC and NO_x emissions in photochemical models are usually illustrated as isopleth plots. Ozone isopleths are commonly seen as a set of parallel curves, see Fig. 5 for one such isopleths diagram constructed from a European photochemical trajectory model [23]. It is not at all common for the isopleths to become closer together as would be implied by the ozone responses to

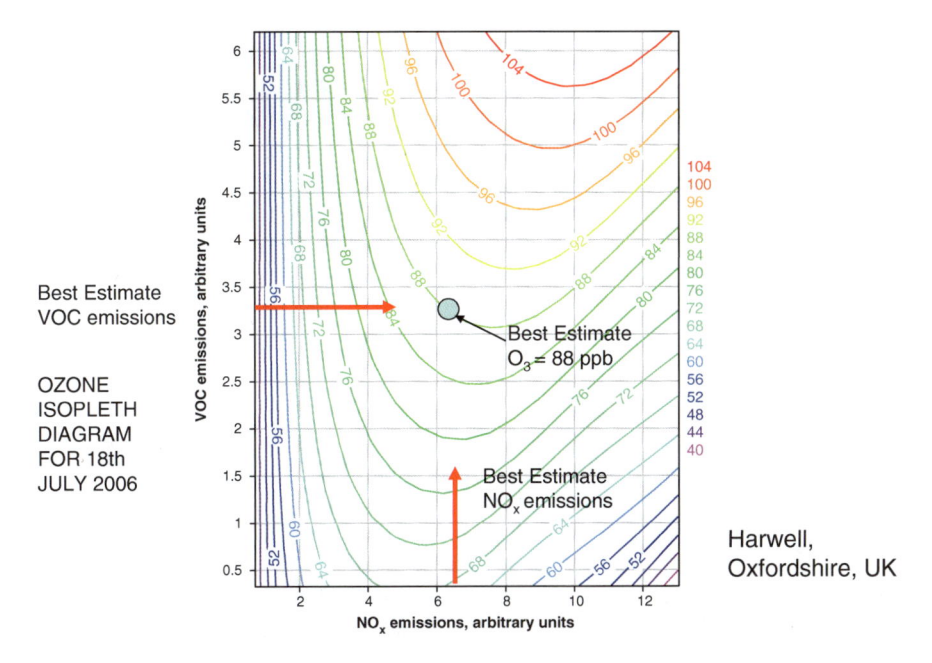

Fig. 5 Ozone isopleth diagram for 18th July 2006 at the EMEP GB0036R Harwell station plotted from the results of VOC and NO_x emission sensitivity experiments performed with a Photochemical Trajectory Model [23]

VOC and NO_x emission reductions increasing with increasing ozone levels and decreasing with decreasing ozone levels.

There must be a further reason why the apparent ozone trends decline with decreasing initial ozone level in Fig. 4. If the experience of the ozone isopleth diagram is accepted, then ozone trends should be independent of the initial ozone level. Other factors must be acting to offset the influence of the regional NO_x and VOC emission reductions at the stations with the lowest initial ozone levels. It is likely that the rising hemispheric ozone levels are this offsetting influence. The regional photochemical ozone production is superimposed on top of the hemispheric ozone level. That is to say, the air masses that cross the North Atlantic Ocean and arrive at the western seaboard of Europe already contain ozone and regional photochemical ozone production adds further to this as the air masses travel eastwards into continental Europe. Any rise in ozone level in these air masses due to hemispheric ozone increase will offset any reduction in episodic peak ozone brought about be regional-scale NO_x and VOC emission reductions. It is apparent that the stations which show the least ozone trends in Fig. 4 are those most likely to be influenced by any growth in hemispheric ozone levels.

5 Discussion and Conclusions

This chapter provides a survey of the episodic peak ozone levels across Europe over the 1980–2009 period as recorded by the EMEP ozone monitoring network. The database of annual highest 8-h average ozone levels contains over 2,291 entries for 178 monitoring stations. All monitoring stations in almost all years show exceedance of the WHO air quality guideline of 50 ppb daily maximum 8-h average ozone level.

The spatial distribution of the decadal average episodic peak ozone levels shows a maximum in central Europe covering much of Belgium, Germany, Switzerland, Slovenia, and northern Italy. Decadal average annual maximum 8-h average ozone levels in this maximum region approach and exceed 90 ppb, just under twice the WHO air quality guideline. Surrounding this maximum, there is a belt of countries including the UK, Netherlands, Czech Republic, France, Austria, Poland and Spain where decadal average annual maximum 8-h average ozone levels fall in the range 80–95 ppb. Outside of this belt, there are some monitoring stations on the fringes of Europe where decadal average annual maximum 8-h average ozone levels still exceed 60 ppb and are hence elevated above the WHO air quality guideline.

The origin of these episodic peak ozone levels is unquestionably photochemical ozone formation from man-made NO_x and VOC precursors. Efficient long-range transboundary transport ensures that these episodic peak levels are observed at even the most remote monitoring stations in the EMEP monitoring network. Equally well, efficient transboundary transport ensures that responses to reductions in man-made NO_x and VOC precursor emissions should be felt across Europe in terms of reductions in episodic peak ozone levels.

These episodic peak ozone levels have unquestionably responded to European air quality policy initiatives that have secured reductions in man-made NO_x and VOC precursor emissions. There are two monitoring stations with long-running time series covering the three decades from the 1980s to the 2000s. These stations show highly statistically significant downwards trends of between -1.2 ± 0.7 ppb per year and -1.4 ppb per year. Examination of residuals shows no evidence of any let up in these downwards trends during the 1990s and 2000s. These downwards trends are fully consistent with expectations based on computer modelling studies [23] based on published European emission inventories for man-made NO_x and VOCs over the 1990–2009 period. This modelling focuses on the GB0036 Harwell monitoring station in the southern United Kingdom. However, the results should be similar for other monitoring stations in northwest Europe.

There are 40 monitoring stations that have suitable monitoring records for trend analyses covering the 1990–2009 period and a further 29 over the slightly reduced 1993–2009 period. Statistically significant ($\alpha \leq 0.1$) downwards trends were found at 23 out of the 69 stations. Downwards trends were larger for those stations with higher initial episodic peak ozone levels in 1990 and decline to zero for those stations with the lowest initial levels. This behaviour is best explained by the resultant of two opposing influences. There is a downwards trend associated with regional NO$_x$ and VOC emission reductions which is offset by increasing hemispheric ozone levels.

The decade of the 2000s saw a huge expansion in the EMEP ozone monitoring network. However, it was a decade which has been characterised by huge year-on-year variability. The years 2003 and 2006 were highly photochemically active whilst 2008 and 2009 were highly inactive. As a result this year-on-year variability has tended to mask any clear-cut appearance of trends. Of the 78 monitoring stations with complete records for the 2000s, only 13 stations showed statistically significant ($\alpha \leq 0.1$) downwards trends. They tended to be the stations with the highest levels as shown by the trend analysis for the 1990–2009 period.

Downwards trends are continuing through the 2000–2009 period from continuing reductions in regional man-made NO$_x$ and VOC emissions. However, even after three decades of downwards trends, episodic peak ozone levels still exceed the WHO air quality guideline. There is much still to achieve with reductions in man-made NO$_x$ and VOC emissions if the WHO air quality guideline is to be archived across Europe. Indeed, in the last year of this assessment, 2009, all 126 monitoring stations still recorded exceedance of the WHO air quality guideline and no stations were in attainment.

So far this survey has had little to say about episodic peak ozone levels in urban areas, where the majority of the population live. This is because urban ozone levels are always lower than those in the rural areas surrounding them. Urban areas contain ozone sinks resulting from the rapid reaction between ozone brought in from the surrounding rural areas and local motor vehicle NO$_x$ sources. Over the last three decades, episodic peak ozone levels in rural areas have been declining as regional man-made NO$_x$ and VOC emissions have been controlled. At the same time vehicular NO$_x$ emissions have been reduced so that urban ozone levels have been rising towards rural levels. Spatial gradients and urban-rural contrasts have thus been eroded. Urban ozone is therefore increasing due to recent decreases in NO emissions resulting from changes in vehicle fleet composition and engine characteristics which, in turn, have decreased urban ozone scavenging.

Finally, attention is given briefly to the most remote monitoring stations in the EMEP ozone monitoring network. These are the WMO GAW monitoring stations at IE0031R Mace Head, on the North Atlantic Ocean coastline of Ireland, AT0034G Sonnblick at an altitude of 3,106 m, CH0001G Jungfraujoch at an altitude of 3,578 m and NO0042G Zeppelinfjell on Spitsbergen, an elevated station in the Arctic. The 2000–2009 decadal average annual 8-h maximum ozone levels at the GAW stations are in the range 55–80 ppb and are all above the 50 ppb WHO air quality guideline. Whilst this is clear evidence of the efficiency of long-range transboundary ozone transport, it is also evidence of the importance of the increasing hemispheric ozone levels because of growing man-made emissions of tropospheric ozone precursors across the northern hemisphere. Further action is required to reduce European man-made emissions of NO$_x$ and VOCs if the long-term air quality policy aim of reaching internationally accepted air quality guidelines is to be achieved across the whole of Europe. However, if this further European policy action is not coordinated with policy actions across the northern hemisphere to control tropospheric ozone precursors, then it will not achieve the stated aim of meeting internationally accepted air quality guidelines for ozone.

Acknowledgements Without the patient help of the EMEP-CCC compiling and recasting the EMEP rural ozone monitoring network data, this assessment could not have been constructed. Thanks are due to all involved with the EMEP rural ozone monitoring network for contributing their data over the three decades. The help and support of Atmosphere and Local Environment, Department for Food and Rural Affairs in the United Kingdom through contract AQ0704 is much appreciated.

References

1. Derwent RG, Stewart HNM (1973) Elevated ozone levels in the air of central London. Nature 241(5388):342–343
2. Guicherit R, van Dop H (1977) Photochemical production of ozone in western Europe (1971–1978) and its relation to meteorology. Atmos Environ 11:145–155
3. Cox RA, Eggleton AEJ, Derwent RG, Lovelock JE, Pack DH (1977) Long-range transport of photochemical ozone in north-western Europe. Nature 255(5504):118–121
4. European Environment Agency (2007) Air pollution by ozone in Europe in summer 2006. EEA Technical Report No. 5, Copenhagen, Denmark
5. Derwent RG (2008) New directions: prospects for regional ozone in north-west Europe. Atmos Environ 42:1958–1960
6. Derwent RG, Simmonds PG, Manning AJ, Spain TG (2007) Trends over a 20-year period from 1987–2007 in surface ozone at the atmospheric research station, Mace Head, Ireland. Atmos Environ 41:9091–9098
7. United Kingdom Photochemical Oxidants Review Group (1993) Ozone in the United Kingdom. Department of the Environment, London, United Kingdom
8. HMSO (1994) Ozone. Expert Panel on Air Quality Standards. Department of the Environment, London, United Kingdom
9. World Health Organization (2006) Air quality guidelines. Global update 2005. WHO Regional Office for Europe, Copenhagen, Denmark
10. Derwent RG, Grennfelt P, Hov O (1991) Photochemical oxidants in the atmosphere. Nordic Council of Ministers Report Nord 1991:7, Copenhagen, Denmark
11. UNECE (1988) Protocol to the 1979 Convention on long-range transboundary air pollution concerning the control of emissions of nitrogen oxides or their transboundary fluxes. http://www.unece.org/env/lrtap/nitr_h1.html
12. UNECE (1991) Protocol to the 1979 Convention on long-range transboundary air pollution concerning the control of emissions of volatile organic compounds or their transboundary fluxes. http://www.unece.org/env/lrtap/vola_h1.html
13. OJ (1988) Council Directive 88/76/EEC of 3 December 1987 amending Directive 70/220/EEC on the approximation of the laws of the Member States relating to measures to be taken against air pollution by gases from the engines of motor vehicles. Official Journal L036 09.02.1988: 1–32
14. UNECE (1999) Protocol to the 1979 Convention on long-range transboundary air pollution to abate acidification, eutrophication and ground-level ozone. http://www.unece.org/env/lrtap/multi_h1.html
15. OJ (2001) Directive 2001/81/EC of the European parliament and of the Council of 23 October 2001 on national emissions ceilings for certain atmospheric pollutants. Official Journal L 309 27.11.2001: 22–30

16. OJ (1999) Council Directive 1999/13/EC of 11 March 1999 on the limitation of emissions of volatile organic compounds due to the use of organic solvents in certain activities and installations. Official Journal L 13 29.03.1999: 1–25
17. Commission of the European Communities (2005) Proposal for a Directive of the European Parliament and of the Council on ambient air quality and cleaner air for Europe. Commission of the European Communities, COM (2005) 447 final, Brussels, Belgium
18. Hjellebrekke A-G, Solberg S, Fjaeraa AM (2011) Ozone measurements 2009. EMEP/CCC-Report 2/2011, Chemical Coordinating Centre of EMEP, Norwegian Institute for Air Research, Kjeller, Norway
19. European Topic Centre on Air and Climate Change (2007) Air pollution by ozone in summer 2006. EEA Technical Report No. 5, European Environment Agency, Copenhagen, Denmark
20. Grennfelt P, Schjoldager J (1984) Photochemical oxidants in the troposphere: a mounting menace. Ambio 13:61–67
21. Aas W, Hjellbrekke A-G, Schaug J (2000) Data quality 1998, quality assurance and field comparisons. EMEP/CCC-Report 6/2000, Chemical Coordinating Centre of EMEP, Norwegian Institute for Air Research, Kjeller, Norway
22. Salmi T, Maatta A, Antitila P, Ruoho-Airola T, Amnell T (2002) Detecting trends of annual values of atmospheric pollutants by the Mann-Kendall test and Sen's slope estimates – The EXCEL template application MAKESENS. Publications on Air Quality No. 31, Finnish Meteorological Institute, Helsinki, Finland
23. Derwent RG, Witham CS, Utembe SR, Jenkin ME, Passant NR (2010) Ozone in central England: the impact of 20 years of precursor emission controls in Europe. Environ Sci Policy 13:195–2004

Persistent Organic Pollutants in the European Atmosphere

Barend L. van Drooge and Joan O. Grimalt

Abstract Since the beginning of the last century, European air has received increasing amounts of organic compounds. Some of them were released as consequence of combustion processes and others were synthesised for specific applications in industry or agriculture. A group of these organic compounds are semi-volatile, resistant to degradation processes, bioaccumulative and toxic. Because of these properties, they are grouped under the general name of persistent organic pollutants (POPs). These physical–chemical properties are found in polycyclic aromatic hydrocarbons or in organic molecules with 5–12 carbon atoms and a high degree of halogen substituents. The present chapter is devoted to describe these two groups of compounds. However, in other review papers, only the organohalogen compounds are considered under the POP concept.

The properties of semi-volatility, bioaccumulation and persistent to degradation processes provide a potential for long-range atmospheric transport of these compounds. They may therefore be deleterious for the ecosystems and human health even in sites located far away from the areas of production and application. Furthermore, nearly all organohalogen POPs are man-made. In nature, there are no compounds with similar chemical structure (or these are in very low concentrations). Thus, human and animal evolution has not had the opportunities to generate specific metabolic ways for their elimination. Thus, they accumulate and remain in the body throughout all life and during all life stages (including in utero development). By the end of the twentieth century, international protocols have been established to reduce the emissions and impact of POPs into the environment and humans.

The following chapter describes the changes in concentration of POPs in ambient air in Europe through time, with a focus on current concentrations. It also

B.L. van Drooge (✉) and J.O. Grimalt
Institute of Environmental Assessment and Water Research, IDÆA-CSIC, Jordi Girona 18, 08034 Barcelona, Catalonia, Spain
e-mail: barend.vandrooge@idaea.csic.es

describes the differences in physical–chemical properties of these pollutants and relates them with the mechanisms of atmospheric transport. Strategies to reduce the ambient air concentrations are discussed. Information is also given on future perspectives in view of "emerging" pollutants and emission sources.

Keywords Gas-to-particle-phase partitioning, Long-range atmospheric transport, Organochlorines, Persistent organic pollutants, Polycyclic aromatic hydrocarbons

Contents

1 Introduction	76
2 Objectives	78
2.1 Organohalogen Compounds	78
2.2 Polycyclic Aromatic Hydrocarbons	81
3 Atmospheric Transport of POP	82
3.1 Gas–Particle Exchange	84
3.2 Atmospheric Deposition	86
4 Historical Trends of POP Atmospheric Pollution in the European Environment	88
5 Atmospheric POP Levels in Europe in the Twenty-First Century	91
6 Future Perspectives	96
References	96

1 Introduction

Since the beginning of human development, anthropogenic activities, such as hunting, wood burning and agriculture, have left their mark in the environment. At present, the increase and intensification of economic progress in our industrialised society, with its unlimited demand and use of energy and manufactured products, have largely raised the environmental impact of these activities. Furthermore, the environmental effects of many chemicals are not restricted to the sites of production and/or application. They may impact distant ecosystems as consequence of long-range transport. When this process is influencing most of the planet, it involves global changes. These changes relate to the presence of toxic compounds in nearby and remote ecosystems like the Arctic regions and European high mountain areas [1, 2] where POPs are present in all environmental compartments such as water, soils and atmosphere, as well as in the organisms living in them.

POPs are chemical compounds responsible for many of these problems. This group of pollutants includes polycyclic aromatic hydrocarbons (PAHs) and organohalogen compounds (OC) such as polychlorobiphenyls (PCBs), hexachlorocyclohexanes (HCHs), hexachlorobenzene (HCB), polybromodiphenyl ethers (PBDEs) and others. After emission to the atmosphere, POPs may undergo long-range atmospheric transport. The environmental effects caused by these compounds result from their toxicity, semi-volatility and resistance against physical, chemical or biological degradation (Sect. 2). Most of these compounds, except PAH, are man-made, or xenobiotic, and were applied in agriculture for crop protection (pesticides), control against diseases such as malaria or protection of materials in industrial applications, such as flame retardants or surface coatings.

Their mass production and uses started in the 1940s when they entered into the environment in large amounts. In the case of PAH, the intensification of industrial applications and human development in the twentieth century largely increased their emissions to the atmosphere in addition to the natural background of spontaneous fires. Nevertheless, already in the beginning of the 1960s, Rachel Carson wrote her inspiring book "Silent Spring" which expressed the concern for the unlimited use of these compounds and their possible consequences [3]. Subsequent research proved quite well most of the worries expressed in this book. The ubiquitous presence of PCBs in the environments was first recognised in the late 1960s when they were found in pike from Swedish lakes [4]. In the late 1970s, the negative effects of OCs were generally acknowledged.

In the last decades, many POPs have been banned in most industrialised countries. The United Nations elaborated a programme on POPs, which was also signed by the European Union in Stockholm (Sweden) in 2001 (http://www.chem.unep.ch/pops/; Bulletin EU 5-2001: 1.4.41). In this programme, countries compelled themselves to reduce or eliminate the production and use of POPs and to investigate their effects in the environment and humans. Thus, in about 60 years after the initial applications of these compounds, strict regulations had to be taken to decrease the impact of POPs in the environment. Accordingly, concentration decreases have been observed since the 1990s, namely, in potential emission source areas where these chemicals were produced and used [5, 6]. The observed changes have been reviewed recently [7]. Nevertheless, POPs are still omnipresent in the environment, and the observed decreases in remote areas, such as the arctic region, are small or hardly detectable [8, 9]. POP recycling among environmental compartments has an important role in the stabilisation of environmental concentrations [10]. The ubiquitous presence of POP in the environment, even at low concentrations, is of concern since a large proportion of these compounds have toxic effects [11, 12].

The POP protocol elaborated on occasion of the Stockholm Convention in 2001 contained 12 compounds, known as the "dirty dozen": aldrin, chlordane, DDT, dieldrin, endrin, heptachlor, HCB, mirex, PCBs, polychlorodibenzo-p-dioxins, polychlorodibenzofurans and toxaphene. After several years of implementation other compounds, were also included: HCHs, PAHs, certain brominated flame retardants, perfluorooctanesulfonic acid (PFOS) and pentachlorobenzene (http://www.chem.unep.ch/pops/). This protocol is aimed to control, reduce and/or eliminate discharges, emissions and spills of POPs into the environment.

In the present chapter, the role of the atmosphere in the distribution of POP in the European environments will be considered (Sect. 3). Data on atmospheric concentration and deposition collected between 2000 and 2009 shows that these compounds are still impacting the European ecosystems. Complementary data from sediment cores allows reconstruction of the historical input of these compounds into the European environments (Sect. 4). Both types of data sets are used to describe the actual status of the atmospheric contamination in the European continent by these compounds and to anticipate future perspectives (Sects. 5 and 6).

2 Objectives

1. To describe the atmospheric background levels of POPs over Europe
2. To identify the processes that are responsible for the atmospheric transport and fate of POP in the European atmosphere
3. To describe the state of the art of POP contamination at the European scale
4. To describe the future trends of POP contamination in the European atmosphere

2.1 Organohalogen Compounds

OC are man-made chemical substances that encompass a wide range of compounds with different structures and applications. They are known for their high chemical stability, which stems from their large proportion of halogen substituents, either chlorine, bromine or others. Some of them were synthesised for use as pesticides and others for industrial applications.

The OC pesticides were initially produced to protect crops and human beings against plague organisms. They are generally incorporated into the environment as consequence of their use in agriculture. Nevertheless, they can also enter into the environment due to bad storage practices.

2.1.1 Hexachlorocyclohexanes

These compounds were originally used as mixtures of different isomers (Fig. 1). These mixtures contained five isomers, α-, β-, γ-, δ-, ε-HCH, of which γ-HCH was the one having insecticide properties. The half-life in biological tissue and other environmental compartments of this isomer is relatively short in comparison to other OCs. γ-HCH and its metabolites have relatively high water solubility, so they can be rather easily washed out from the atmosphere by rain [13].

The first production of this group of compounds started in 1943. The production of lindane (γ-HCH) in developed countries has been estimated to be 720,000 tonnes. The global production of technical-grade HCHs over the period between 1948 and 1997 has been estimated to be 10 million tonnes. Today, the production and use of HCHs have been strongly reduced due to the international restriction. The principal emission sources of this group of compounds are shown in Fig. 3. Their physical–chemical properties are summarised in Table 1.

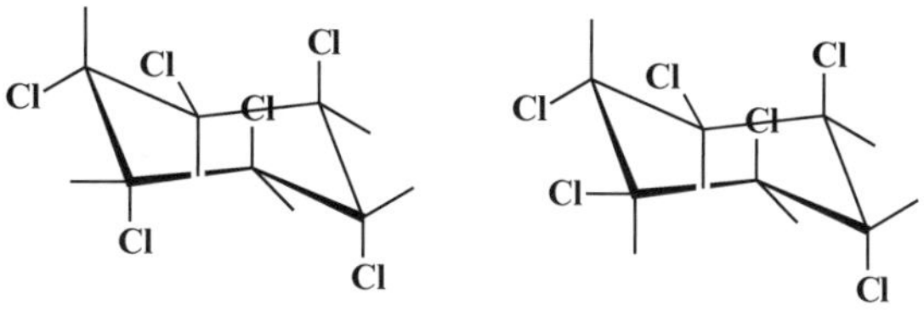

Fig. 1 Structures of two hexachlorocyclohexane isomers (*1*) α-HCH, (*2*) γ-HCH

Table 1 Selected physico-chemical properties of studied POPs at 278 K [14–20]

POPs	Mol Wt.[a]	P_L^b	log K_{ow}^c	H^e
Fluorene	166	0.72	4.2	7.9
Phenanthrene	178	0.11	4.6	3.2
Anthracene	178	7.8×10^{-2}	4.5	4.0
Fluoranthene	202	8.7×10^{-3}	5.2	0.6
Pyrene	202	1.2×10^{-2}	5.2	0.9
Benz[*a*]anthracene	228	6.1×10^{-4}	5.9	0.6
Chrysene	228	1.1×10^{-4}	5.9	0.2
Benzo[*b+j*]fluoranthene	252		5.8	0.15
Benzo[*k*]fluoranthene	252	4.1×10^{-6}	6.0	0.11
Benzo[*a*]pyrene	252	2.1×10^{-5}	6.0	0.045
Indeno[123-*cd*]pyrene	276			
Dibenz[*ah*]anthracene	278	9.2×10^{-8}	6.8	0.069
Benzo[*ghi*]perylene	276	1.4×10^{-8}	6.5	0.076
HCB	285	3.1×10^{-3}	5.5	53
α-HCH	181	7.3×10^{-3}	3.9	0.10
γ-HCH	181	1.9×10^{-3}	3.9	0.061
PCB18	257	8.0×10^{-2}	5.6	32
PCB28	257	2.0×10^{-2}	5.8	29
PCB52	292	1.0×10^{-2}	6.1	32
PCB70	292	6.0×10^{-3}	6.3	17
PCB101	326	3.0×10^{-3}	6.4	25
PCB110	326	2.0×10^{-3}	6.3	20
PCB138	361	5.0×10^{-4}	7.0	13
PCB153	361	2.0×10^{-4}	6.9	17
PCB180	395	1.0×10^{-5}	6.9	11
PCB194	430	1.0×10^{-5}	7.6	29

[a]Molecular weight (amu)
[b]Sub-cooled liquid vapour pressure (Pa)
[c]Logarithm of the octanol–water partition coefficients
[d]Henry's law constants (Pa m^3/mol)

2.1.2 Hexachlorobenzene

HCB (Fig. 2) was used mainly as fungicide in wood and seed treatment, but it is also a by-product of several chlorination procedures for the production of organochlorine solvents, plastics and in the secondary aluminium industry (Fig. 3). A total

Fig. 2 Molecular structure of HCB

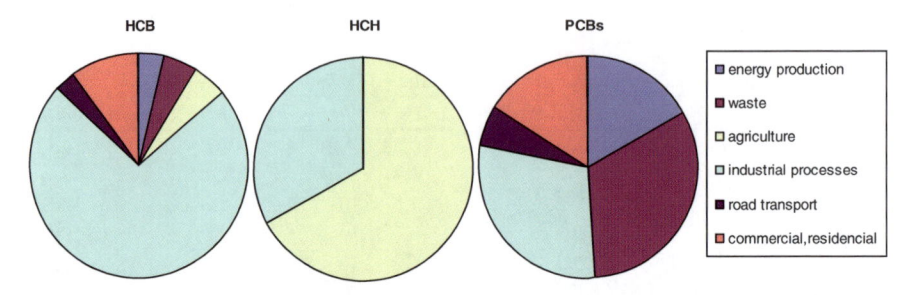

Fig. 3 European Union emissions of HCB, HCHs and PCBs (adapted from [7])

global emission of 23,000 kg per year has been estimated with a range between 12,000 and 92,000 kg per year [21].

This compound is poorly soluble in water and evaporates rapidly after release into the atmosphere. Due to its relatively high vapour pressure, it is usually found in the gas phase, not in the aerosol fraction. Today, its production is banned. However, it is present in many products as an impurity and can be released unintentionally. HCB has shown to be toxic for men, when in Turkish Kurdistan, between 1955 and 1959, it caused an outbreak of *porphyria cutanea tarda* after intake of contaminated food. A relationship between thyroid cancer increase and chronic exposure to HCB has been identified [22].

2.1.3 Polychlorobiphenyls

PCBs are industrial compounds used as industrial, dielectric and heat transfer fluids, organic solvents, flame retardants, plasticizers, sealant and surface coatings. They may also be released to the atmosphere by waste incineration (Fig. 3). The worldwide production of this compound has been 1.3 million tonnes, of which 97% in the northern hemisphere [23]. The amount of chlorine atoms in the biphenyl mixtures is related to the duration and temperature of the chlorination process. The commercial mixtures were distributed under names such as Aroclor (Monsanto, USA) or Clophen (Bayer, EU). The chlorine atoms can substitute the *para*, *meta* and/or *ortho* positions of the biphenyls. There are 209 possible congeners. PCBs can be divided into nine isomeric groups and one decachlorobiphenyl, all with an empirical formula of $C_{12}H_{10}$-nCl_n ($n = 1$–10) (Fig. 4).

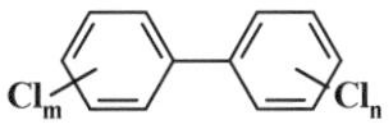

Fig. 4 General molecular structures of PCBs

PCBs were found for the first time in the environment by Sören Jensen, who found these contaminants in pike from Swedish waters [4]. After few years of research, the adverse effects of these compounds for organisms were identified and led to restrictions in their production and use. Finally (late 1970s), they were banned in most industrialised countries [24, 25]. The position of the chlorine atoms in the molecule influences the toxicity properties of the PCBs. They are more toxic when the *ortho* positions of the molecule are not substituted, like in the case of PCB-77 (3,3′,4,4′-tetraCB), PCB-126 (3,3′,4,4′5-pentaCB) and PCB-169 (3,3′,4,4′,5,5′-hexaCB). In these conditions, the two phenyl groups can rotate freely, and the molecule is stabilised as a flat structure (planar or coplanar PCBs) due to the interaction between the π orbitals of both biphenyls. *Mono–ortho* and *di–ortho* PCBs (with the 2 and 2′ positions substituted) are less toxic [25] than the unsubstituted *ortho* isomers. Reported biological effects on terrestrial animals exposed to PCBs involve liver damage, dermal disorders (chloracne), reproductive toxicity, thymic anthropy, body weight loss, immunotoxicity, teratogenicity and induction of several cytochrome P-450 (enzyme) isosystems [25]. Although the use of these products is prohibited for many years now, they still cause toxic effects at trace levels, as it has been shown in a study on Dutch school children [12].

PCBs are discharged into the environment by leaking from hydraulic systems, diffusion from coatings and during production, waste incineration or waste disposal. Atmospheric PCBs are predominately presented in the gas phase. However, significant fractions of the more chlorinated congeners are also presented in the particulate phase [26].

2.2 Polycyclic Aromatic Hydrocarbons

PAHs are constituted of C and H atoms in the form of fused benzene rings. The atmospheric PAH mixtures are essentially generated in incomplete combustion processes involving fossil fuels or any other organic materials, such as wood. These sources are numerous (Fig. 5). At present, residential emissions from wood and coal combustion for domestic heating and vehicular exhaust emissions are important sources. The composition of the PAH mixture depends on the precursor organic material and combustion conditions [27, 28]. Diagenetic processes may also produce PAHs, but their significance as sources for the atmosphere is small.

Most PAHs enter into the environment by atmospheric emissions. They may therefore be atmospherically transported over long distances, especially when they are bound to aerosols. In this way, they have become ubiquitous contaminants reaching remote areas [29]. There is a direct relationship between pyrolytic PAH

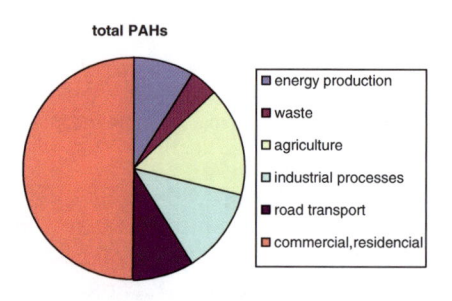

Fig. 5 European emissions of PAHs in 2009 by sector group (adapted from [7])

total PAHs
- energy production
- waste
- agriculture
- industrial processes
- road transport
- commercial, residencial

and black carbon (BC) particles. BC is produced by incomplete combustion in fossil fuel and biomass burning. Like PAHs, BC occurs ubiquitously and can account for about half of the carbonaceous particles in the atmosphere. Structurally, BC may be essentially differentiated between soot carbon (small particle size; SC) and char/charcoal. The former originates during the condensation of hot combustion gases involving free radical reactions of acetylene species leading into PAH, macro PAH and SC. Conversely, charcoal originates from incomplete combustion of plant tissue and diagenesis [30] and always contains a core of unburned biomass material. During the past decades, the study of SC as a strong sorption matrix has received increasing attention [30–32]. Adsorption onto SC has been shown to be significant for the overall atmospheric transport of PAH to European alpine areas [33, 34].

Some PAH like benzo[a]pyrene and their metabolites are teratogenic and mutagenic [27]. In the late 1970s, the United States Environmental Protection Agency (USEPA) listed 16 PAH as "priority pollutants" (Fig. 6). This list was later adopted by the European Union. Moreover, the CAFE directive of the European Union establishes an annual limit value of 1 ng/m^3 for the occurrence of benzo[a]pyrene in the atmosphere (Council Directive 2004/107/EC *Official Journal*, L 023, 26/01/2005; pp. 3–6).

In general, fluoranthene and higher molecular weight PAH tend to bound to atmospheric particles, while the lower molecular weight PAH is present in the gas phase [26, 35, 36]. Thus, the lower molecular weight PAH is more susceptible to direct or indirect photodegradation than the particle-bound higher molecular weight PAH [37, 38].

3 Atmospheric Transport of POP

The presence of POPs in remote areas demonstrates the importance of long-range atmospheric transport in the environmental distribution of these compounds. The atmosphere, due to its large volume and elevated speed at which it can transport air masses over long distances, is the most important pathway for distribution of POPs into the environment [39, 40]. POPs have vapour pressures (P_L) between 10^{-2} and

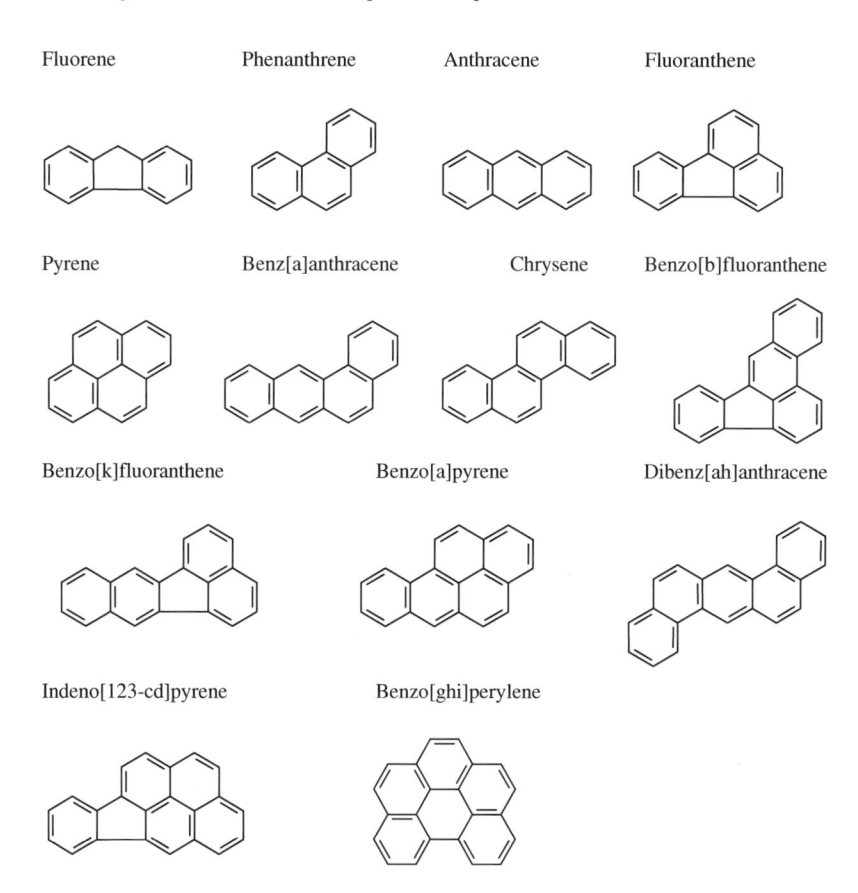

Persistent Organic Pollutants in the European Atmosphere

Fig. 6 Molecular structures of PAHs

10^{-5} Pa (Table 1) which are in the range of phase changes at ambient temperatures. In this way, POPs tend to evaporate with rising ambient temperatures and condense, or adsorb to surfaces, at decreasing temperatures. These temperature-dependent processes have been described using the concept of "global distillation" taking as example a gas chromatography process [41] (Fig. 7). Other concepts used to define this process have been "global" or "latitudinal fractionation," "cold condensation" or "selective trapping" [2, 29, 42, 43]. These effects have allowed to explaining the changes in POP composition in different environmental matrices in relation with latitude [43–46] and altitude [2, 44, 47, 48].

Atmospheric POPs are distributed between the gas and particulate phases and, as mentioned above, the distribution rates are temperature-dependent [35]. However, other atmospheric processes such as photochemical oxidation and deposition are also relevant for the ultimate value of POPs in the air compartment (Fig. 7).

Once the compounds have been deposited on land or water surfaces, they can be incorporated into these compartments and the organisms living in them (Fig. 7). Soils and sediments can be considered as final POP sinks. However, the retention

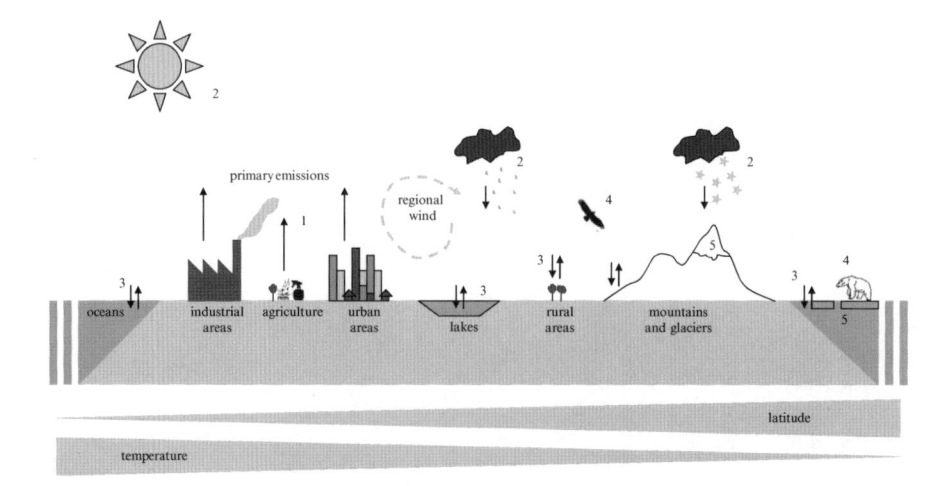

Fig. 7 Conceptual representation of processes influencing the atmospheric transport and fate of POPs. (*1*) Primary emissions of POPs to the atmosphere, (*2*) atmospheric deposition and photochemical degradation/transformation, (*3*) re-volatilisation from secondary sources in the different environmental compartments and burial in sediments, (*4*) bioaccumulation and biotic transport, (*5*) accumulation in glaciers and ice caps, with probable releases due to melting

capacity of soils and lake sediments depends on several factors, such as its organic content, ambient temperatures, biological activity, physical disturbance and chemical transformation [6, 49–53].

3.1 Gas–Particle Exchange

The exchange of chemical compounds from the gas phase to a surface, e.g. atmospheric particles, soil, water, vegetation or other surfaces, is controlled by the affinity of the compound to this surface. The ratio of vapour pressure to water solubility can be used as indicator between levels in the atmosphere and water surface (Henry's law; H constant). In many model calculations, the ratio between POP levels in octanol and water, the octanol–water partitioning coefficient (K_{ow}), is used as reference for the distribution of POP in organic material [14]. Consequently, the expression H/RT (C_{air}/C_{water}) and K_{ow} ($C_{octanol}/C_{water}$) provide the octanol–air partitioning coefficient (K_{oa}):

$$K_{oa} = K_{ow}\, RT/H$$

which can be used as the indicator for the distribution of POP between the atmosphere and surfaces. Since H is temperature-dependent, K_{oa} is temperature-dependent as well [54, 55].

Vegetation has been used in the past as indicator of the degree of atmospheric contamination on small and global scale [44, 54, 56, 57]. Although important differences between chemical species have been observed, in general the lipophilic organic pollutants (log $K_{ow} > 4$) are taken up from the atmosphere via the cuticle waxes or by the stomata of the leaf, where they tend to accumulate. Uptake of these compounds from the soil via the roots is not significant; neither is POP transported to other parts of the plant. Due to the large surface area of leaves, atmospheric variability of POP concentrations and ambient temperatures affect directly the concentrations found in the leaves of plant species [56]. Leaves can therefore be useful as bioindicators for the atmospheric quality in terms of POP levels.

The significance of soot carbon for atmospheric transport and particulate–gas partitioning of PAH has been assessed in several studies [30–34, 52]. This exchange is based on the particulate–gas partitioning coefficient K_p

$$K_p = C_p/(C_g \times \text{TSP}),$$

where C_p and C_g are the concentrations in the particulate and gas phase, respectively.

Usually, there is a significant linear fit between K_p and the sub-cooled liquid vapour pressure (P_L; [58]) of the PAHs. Organic matter plays an important role in the particle/gas partitioning of these compounds ([58] and references therein), which can be described by the following equation:

$$\log K_p = \log K_{oa} + \log f_{OM} - 11.91, \quad (1)$$

where K_{oa} is the temperature corrected octanol–air partitioning coefficient. $\log K_{oa} = \log K_{ow} \times RT/H$, where K_{ow} is the octanol–water coefficient, R is the gas constant, T the mean air temperature and H the temperature corrected Henry's law constant.

However, when applying the fraction of organic matter (f_{OM}) to the model calculations, there is often an underestimation of the predicted K_p, especially in remote areas. This underestimation could only be corrected by using unrealistically high f_{OM} values. Notwithstanding, a general agreement between predicted and measured K_p is observed when both organic matter absorption and soot-carbon adsorption are included in the model (see equation below):

$$K_P = \frac{f_{OM} MW_{OCT} \gamma_{OCT}}{\rho_{OCT} MW_{om} \gamma_{OCT} \times 10^{12}} \cdot K_{oa} + f_{EC} \frac{a_{ec}}{a_{ac} \times 10^{12}} K_{sa},$$

where the expression $\frac{f_{OM} MW_{OCT} \gamma_{OCT}}{\rho_{OCT} MW_{om} \gamma_{OCT} \times 10^{12}} \cdot K_{oa}$ is the relationship between K_p and K_{oa} as described in its simplified form in Eq. (1), assuming γ_{OCT}/γ_{OM}, MW_{OCT}/MW_{OM} to be equal to 1 and ρ_{OCT} to be 0.820 kg/L. f_{EC} is the fraction of elemental carbon in atmospheric particles, which was estimated to be less than 5% in rural and remote sites [59], a_{ac} is the surface area of active carbon, a_{ec} is the specific surface of elemental carbon and 10^{-12} is a factor for unit correction. Assuming that the ratios γ_{OCT}/γ_{OM}, MW_{OCT}/MW_{OM} and a_{ec}/a_{ac} are equal to 1, it is possible to predict K_p values from f_{OM} and f_{EC} [30–34, 52].

Fig. 8 log K_{ow} and log K_{aw} of different organic molecules belonging to the group of POPs. The properties of these compounds make them suitable for long-range atmospheric transport (adapted from [60])

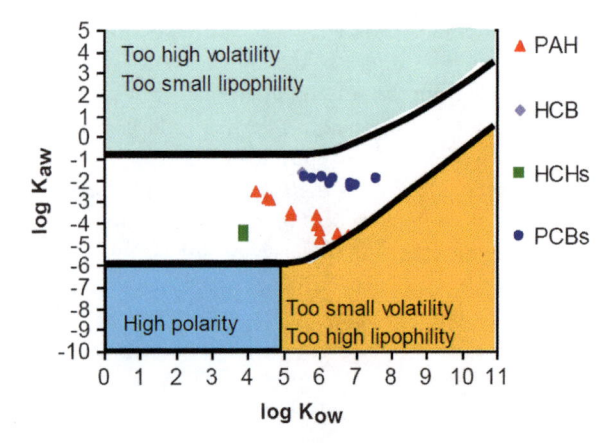

In general, POPs are volatile enough to be present in the atmosphere at ambient temperatures and travel over long distances. However, they are not too volatile, so they can be absorbed to surfaces, particularly if those are organic-rich, and be incorporated to other environmental compartments. Their semi-volatility is expressed by the partitioning coefficient air–water, K_{aw}. Moreover, they have higher affinity for organic matter than water, which is expressed by the partition coefficient octanol–water, K_{ow}, and this delays the washout from the atmosphere by rain. The K_{ow} of the different POPs versus their K_{aw} is shown in Fig. 8. POPs are found in an area of intermediate K_{aw} and K_{ow}. Any other compound with similar K_{aw} and K_{ow} values is a potential POP.

3.2 Atmospheric Deposition

Deposition of POPs may take place by (1) snow and rain scavenging of gases and aerosols (wet deposition), (2) dry particle deposition and (3) gas exchange with surfaces [61]. The distributions of POPs between the gas and particle phases depend on their physical–chemical properties (Fig. 8) as well as the environmental conditions in the atmosphere, such as temperature, amounts and composition of particles [26, 35].

3.2.1 Wet Deposition

Wet deposition involves the adsorption of gases and aerosols on surfaces after removal by rain, snowfall, etc. Water solubility and vapour pressure are determinant for this process. Compounds with low H constants are washed out from the atmosphere faster than those with high H. The efficiency in which rain (C_{rain}) or snow removes a gaseous compound from the atmosphere ($C_{\text{air gas}}$) can be defined from the washout ratio of the gas phase (W_g) [61]:

$$W_g = C_{\text{rain}}/C_{\text{air gas}} = RT/H \text{ (mass/volume)}.$$

According to this equation, compounds with low H tend to be washed out faster than compounds with high H.

The washout factor of the particulate phase (W_p) is important in rain and snow scavenging of particle-bound POP ($C_{\text{air particulate}}$) [61]:

$$W_p = C_{\text{rain}}/C_{\text{air part}} \text{ (mass/volume)}.$$

The W_p is however very variable and depends on the size of the particle, its structure and chemical composition and on the meteorological conditions.

3.2.2 Dry Deposition

The velocity of particle dry deposition, and that of the particulate-bounded compounds, depends largely on the size and composition of the aerosol. For large particles (>25 μm), the deposition is mainly governed by gravity. Aerosols of this size are deposited near the emission sources. Smaller aerosols between 0.1 and 10 μm are suitable to long-range atmospheric transport, involving sometimes more than hundreds of kilometres. Invasions of Saharan dust to the North Atlantic and Europe mainly consist of aerosols of these sizes (mostly <2.5 μm). Particles smaller than 0.1 μm depend on molecular diffusion [61]. POPs have higher affinity to particles with high organic carbon content due to their relative high lipophilicity (log $K_{\text{ow}} > 4$).

The particulate flux to a surface ($F_{\text{d part}}$) and the particle-bound atmospheric concentration of a compound ($C_{\text{air part}}$) at a given height are related to the dry deposition velocity ($V_{\text{d part}}$):

$$V_{\text{d part}} = F_{\text{d part}}/C_{\text{air part}}.$$

This velocity is very variable and depends, for example, on the meteorological conditions, particle size and composition and surface properties.

The lower molecular weight compounds are in general more abundant in the wet deposition than the heavier molecular weight compounds (Fig. 9). The former also occur predominantly in the atmospheric gas phase, while the latter are particle-bound in larger proportion [35]. The transfer of gas phase POP to rain and/or snowflakes is very effective [62–64]. Wet deposition accounts for the predominant incorporation pathway of gas phase and particulate phase POPs. But the efficiency depends on the quantity of precipitation, which may show high variability between different sites and seasons.

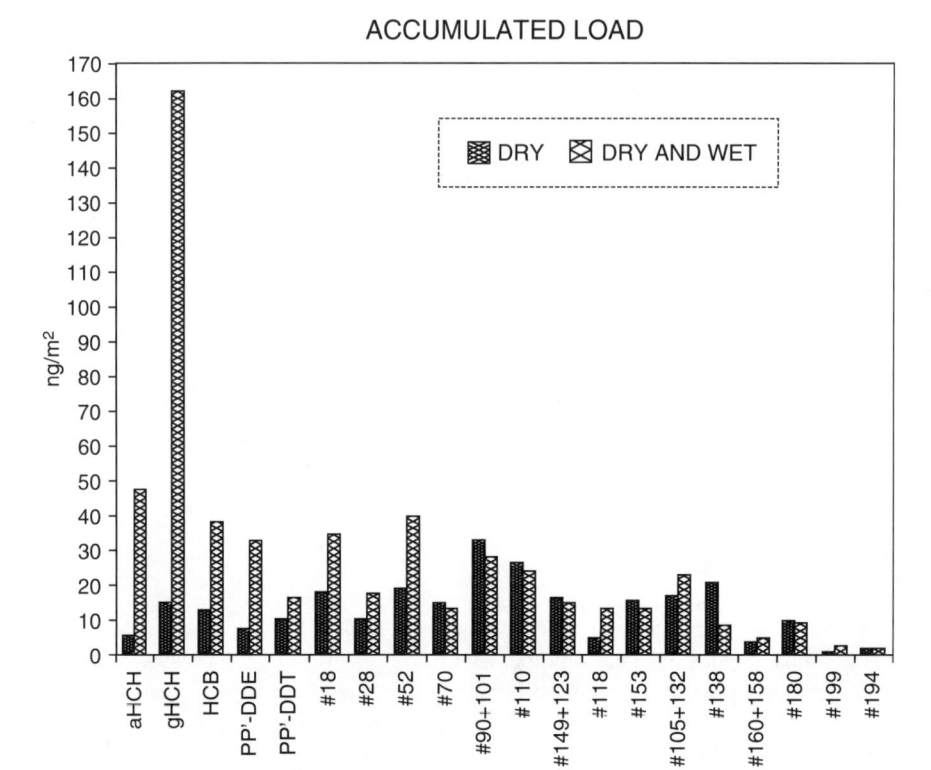

Fig. 9 Accumulated load of organochlorine compounds in deposition samples collected between 20 May 1999 and 19 May 2000 at Teide (2,367 m; Tenerife) [13]

4 Historical Trends of POP Atmospheric Pollution in the European Environment

Data on past atmospheric concentrations of POPs is scarce. However, alpine lake ecosystems can be used as sentinel ecosystems to reconstruct the atmospheric pollution load over the European continent [50, 51, 65, 66].

Alpine lakes essentially receive pollutant inputs from regional or long-range atmospheric transport [50, 51]. Depending on the physical–chemical properties of the compound and the climate conditions, the compounds enter these lakes through gas exchange, dissolved in rain drops, snowflakes or adsorbed/absorbed to aerosols. Once the POP has entered into the lake system, interactions take place between the compounds, particles, water and organisms. The strength of the interactions depends on the water solubility of the individual POPs, or their affinity to organic material, as well as on the climate conditions, e.g. water temperature [67–71]. Thus, POPs with relatively high solubility, like HCHs, mostly occur in the dissolved phase, while higher molecular weight PCBs and PAHs exhibit in general higher association to the particles in the water column [72–74].

POPs can be absorbed actively or passively by zooplankton and phytoplankton species, or by higher organisms, such as fish, via ingestion, respiration and diffuse absorption. Some compounds may accumulate in body tissue, while others are transformed into other compounds by the metabolism of the organism or excreted to the water column in its original form [75–77]. As consequence of particle deposition, POPs will eventually end up on the bottom of the lake and be incorporated into the sediment were they are generally well preserved due to the low biological activities in these lakes [50, 51].

The above mentioned processes result in some differences in POP composition between the atmosphere and the sediment. For example, the relative composition of PCB congeners in sediment is predominated by the compounds of higher molecular weight, while the lower molecular weight congeners are in higher proportion in the atmosphere. This difference is temperature-dependent [53]. Overall, specific relationship can be established between ambient air temperatures of high mountain lakes and sedimentary PCB concentrations in the European alpine areas [2].

On the other hand, comparison of the average PAH profiles in the sediments (Fig. 10) with snow deposition [62] and atmospheric PAH gas and particulate phases distribution [34] shows a good agreement between the PAH composition in the sediments and in the atmospheric particulate phase ($r^2 = 0.94; p < 0.01$) and snow deposition ($r^2 = 0.92; p < 0.01$).

Studies on the downcore concentrations of POPs in alpine lakes allow to reconstructing the historical trends of the incorporation of these compounds to the European ecosystems. The study of these high mountain lakes provides a reference for the overall background load of these compounds. As shown in Fig. 11, there is a rather good agreement between a sediment core from a high mountain lake from the Tatra mountains and a peat core from northwestern UK. Both records show a steep increase of OCs in the 1950s and peak concentrations between 1960 and 1990. After 1990, the sediment concentrations in the alpine lakes started to decrease. Conversely, PAH show different profiles in the alpine sediments from eastern Europe and in the peat samples from northwestern UK. In the former case, the concentrations increased after 1950 and started to decrease after 1990, whereas in the latter these compounds started to increase in the nineteenth century (Fig. 11). These differences may reflect the later industrial development in eastern than in western Europe. Thus, in the UK, the PAH concentration increased after 1820 with a peak concentration around 1925. This onset corresponds to the start of the industrial revolution in the UK. After 1960, the concentration decreased to low concentrations in the 1980s, which could reflect lower inputs from heavy industry and changes in combustion fuels for domestic heating as consequence of air quality legislation [6]. A description of the historical accumulation of PAHs over Europe based on the study of sediment cores collected in diverse alpine lakes is available in Fernández et al. [49].

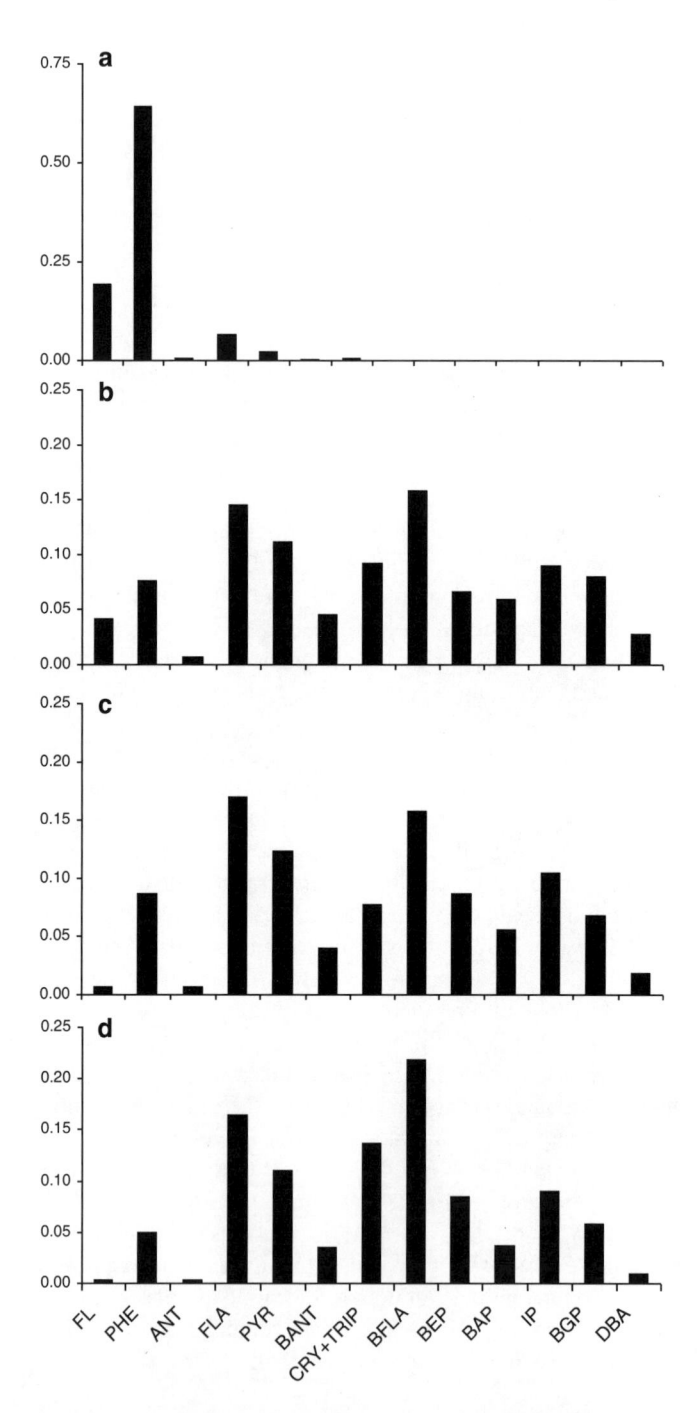

Fig. 10 PAH profiles in environmental compartments from the High Tatras; (**a**) air gas phase [34]; (**b**) air particulate phase [34]; (**c**) snow deposition [62]; (**d**) sediment [66]. *FL* fluorene, *PHE*

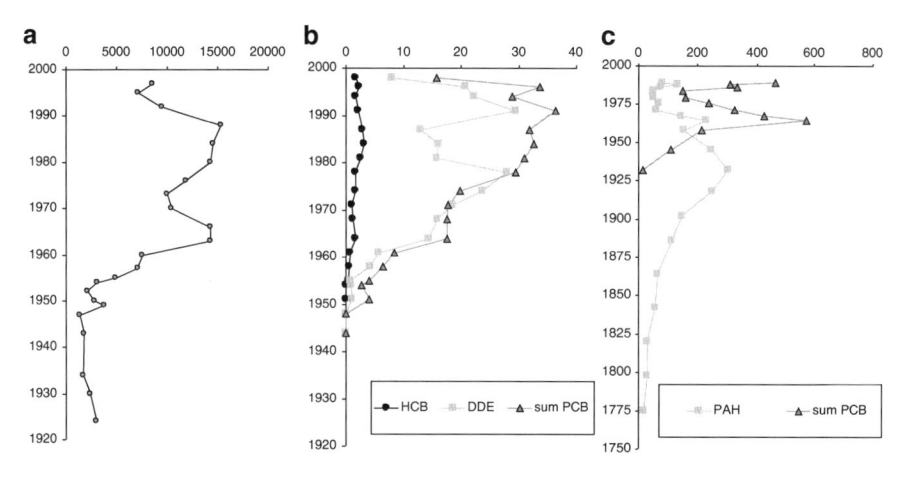

Fig. 11 Downcore concentrations (ng/g) of ∑PAH (**a**) and OCs (**b**) in a sediment core from the High Tatras (46.163°N; 20.106°E) (adapted from [50, 51]) (**c**) concentrations of ∑PAH (mg/g) and ∑PCB (µg/g) from a peat core in the flaxmere bog in northwestern UK (adapted from [6])

5 Atmospheric POP Levels in Europe in the Twenty-First Century

Since the beginning of the twenty-first century, atmospheric POP concentrations in the European Union are measured regularly in different sites of Europe as part of scientific projects or regular atmospheric monitoring programmes, such as the European Monitoring and Evaluation Programme (EMEP) driven under the Convention on Long-Range Transboundary Air Pollution of the United Nations (UN-LRTAP). This monitoring programme is similar to the one run in the Arctic region [Arctic Monitoring and Assessment Programme (AMAP)] or other programmes existing at regional or national scale, such as TOMPS in UK, NJADN in New Jersey, USA and CBADS in Chesapeake Bay, USA. Moreover, national governmental institutions report to the European Commission their national estimates on POP emissions, which are published annually by the European Environmental Agency.

Obviously, uncertainties are inherent to measurements of atmospheric POPs and emission estimates. Furthermore, the coverage of monitoring stations may not encompass all different European environments [78, 79]. Nevertheless, the published data gives an appraisal of the atmospheric POP concentrations over the European continent. As an example, the year-round trend of PCB concentration and

---

Fig. 10 (continued) phenanthrene, *ANT* anthracene, *FLA* fluoranthene, *PYR* pyrene, *BANT* benz[*a*]anthracene, *CRY* + *TRIP* chrysene + triphenylene, *BBJFLA* benzo[*b*+*j*]fluoranthene, *BKFLA* benzo[*k*]fluoranthene, *BEP* benzo[*e*]pyrene, *BAP* benzo[*a*]pyrene, *PER* perylene, *IP* indeno[123-*cd*]pyrene, *BGP* benzo[*ghi*]perylene, *DBA* dibenz[*ah*]anthracene

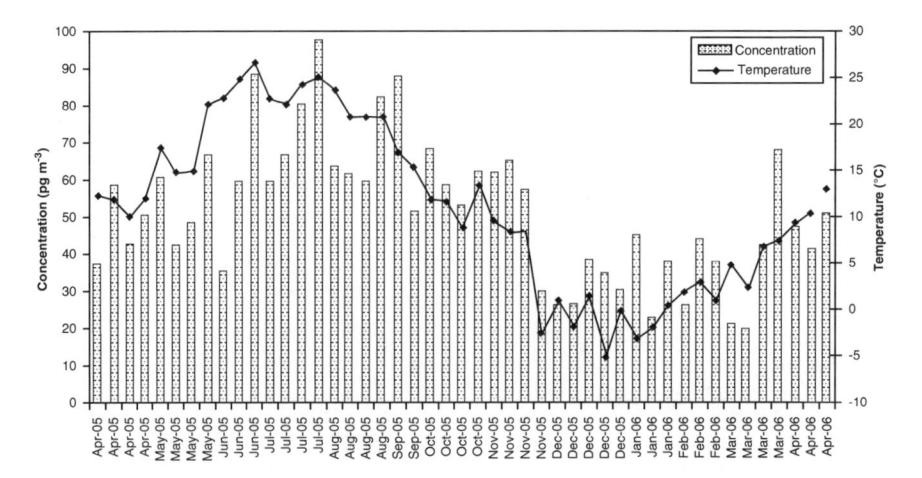

Fig. 12 Weekly air concentration (gas + particle phase) of the summed seven ICES PCB congeners and ambient air temperature from April 2005 to 2006 in the semi-rural area of Ispra, northern Italy [80]

ambient air temperatures for a semi-rural site in northern Italy is shown in Fig. 12 [80]. Summertime concentrations are about twice as high as wintertime concentrations, due to the influence of temperature (e.g. ambient air PCB concentrations increase with increasing temperature due to volatilization from soil or water). On the other hand, PAHs in the same site show much higher concentrations in cold than in warmer periods (Fig. 13), which is likely due to increases of biomass combustion emissions for domestic heating and litter removal in fields and gardens [81].

Reports of the European Environmental Agency (e.g. [7]) have shown several national emission inventories and have described the temporal trend of POP emissions between 1990 and 2009 (Fig. 14). During this period, the reported POP emissions have decreased considerably to levels that in 2009 were less than half of those in 1990. In the case of PAH, the mean reduction was 61%, and this was due to the decreased residential use of coal for domestic heating, to the improvement in the technologies for metal refining and smelting and to regulations on emissions for road transport. Since 2003, the emissions of these compounds do not show variations. The average HCB emissions decreased 98% in Europe as consequence of regulations in the chemical industry, although the secondary aluminium industry is responsible for a relative increase of HCB emissions. HCH emissions decreased 80% by changes in agriculture uses. PCB emissions decreased about 75% due to changes in electronic equipment, while there was a relative emission increase as consequence of industrial waste incineration.

POP concentrations in the European atmosphere can be estimated from these national emission inventories by using model calculations, such as the MSCE–POP model. This is a three-dimensional Eulerian multi-compartment model operating within the geographical scope of EMEP region with a spatial resolution of 50 km × 50

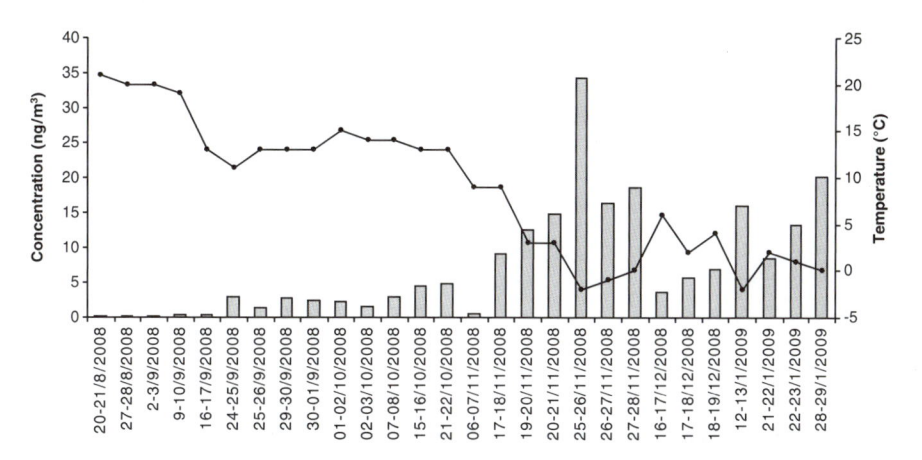

Fig. 13 ∑PAH particle phase concentrations (ng/m^3; *left* y-axis) and ambient air temperature from summer 2008 to winter 2009 (°C; *right* y-axis) in the semi-rural area of Ispra, northern Italy [81]

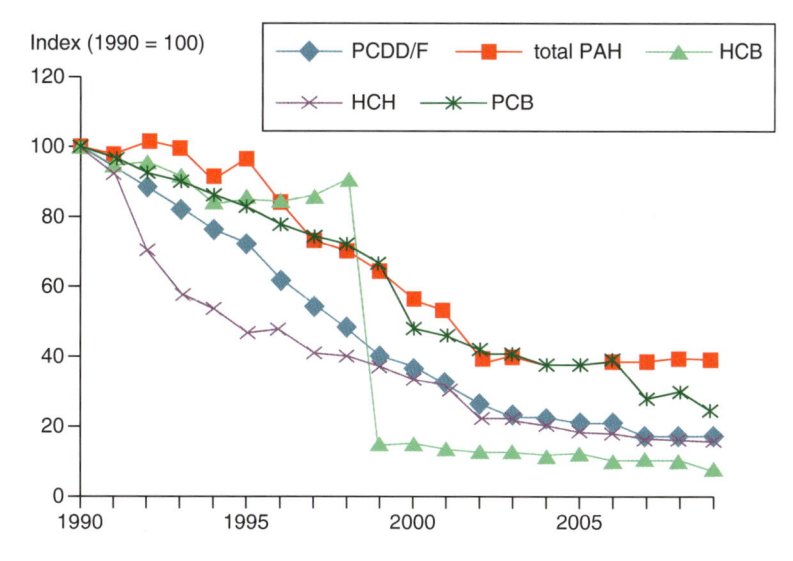

Fig. 14 Emission trends of several POPs between 1990 and 2009 [7]

km [86]. Examples of these model calculations of the atmospheric emissions and concentrations of PCB153 and benzo[*a*]pyrene are shown in Figs. 15 and 16. They show that both combustion-related compounds, such as PAH and industrial products, such as PCBs, have larger emissions in urbanised than in rural areas. However, in some semi-rural areas, such as northern Italy or southern Poland, large PAH emissions are reported, mainly due to induced biomass and coal combustion for domestic heating and specific meteorological conditions in winter (e.g. stagnant atmospheric conditions) [81, 82]. These model calculations predict ambient air concentrations on the European

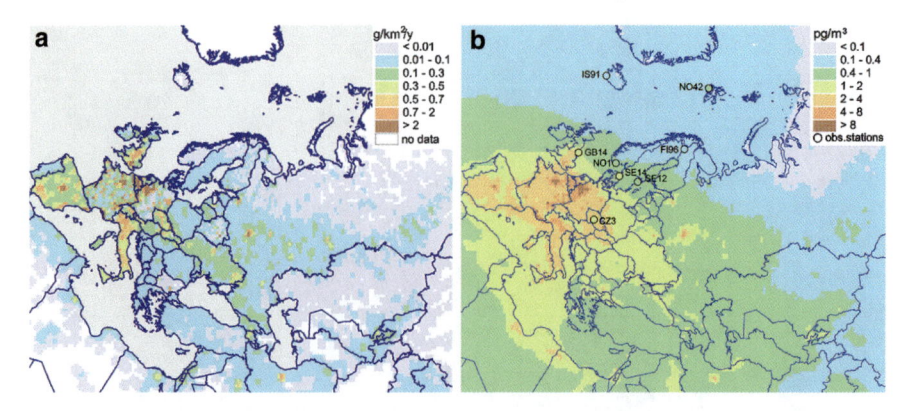

Fig. 15 Estimated environmental distributions of PCB 153: (**a**) emissions to the atmosphere; (**b**) ambient air concentrations over the European continent (Gusev et al. 2007)

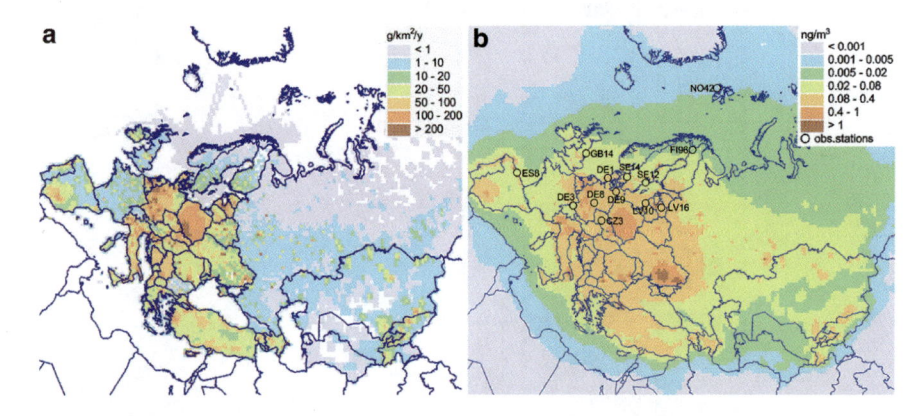

Fig. 16 Estimated environmental distributions of benzo[*a*]pyrene: emissions to the atmosphere (**a**) and ambient air concentrations over Europe (**b**) (Gusev et al. 2007)

continent between 1–8 pg/m^3 for PCB153 and 0.005–1 ng/m^3 for benzo[*a*]pyrene which are in agreement with the ambient air POP measurements on the European continent (Figs. 15b and 16b).

Results from the EMEP observation stations that annually report the monthly POP concentrations [EMEP/ccc-reports (http://www.nilu.no/projects/ccc/reports)] are shown in Figs. 15b and 16b. These stations are located in central and northern Europe, so temporal information on POP concentrations in southern Europe, such as the Mediterranean, is missing. The present network of monitoring stations shows that in the period between 2002 and 2009, the lowest concentrations were observed at high latitudes in Spitsbergen, in the Arctic sea and that they increased with decreasing latitudes, with highest concentration in central Europe (e.g. Czech Republic).

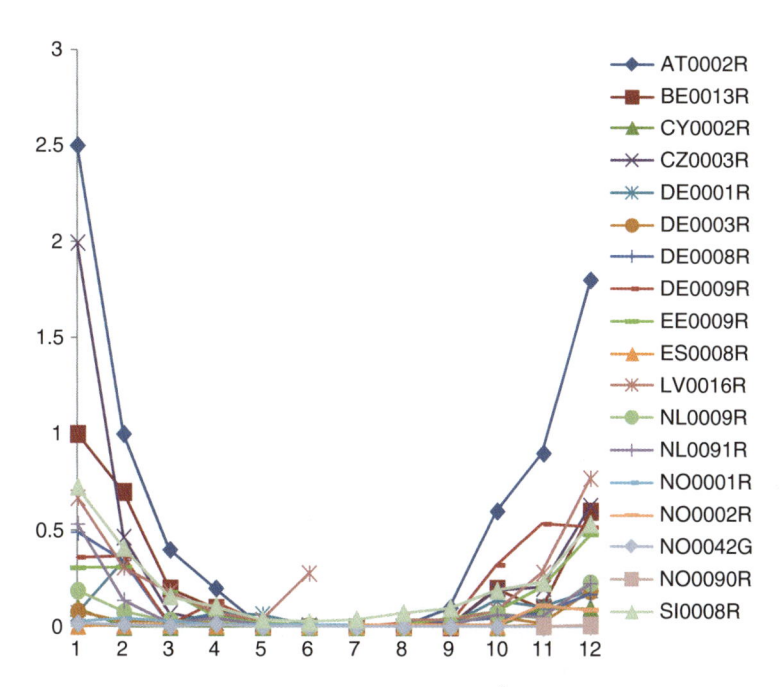

Fig. 17 Monthly concentrations of benzo[a]pyrene (ng/m^3) in the ambient air of the EMEP stations in 2009 (1 = January; 12 = December) (EMEP/ccc-report, 2011)

The temporal evolution of the PCB levels in central Europe involved a decrease of 90% within 4 years (between 2002 and 2006) to more stable levels between 2006 and 2009. The latest concentrations were comparable to those observed in 2009 in Germany (EMEP/ccc-report, 2011). A well-defined reduction in PCB air concentrations was observed in the Czech Republic during this second period in which all other stations showed more or less uniform concentrations.

Other POPs, such as γ-HCH, showed very significant reductions in the ambient air concentrations, which was in agreement with the reductions of emissions from agriculture. The concentrations of these compounds in northern Europe are at present between 1 and 4 pg/m^3 and around 20 pg/m^3 in central Europe. On the other hand, HCB did not show any decrease in atmospheric levels and remained around 50 pg/m^3 on the European continent and at 5 pg/m^3 in Iceland.

Ambient air concentrations of combustion products, such as PAH, showed more or less constant values in the period between 2002 and 2009; only one site in northern Finland showed a significant decrease. These results are in agreement with the reported PAH emissions, which were also constant in this period. The concentrations of benzo[a]pyrene were lowest in Spitsbergen (0.003 ng/m^3) and highest in the Czech Republic (0.36 ng/m^3). As mentioned above, the PAH concentrations showed strong seasonal variations. This is again illustrated in Fig. 17, showing high increases in the cold months due to the use of fossil fuels for heating. These observed concentrations are in agreement with the emission inventories reporting that residential emission sources are responsible for almost half of the PAH emissions on the European continent.

6 Future Perspectives

Under the Stockholm Convention and other agreements, the European countries compelled themselves to eliminate or reduce the production, use and emissions of POPs and to decrease human and ecosystem exposure. Now, one aspect under discussion is the extension of the pollutants with POP properties to new compounds needing environmental monitoring. Several candidates have been intensively studied in the recent years. One obvious requirement concerns the proven toxicity effects of the molecules under study.

The emission reports and air concentration measurements indicate that the atmospheric concentrations of most of the well-studied POP, e.g. PCBs, HCH and HCB in Europe, declined between 1990 and 2000 but decreased more slowly during the last decade. The overall reduction results from the international actions undertaken in the last decades for the elimination/reduction of production and use of these compounds. However, there are still primary emissions from diffuse sources, such as old stockpiles or old equipment, which are often hard to target. Model calculation using different policy scenarios show that between now and 2020, little decrease on the emissions should be expected for PCBs, HCB, HCHs and PAHs [79] because most of the reduction programmes have already been completed. The ubiquitous presence of these persistent compounds in the European ecosystems involve that soils, vegetation and water bodies constitute secondary emission sources for POP re-emissions to the atmosphere [10]. This scenario outlines the need for a comprehensive assessment of the dynamics of POPs in the European ecosystems, particularly in the mountain areas in which freshwater is primarily accumulated. The interdependences between long-range transport, degassing and deposition involve a background contamination of these water bodies at the onset of water accumulation in lakes and, subsequently, when transferred to rivers.

In this context, the progressive warming associated to human-driven climate change may also add to the redistribution of POPs in the European environment. At increasing ambient air temperatures, higher degassing of POP flux towards the atmosphere in high mountain regions is expected [83]. The increase of ambient air temperatures in the Arctic region has also been related to the re-volatilisation of POP to the atmosphere [84]. In the European continent, the accelerated retreat of glaciers due to higher temperature has also led to an induced release of POP to meltwater that is reflected in the increasing POP concentrations in sediments from lakes that are fed by this water. These changes have not been observed in lakes without glacier input [85].

References

1. AMAP (1998) Arctic pollution issues: a state of the arctic environment report. Arctic Monitoring and Assessment Programme, Oslo
2. Grimalt JO, Fernández P, Berdie L, Vilanova RM, Catalan J, Psenner R, Hofer R, Appleby PG, Rosseland BO, Lien L, Massabuau LC, Batterbee RW (2001) Selective trapping of organochlorine compounds in mountain lakes of temperate areas. Environ Sci Technol 35:2690–2697

3. Carson RL (1962) Silent spring. Houghton Mifflin Company, NY
4. Jensen S (1966) Report of a new chemical hazard. New Sci 32:612
5. Kjeller L-O, Rappe C (1995) Time trends in levels, patterns, and profiles for polychlorinated dibenzo-p-dioxins, dibenzofurans, and biphenyls in a sediment core from the Baltic proper. Environ Sci Technol 29:346–355
6. Sanders G, Jones KC, Hamilton-Taylor J (1995) PCB and PAH fluxes to a dated UK peat core. Environ Pollut 89:17–25
7. EEA (2011) European Union emission inventory report 1990–2009 under the UNECE convention on long-range transboundary air pollution (LRTAP). European Environmental Agency, Technical Report N°9
8. Hung H, Halsall CJ, Blanchard P, Li HH, Fellin P, Stern G, Rosenberg B (2001) Are PCBs in the Canadian Arctic atmosphere declining? Evidence from 5 years of monitoring. Environ Sci Technol 35:1303–1311
9. Hung H et al (2010) Atmospheric monitoring of organic pollutants in the Arctic under the Arctic Monitoring and Assessment Programme (AMAP): 1993–2006. Sci Total Environ 408:2854–2873
10. Nizzetto L et al (2010) Past, present, and future controls on levels of persistent organic pollutants in the global environment. Environ Sci Technol 44:6526–6531
11. Porta M, Malats N, Jariod M, Grimlat JO, Rifà J, Carrato A, Guarner L, Santiago-Silva M, Corominas JM, Andreu M, Real FX (1999) Serum levels of organochlorine compounds and K-ras mutations in exocrine pancreatic cancer. Lancet 354:2125–2129
12. Vreugdenhil HJI, Slijper FME, Mulder PGH, Weisglas-Kuperus N (2002) Environ Health Perspect 110:593–598
13. van Drooge BL, Grimlat JO, Torres C, Cuevas E (2001) Deposition of semi-volatile organochlorine compounds in the free troposphere of the eastern Atlantic Ocean. Mar Pollut Bull 42(8):628–634
14. Mackay D, Shui WY, Ma KC (1992) Illustrated handbook of physical-chemical properties and environmental fate for organic chemicals, vol I and II. Lewis Publishers, London
15. Cotham WE, Bidleman TF (1989) Degradation of malathion, endosulphane and fenvalerate in seawater and seawater/sediment microcosms. J Agric Food Chem 37:824–828
16. Ballschmiter K, Wittlinger R (1991) Interhemisphere exchange of hexachlorocyclohexanes, hexachlorobenzene, polychlorobiphenyls, and 1,1,1-trichloro-2,2-bis(p-chlorophenyl)ethane in the lower troposphere. Environ Sci Technol 25:1103–1111
17. Fischer RC, Kramer W, Ballschmiter K (1991) Hexachlorocyclohexane isomers as markers in the water flow of the Atlantic Ocean. Chemosphere 23(7):889–900
18. ten Hulscher TEM, van der Velde LE, Bruggeman WA (1992) Temperature dependence of Henry's law constants for selected chlorobenzenes, polychlorinated biphenyls and polycyclic aromatic hydrocarbons. Environ Toxicol Chem 11:1595–1603
19. Dunnivant FM, Elzerman AW, Jurs PC, Hasan MN (1992) Quantitative structure–property relationships for aqueous solubilities and Henry's law constants of polychlorinated biphenyls. Environ Sci Technol 26(8):1567–1573
20. Hargrave BT, Barrie LA, Bidleman TF, Welch HE (1997) Seasonality in exchange of organochlorines between Arctic air and seawater. Environ Sci Technol 31:3258–3266
21. Bailey RE (2001) Global hexachlorobenzene emissions. Chemosphere 43:167–182
22. Grimalt JO, Sunyer J, Moreno V, Amaral O, Sala M, Rosell A, Albaiges J (1994) Risk excess of soft-tissue sarcoma and thyroid cancer in a community exposed to airborne organochlorinated compound mixtures with a high hexabenzene content. Int J Cancer 56:200–203
23. Breivik K, Sweetman A, Pacyna JM, Jones KC (2002) Towards a global historical emission inventory for selected PCB congeners – a mass balance approach. 2. Emissions. Sci Total Environ 290:199–224
24. Safe S (1991) Polychlorinated dibenzo-p-dioxins and related compounds: sources, environmental distribution and risk assessment. Environ Carcinog Ecotoxicol Rev 9:261–302

25. de Voogd P, Wells DE, Reutergardh L, Brinkman UATh (1990) Biological activity, determination and occurrence of planar, mono-, di-ortho PCBs. Int J Anal Chem 40:1–46
26. Pankow JF (1987) Review and comparative analysis of the theories on partitioning between gas and aerosol particulate phases in the atmosphere. Atmos Environ 21:2275–2283
27. Howsam M, Jones KC (1998) Sources of PAHs in the environment. In: Neilson AH (ed) The handbook of environmental chemistry, vol 3, Part I, PAHs and related compounds. Springer-Verlag
28. Galarneau E (2008) Source specificity and atmospheric processing of airborne PAHs: implications for source apportionment. Atmos Environ 42:8139–8149
29. Wania F, Mackay D (1996) Tracking the distribution of persistent organic pollutants. Environ Sci Technol 30:390–396
30. Gustafsson Ø, Bucheli TD, Kukulska Z, Andersson M, Largeau C, Rouzaud J-N, Reddy CM, Eglinton TI (2001) Global Biogeochem Cycles 15:881–890
31. Bucheli TD, Gustafsson Ø (2000) Quantification of the soot-water distribution coefficient of PAHs provides mechanistic basis for enhanced sorption observations. Environ Sci Technol 34:5144–5151
32. Dachs J, Einsenreich S (2000) Adsorption onto aerosol soot carbon dominates gas-particle partitioning of polycyclic aromatic hydrocarbons. Environ Sci Technol 34:3690–3697
33. Fernández P, Grimalt JO, Vilanova RM (2002) Atmospheric gas-particle partitioning of polycyclic aromatic hydrocarbons in high mountain regions of Europe. Environ Sci Technol 36:1162–1168
34. van Drooge BL, Fernández P, Grimalt JO, Stuchlík E, Torres-García CJ, Cuevas E (2010) Atmospheric polycyclic aromatic hydrocarbons in remote European and Atlantic sites located above the boundary mixing layer. Environ Sci Pollut Res 17:1207–1216
35. Pankow JF, Bidleman TF (1992) Interdependence of the slopes and intercepts from log–log correlations of measured gas-particle partitioning and vapor pressure-I. Theory and analysis of available data. Atmos Environ 26A(6):1071–1080
36. Rosell A, Grimalt JO, Rosell MG, Guardino X, Albaiges J (1991) The composition of volatile and particulate hydrocarbons in urban air. Fresenius J Anal Chem 339:689–698
37. Brubaker WW, Hites RA (1998) OH reaction kinetics of polycyclic aromatic hydrocarbons and polychlorinated dibenzo-p-dioxins and dibenzofurans. J Phys Chem 102:915–921
38. Esteve W, Budzinski H, Villenave E (2006) Relative rate constants for the heterogeneous reactions of NO_2 and OH radicals with polycyclic aromatic hydrocarbons adsorbed on carbonaceous particles. Part 2: PAHs adsorbed on diesel particulate exhaust SRM 1650a. Atmos Environ 40:201–211
39. Ballschmiter K (1992) Transport and fate of organic compounds in the global environment. Angew Chem 31(5):487–515
40. Duce RA, Liss PS et al (1991) The atmospheric input of trace species to the world ocean. Global Biogeochem Cycles 5:193–259
41. Goldberg ED (1975) Synthetic organohalides in the sea. Proc R Soc Lond B 189:277–289
42. Wania F, Mackay D (1993) Global fractionation and cold condensation of low volatility organochlorine compounds in polar regions. Ambio 22(1):10–18
43. Argell C, Okla L, Larsson P, Backe C, Wania F (1999) Evidence of latitudinal fractionation of polychlorinated biphenyl congeners along the Baltic Sea region. Environ Sci Technol 33(8):1149–1156
44. Calamari D, Bacci E, Focardi S, Gaggi C, Morosini M, Vighi M (1991) Role of plant biomass in the global environment partitioning of chlorinated hydrocarbons. Environ Sci Technol 25:1489–1495
45. Ockenden WA, Sweetman AJ, Prest HF, Steinness E, Jones KC (1998) Towards an understanding of the global atmospheric distribution of persistent organic pollutants: the use of semipermeable membrane devices as time-integrated passive samplers. Environ Sci Technol 32:2795–2803
46. Kalantzi OI, Alock RE, Johnston PA, Santillo D, Stringer RL, Thomas GO, Jones KC (2001) The global distribution of PCBs and organochlorine pesticides in butter. Environ Sci Technol 35:1013–1018

47. Galassi S, Valsecchi S, Tartari GA (1997) The distribution of PCB's and chlorinated pesticides in two connected Himalayan lakes. Water Air Soil Pollut 99:717–725
48. Blais JM, Schindler DW, Muir DCG, Kimpe LE, Donald DB, Rosenberg B (1998) Accumulation of persistent organochlorine compounds in mountains of western Canada. Nature 395:585–588
49. Fernández P, Vilanova RM, Martinez C, Appleby P, Grimalt JO (2000) The historical record of atmospheric pyrolytic pollution over Europe registered in the sedimentary PAH from remote mountain lakes. Environ Sci Technol 34:1609–1913
50. Grimalt JO, van Drooge BL, Ribes A, Fernández P, Appleby P (2004) Polycyclic aromatic hydrocarbon composition in soils and sediments of high altitude lakes. Environ Pollut 131:13–24
51. Gimalt JO, van Drooge BL, Ribes A, Vilanova RM, Fernández P, Appleby P (2004) Persistent organochlorine compounds in soils and sediments of European high altitude mountain lakes. Chemosphere 54:1549–1561
52. Ribes S, Van Drooge B, Dachs J, Gustafsson Ø, Grimalt JO (2003) Influence of soot carbon on the soil–air partitioning of polycyclic aromatic hydrocarbons. Environ Sci Technol 37:2675–2680
53. Meijer SN, Ockenden WA, Sweetman A, Breivik K, Grimalt JO, Jones KC (2003) Global distribution and budget of PCBs and HCB in background surface soils: implications for sources and environmental processes. Environ Sci Technol 37:667–672
54. Kömp P, McLachlan MS (1997) Interspecies variability of the plant/air partitioning of polychlorinated biphenyls. Environ Sci Technol 31:2944–2948
55. Kömp P, McLachlan MS (1997) Octanol/air partitioning of polychlorinated biphenyls. Environ Toxicol Chem 16(12):2433–2437
56. Simonich SL, Hites RA (1995) Organic pollutant accumulation in vegetation. Environ Sci Technol 29:2905–2914
57. Grimalt JO, van Drooge BL (2006) Polychlorinated biphenyls in mountain pine (*Pinus uncinata*) needles from Central Pyrenean high mountains (Catalonia, Spain). Ecotoxicol Environ Saf 63:61–67
58. Lohmann R, Lammel G (2004) Adsorptive and absorptive contributions to the gas-particle partitioning of polycyclic aromatic hydrocarbons: state of knowledge and recommended parametrization for modelling. Environ Sci Technol 38:3793–3803
59. Seinfeld JH, Pandis SN (2006) Atmospheric chemistry and physics – from air pollution to climate change. John Wiley & Son
60. Macleod M, Mackay D (2004) Modeling transport and deposition of contaminants to ecosystems of concern: a case study for the Laurentian Great Lakes. Environ Pollut 128:241–250
61. Cotham WE, Bidleman TF (1991) Estimating the atmospheric deposition of organochlorine contaminants to the arctic. Chemosphere 22(1–2):165–188
62. Carrera G, Fernandez P, Vilanova R, Grimalt JO (2001) Persistent organic pollutants in snow from European high mountain areas. Atmos Environ 35:245–254
63. Carrera G, Fernandez P, Grimalt JO, Ventura M, Camarero L, Catalan J, Nickus U, Thies H, Psenner R (2002) Atmospheric deposition of organochlorine compounds to remote high mountain lakes of Europe. Environ Sci Technol 36:2581–2588
64. Lei YD, Wania F (2004) Is rain or snow a more efficient scavenger of organic chemicals? Atmos Environ 38:3557–3571
65. Fernández P, Grimalt JO (2003) On the global distribution of persistent organic pollutants. Chimia 57:514–521
66. van Drooge BL, López J, Fernández P, Grimalt JO, Stuchlík E (2011) Polycyclic aromatic hydrocarbons in lake sediments from the High Tatras. Environ Pollut 159:1234–1240
67. Karickhof SW, Brown DS, Scott TA (1979) Sorption of hydrophobic pollutants on natural sediments. Water Res 13:241–248
68. Chiou CT, Porter PE, Schmedding DW (1983) Partitioning equilibria of nonionic organic compounds between soil organic matter and water. Environ Sci Technol 17:227–231

69. Mackay D, Powers B (1987) Sorption of hydrophobic chemicals from water: a hypothesis for the mechanism of the particle concentration effect. Chemosphere 16:745–757
70. Baker JE, Eisenreich SJ, Eadie BJ (1991) Sediment trap fluxes and benthic recycling of organic carbon, polycyclic aromatic hydrocarbons, and polychlorobiphenyls in Lake Superior. Environ Sci Technol 25:500–509
71. Swackhamer DL, Skoglund RS (1993) Bioaccumulation of PCBs by Algae: kinetics versus equilibrium. Environ Toxicol Chem 12:831–838
72. Vilanova RM, Fernandez P, Grimalt JO (2001) Polychlorinated biphenyls partitioning in the waters of a remote mountain lake. Sci Total Environ 279:51–62
73. Vilanova RM, Fernandez P, Martinez C, Grimalt JO (2001) Polycyclic aromatic hydrocarbons in remote mountain lake waters. Water Res 35:3916–3926
74. Vilanova R, Fernandez P, Martinez C, Grimalt JO (2001) Organochlorine pollutants in remote mountain lake waters. J Environ Qual 30:1286–1295
75. Barron M (1990) Bioconcentration. Environ Sci Technol 24:1612–1618
76. Clark KE, Gobas FAPC, Mackay D (1990) Model of organic chemical uptake and clearance by fish from food and water. Environ Sci Technol 24:1203–1213
77. Vives I, Grimalt JO, Catalan J, Rosseland BO, Batterbee RW (2004) Influence of altitude and age in the accumulation of organochlorine compounds in fish from high mountain lakes. Environ Sci Technol 38:690–698
78. Breivik K, Vestreng V, Rozovskaya O, Payna JM (2006) Atmospheric emissions of some POPs in Europe: a discussion of existing inventories and data records. Environ Sci Policy 41:9245–9261
79. Denier van der Gon H, van het Bolscher M, Visschedijk A, Zandveld P (2007) Emissions of persistent organic pollutants and eight candidate POPs from UNECE-Europe in 2000, 2010 and 2020 and the emission reduction resulting from the implementation of the UNECE POP protocol. Atmos Environ 41:9245–9261
80. Castro-Jiménez J, Dueri S, Eisenreich SJ, Mariani G, Skejo H, Umlauf G, Zaldívar JM (2009) Polychlorinated biphenyls (PCBs) in the atmosphere of sub-alpine northern Italy. Environ Pollut 157:1024–1032
81. van Drooge BL, Pérez-Ballesta P (2009) Seasonal and daily source apportionment of polycyclic aromatic hydrocarbon concentrations in PM_{10} in a semirural European area. Environ Sci Technol 43:7310–7316
82. Umlauf G, Cristoph EH, Eisenrreich SJ, Mariani G, Paradiž B, Vives I (2009) Seasonality of PCDD/Fs in the ambient air of Malopolska Region, southern Poland. Environ Sci Pollut Res. doi:10.1007/s11356-009-0215-4
83. Grimalt JO, Catalan J, Fernandez P, Piña B, Munthe J (2010) Distribution of persistent organic pollutants and mercury in freshwater ecosystems under changing climate conditions. In: Kernan M, Battarbee RW, Moss B (eds) Climate change impacts on freshwater ecosystems. Wiley-Blackwell, Chichester, pp 180–202 (Chapter 8)
84. Ma J, Hung H, Tian C, Kallenborn R (2011) Revolatilization of persistent organic pollutants in the Arctic induced by climate change. Nat Clim Chang 1:255–260
85. Bogdal C, Schmid P, Zennegg M, Anselmetti F, Schreringer M, Hungerbühler K (2009) Blast from the past: melting glaciers as a relevant source for persistent organic pollutants. Environ Sci Technol 43:8173–8177
86. Gusev A, Mantseva E, Rozovskaya O, Shatalov V, Vulykh N, Aas W, Breivik K (2007) Persistent Organic Pollutants in the Environment. EMEP Status Report 3/2007

Wildfires as a Source of Aerosol Particles Transported to the Northern European Regions

Sanna Saarikoski and Risto Hillamo

Abstract Each year large areas of forested land in Europe are burned by more than 50,000 fires. Over the past few years, climatic anomalies in temperature and precipitation have resulted in an increase in fire events. The exceptional fire occurrences in the 2000s and their regional consequences on atmospheric air quality have been observed in the northern European regions. In the last 10 years almost annually the episodes of long-range transported (LRT) biomass smokes from Eastern European fires have been reported, exceptionally intense smoke plumes having been detected in 2002 and 2006. Typically, the smoke episodes occur in spring or/and late summer and they last for few days. As the particulate matter (PM) concentrations are generally quite low in Northern Europe, the LRT smoke plumes increase the PM concentrations in several folds even at the background sites with no local emissions. As a result, there are exceedances in the European Union PM daily limit values, which result in serious health problems. This chapter describes the episodes of wildfire particles observed in Northern Europe in the last 10 years. It discusses the chemical and physical properties of particles, the transformation during the transport as well as the methods to investigate the composition and source areas of smoke plumes.

Keywords Aerosols, Chemical composition, Long-range transport, Wildfires

Contents

1 Introduction 102
2 LRT Wildfire Smoke Episodes Detected in Northern Europe in 2002–2010 103
3 Experimental Methods 107

S. Saarikoski (✉) and R. Hillamo
Finnish Meteorological Institute, Erik Palménin aukio 1, 00560 Helsinki, Finland
e-mail: Sanna.Saarikoski@fmi.fi

3.1 Smoke-Specific Methods 107
3.2 General Characterization of LRT Smoke Particles 108
4 Properties of LRT Smoke Particles 109
  4.1 Physical Properties 109
  4.2 Chemical Composition 111
  4.3 Transformation During Transport 114
5 Modeling 115
6 Health Effects 116
7 Conclusions 117
References 118

1 Introduction

The exceptional fire occurrences in continental Eurasia in the 2000s and their regional consequences on atmospheric air quality have been observed in the northern European regions. Each year about 5,000 km^2 of forested land in Europe is burned by more than 50,000 fires. Wildfires occur in all European countries but they are particularly intensive in the arid southern and eastern regions. Over the past few years, climatic anomalies in temperature and precipitation have resulted in an increase in wildfire events across the boreal Russian Federation. More than about 7 Mha of land are burned in Russia each year through forest fires, peat fires, and agricultural burning [1]. The term wildfires denotes here open fires of various vegetation, e.g., forest, grasslands, agricultural residue, and peat. Therefore, it refers to fires caused by both natural and anthropogenic causes.

Abundance and development of biomass fires have been observed in Eastern Europe every year by satellites. Main fire seasons are spring and late summer. Most fires are caused by the human activity, either intentionally or accidentally. In springtime wildfires are usually as a result of agricultural burning, a traditional cultivation technique to burn the fields before the start of the new growing season. That practice is illegal in the European Union but is still widely used in many European countries. It is quite common that agricultural fires get out of control and extend to nearby forests or human property. The uncontrolled use of fire for clearing forest and woodland for agriculture is estimated to account for up 90% of world's wildfires. Russia was the largest contributor to agricultural burning globally since it produced 31–36% of all agricultural fires in 2001–2003 [2].

The number and timing of fires vary with years, which are mainly caused by differences in meteorological conditions. Prolonged dry and hot periods (e.g., summer 2002 and 2006) are favorable for fire ignition and burning. In general in Europe, the regions most impacted by wildfire emissions, either local or transported, are Eastern and Central Europe as well as Scandinavia [3]. In springtime 5–35% of the European fine fraction particles is attributable to wildfires with the largest contribution observed in April (20–35%).

Wildfires represent a major source of atmospheric particulate matter (PM) especially fine particles with diameter <2.5 μm (PM$_{2.5}$). High ambient fine particle concentrations are associated with serious health problems [4]. At times extensive wildfires can have a larger impact on air quality than other common anthropogenic sources, in both urban and rural areas. Fire emissions are not just local issue since particle emissions can be transported over long distances, leading to regionally elevated loadings.

In Northern Europe the concentrations of particulate matter are typically quite small compared with the more polluted regions in Europe [5, 6]. In Northern Europe annual average $PM_{2.5}$ concentrations are ~10 μg m^{-3} in urban areas, whereas in Eastern and Southern Europe the concentrations are two or three times larger [7]. In areas with low local emissions the composition and the mass concentrations of fine particles are substantially affected by the long-range transport (LRT) from other areas, even from thousands of kilometers away. For example in the urban areas of Helsinki, Finland, 64–76% of the $PM_{2.5}$ mass was estimated to originate from the LRT in 1998–2002 [8]. Regarding special pollution episodes, a total of 27 pollution episodes were attributed to LRT in Southern Finland in 2001–2007 [9]. Half of those episodes was caused by smokes from wildfires in Eastern Europe, and the total duration of the episodes was 26 days in 7 years time period.

This chapter describes the episodes of wildfire particles observed in Northern Europe in the last 10 years. Biomass fires were usually located in Russia or other Eastern European countries but the smokes from the fires were transported to the northern European regions. Transport of smoke plumes to Northern Europe is typically caused by low and high pressure centers located in the northern hemisphere. An anticyclonic system over northeastern Europe, the burning area, and a low pressure center over northern Atlantic can create a strong flow from south to north, from the fire region to Northern Europe [10]. Smoke episodes observed in Northern Europe in late summer 2002 and spring 2006 were exceptionally intense with the smoke being carried to regions as far as the European Arctic [10] and UK [11] causing severe air quality deterioration. This chapter presents the smoke events detected in Northern Europe in 2002–2010, the chemical and physical properties of the observed LRT particles as well as the methods to investigate the composition and source areas of smoke plumes.

2 LRT Wildifire Smoke Episodes Detected in Northern Europe in 2002–2010

The episodes of LRT biomass burning smokes have been detected in the northern European regions almost every year in the last 10 years (Table 1). The episodes were observed most extensively in Finland but also in Sweden, Norway, and Denmark. At the seasonal level, the episodes of biomass burning smokes occurred most likely in spring, from March to May, and in late summer in August and September. Exceptionally in 2010, the smoke plumes were detected in Kuopio, Finland, already at the end of July [32]. The duration of the episodes varied from

Table 1 A summary of forest and wildfire episodes detected in Northern Europe in 2002–2010

Year	Month	Location	Type of site[a]	Reference
2002	Mar	Ähtäri, Hyytiälä, Helsinki, Virolahti, Utö (Finland)	B, U	[12]
	Aug to Sep	Ähtäri, Hyytiälä, Helsinki, Virolahti, Utö, Imatra (Finland)	B, U	[13]
		Helsinki (Finland)	U	[14]
		Several sites at Southern Finland	U	[15]
2004	Apr to May	Helsinki (Finland)	U	[16]
		Hyytiälä (Finland)	B	[17]
2006	Apr to May	Helsinki (Finland)	U	[18–20]
		Kotka (Finland)	U	[21]
		Virolahti (Finland)	B	[22, 23]
		Helsinki and Jokioinen (Finland)	B, U	[24]
		Several sites in Sweden and in Pallas (Finland)	B, U	[25, 26]
		Spitsbergen (Norway)	B	[10, 27, 28]
	Aug	Helsinki (Finland)	U	[18, 20]
		Virolahti (Finland)	B	[22, 23]
2007	Mar	Helsinki (Finland)	U	[9]
2008	Apr	Hyytiälä, Kuopio, Virolahti, Utö (Finland)	B	[29]
2009	Apr	Helsinki (Finland)	U	[30]
	Aug to Sep	Birkenes (Norway), Hyytiälä (Finland), Lille Valby (Denmark), Vavihill (Sweden)	B	[31]
2010	July to Aug	Kuopio (Finland)	B, U	[32, 33]

[a]*B* Background, *U* Urban

year-to-year and site-to-site. Usually elevated concentrations lasted from a day to several days, but for example in August 2006, the average duration of intensive smoke plumes detected in Helsinki was only few hours [18]. The source areas for all the LRT smokes observed in Northern Europe were Russia and other eastern European countries including, e.g., Ukraine, Belarus, and Baltic countries. LRT smoke plumes are usually wide in size and have been observed at both the background and urban sites (Table 1).

Typically, the LRT smoke episodes are first detected by the increase in the PM concentration at the measurement site. If adequate real-time instruments are available, also the changes in the physical properties (e.g., particle size) and chemical composition of particles can be observed. Every PM pollution episode observed in Northern Europe is not associated with LRT smokes. Therefore, the origin of the polluted air masses must be identified, e.g., by using the backward air mass trajectory models (e.g., HYSPLIT provided by the National Oceanic and Atmospheric Administration or FLEXTRA; [34]), that shows the path of air masses before arriving the measurement site. Additionally, Web Fire Mapper (http://maps.geog.umd.edu) shows the location and intensity of fires areas obtained from MODerate-resolution Imaging spectroradiometer (MODIS; [10]) onboard the satellites.

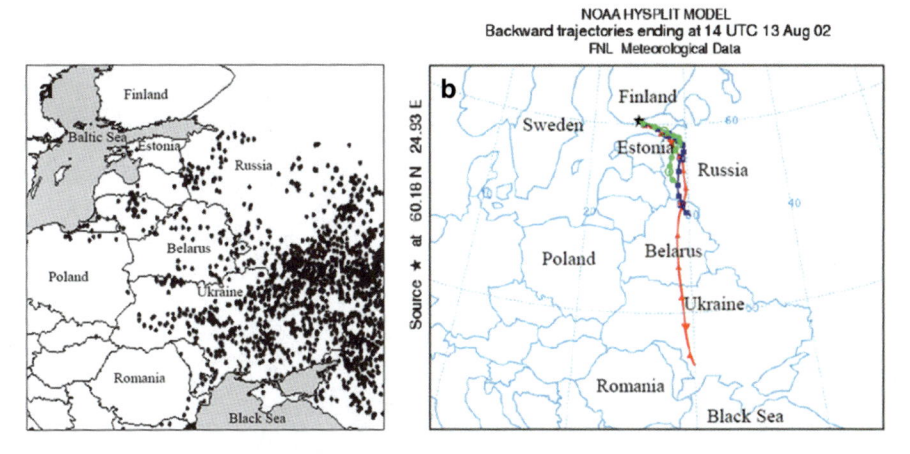

Fig. 1 Fires in Eastern Europe detected by MODIS on 10–15 August 2002 (**a**) and backward air mass trajectories to Helsinki on 13 August 2002 (**b**) [13]

During the last 10 years the strongest episodes caused by the LRT smokes were observed in Northern Europe in August to September 2002 and in April to May 2006. The smoke episode in August and September 2002 was detected in Finland in three parts: first on 12–15 August, second on 26–28 August, and third on 5–6 September 2002 [13]. The episodes were the strongest in Southern Finland, but they were also experienced in the more northern parts of Finland. The third episode on 5–6 September was especially strong in southeast Finland where the maximum hourly PM_{10} concentrations were 200–220 µg m^{-3}. In addition to Northern Europe, these major episodes in late summer 2002 were also observed in the UK [11], but the episodes occurred in the UK almost a week later than in Finland.

August and the beginning of September 2002 were exceptionally warm and dry in Finland and in the neighboring areas. Satellite observations indicated that there were many forest and peat fires in Russia, Ukraine, Belarus, and the Baltic countries (Fig. 1a) leading to the accumulation of large smoke plumes over Eastern Europe that were later transported to Northern Europe (Fig. 1b). Results from Vilnius, Lithuania, showed that the PM_{10} concentrations were elevated significantly close to the fire areas [35]. Twenty-four-hour average PM_{10} rose to above 250 µg m^{-3} at the maximum in Vilnius in August and September 2002. In Vilnius region, more than 976 ha of land were burned in 2 months period so it is possible that smokes from those fires were also transported to Northern Europe.

Another major smoke episode occurred in Northern Europe in late April and in early May 2006. At that period there was a large number of fires in the Baltic countries, Western Russia, Belarus, and the Ukraine. Compared to the spring 2002 when the smoke episode was observed in Finland already in the middle of March (Table 1; [12]), in 2006 there was a lot of snow on the ground in Eastern Europe until the end of March. Therefore, the farmers had to wait with the burning for the unusually late snow melt, causing high frequency of fires at the end of April and beginning of May.

The duration of the smoke episode observed in April to May 2006 was extraordinarily long. For example in Helsinki, elevated $PM_{2.5}$ concentrations lasted almost a period of 2 weeks [19]. There was a long-lasting anticyclonic system over Western Russia which caused the smoke particles to be transported northward of the burning region. During the episode the visibility decreased remarkably and the air was dry. In Helsinki the highest 30-min average $PM_{2.5}$ concentration was 69 µg m^{-3} [19]. Coarse particle concentration ($PM_{2.5-10}$) increased as well but less than $PM_{2.5}$. Besides in Helsinki, the smokes from biomass burning were observed in Southern Finland in Kotka [21] and in Virolahti [22]. In Northern Finland elevated concentrations were reported from Pallas [25].

In Sweden the spring 2006 wildfire episode was detected in two parts: first from 24 April to 3 May and later on 5–9 May [25]. The first part was observed only in the eastern and northern parts of Sweden whereas the second part of the episode was also noticed in Western and Southern Sweden. PM concentrations rose remarkably in Sweden, at stations around Stockholm the PM_{10} concentrations reached the values equal to 70 µg m^{-3} being 5–9 times higher than in the reference period. PM_{10} was also elevated in Denmark from 3 to 8 May with the maximum PM_{10} concentrations of 70–90 µg m^{-3} [11]. In Norway PM_{10} levels showed a slight increase above the baseline from 2 to 12 May but there were no obvious peaks and the concentrations generally remained below 60 µg m^{-3}. However, as aerosols were not removed from the atmosphere to a significant extent by precipitation, a severe air pollution episode was observed in the Norwegian Arctic at Spitsbergen on 27 April and early May 2006 [10]. In addition to Northern Europe, high particle concentrations were also observed in other areas in Europe, e.g., in Lithuania [36], UK, and Germany [11]. At the same time with the smoke episodes, there was also a sudden release of pollen in European countries, mainly from birch and pine trees [11]. Pollen was observed in the smoke plume at Spitsbergen as it was transported across Scandinavian [10].

In contrast to the nearly constantly elevated PM concentrations in spring 2006, in August 2006 there was a series of short-term smoke plumes. In August 2006 the fires were close to Helsinki (~200–300 km from Helsinki; [18]), whereas in spring the smoke particles had been in the atmosphere probably for days. Besides different source areas and transport times, the burning material was assumed to be different in August from spring: in spring the burning material was dry hay and agricultural biomass from previous season, whereas in August mainly bog peat and fresh forest were burning. The average $PM_{2.5}$ mass concentration during the smoke plumes was equal to 45 µg m^{-3} in Helsinki with the highest 1-h average concentration peak being 180 µg m^{-3} on 21 August.

In 2010 smokes from the extensive wildfires in the western part of Russia, near Moscow, were observed in Eastern Finland, Kuopio [32, 33]. Kuopio is around 1,000 km northwest of Moscow and the transport time from the fire area was estimated to be around 1–2 days. Especially on 2 days, on 29 July and 8 August, the influence of wildfire smokes was clear with the highest $PM_{2.5}$ and PM_{10} concentrations around 50 and 60–70 µg m^{-3}, respectively. There was a difference in the vertical location of the plume between the 2 days. In July the plume was located close to the surface, mainly below 2 km, whereas on 8 August the smoke plume had two layers; one between 1 and 2 km and the other between 2.5 and 4 km.

3 Experimental Methods

There are several methods how the physical and chemical properties of the wildfire aerosol particles can be investigated, but only few of them are specific for smoke components. In general, the characteristics of smoke-related aerosols can be examined with ground-based observations, remote sensing instruments, or space born instruments. Ground-based instruments include both direct measurements (real-time methods) and laboratory (filter/impactor-based) methods proving information on the particles close to the ground. Remote sensing instruments, like sun photometer and lidar, reveal column concentration and vertical distribution of PM layers, respectively. Space born instruments comprise of the instruments onboard the satellites orbiting the Earth.

3.1 Smoke-Specific Methods

Biomass burning smoke particles contain some inorganic and organic compounds that are specific for biomass burning smoke. Monosaccharide anhydrides (MAs), levoglucosan and its isomers mannosan and galactosan, are probably the most exclusive tracers for biomass combustion since they are formed solely from the incomplete burning of cellulose and hemicelluloses [37]. MAs are usually determined by using either gas chromatography–mass spectrometer [37], liquid chromatography–mass spectrometer [31], or anion-exchange chromatography mass spectrometer [38]. The disadvantage of using MAs for identifying the smoke episodes is that they are usually analyzed from the filter samples, and the collection time varies typically from a day to several days. Therefore, the sample is often a mixture of smoke particles and the particles that are present in the background air.

In contrast to filter samples, real-time methods enable fast detection of smoke-related particles. An aerodyne high-resolution time-of-flight aerosol mass spectrometer (HR-ToF-AMS) provides information on the whole mass spectrum of the particle mass, which allows the investigation of the selected mass to charge (m/z) ratios specific to biomass burning (m/z 60 and 73; [39]). For example, HR-ToF-AMS enabled to study the biomass smokes with a time resolution of 5 min in April 2009 in Helsinki [30]. Water-soluble ions potassium (K) and oxalate are also often associated with biomass burning smokes [14]. A particle-into-liquid sampler (PILS) combined with an ion chromatograph (IC) allowed to study the evolution of LRT smoke event in spring 2006 with a time resolution of 15 min [19]. However, neither oxalate nor potassium is unique for the biomass burning since they have several other origins.

3.2 General Characterization of LRT Smoke Particles

The most often measured property of the LRT wildfire smokes is the mass concentration of aerosol particles in fine ($PM_{2.5}$) or submicron (PM_1) or in fine and coarse size fraction together (PM_{10}). The mass concentration is usually determined either with a tapered element oscillating microbalance (e.g., [19]), or with the instruments based on β-attenuation (e.g., [13]). In addition to mass concentration, real-time instruments can measure other physical properties of particles, e.g., number concentration of particles in different size fractions using a differential mobility particle sizer [13] or the optical characteristics of smoke by a nephelometer [32]. Ground-based instruments also include remote sensing instruments such as sun photometers [32].

Organic carbon (OC) and elemental carbon (EC) concentrations in smoke aerosol can be measured by using a semicontinuous thermal-optical OC/EC instruments [18]. Water-soluble organic carbon (WSOC) and water-soluble ions can be detected in smoke plumes with a PILS combined with a total organic carbon analyzer [30] and IC [19]. Besides smoke-specific tracers, the HR-ToF-AMS enables to study the concentrations of organic matter (OM), nitrate, ammonium, sulfate, and chloride in smoke particles [30]. Black carbon (BC) can be measured in real-time with several instruments, e.g., with aethalometer [19], multi-angle absorption photometer [29], and particle soot absorption photometer [25].

As already mentioned, smoke particles can also be studied by collecting them on a filter and analyzing that in the laboratory. However, collected smoke particles can transform during the sampling or sample preparation [12]. The advantage of the filter methods is that smoke samples are available for several different chemical analytical techniques. In addition to OC, EC, WSOC, and ions, also the elemental composition of smoke aerosol can be determined from the filter samples [22]. Typically chemical components are analyzed from the bulk samples, but the elemental composition can also be determined for individual smoke particles by scanning electron microscope coupled with an energy-dispersive X-ray microanalyzer [12]. With filter samples the composition of organic fraction in aerosol can be explored in detail. From biomass burning samples, e.g., dicarboxylic acids [14] and polycyclic aromatic hydrocarbons (PAHs; [23]) have been determined.

Most widely used space born measurement for the wildfire studies is MODIS. There are two MODIS's onboard satellites and they cover the Earth's surface every 1–2 days. MODIS gives information on the location of fires and also on the optical properties of aerosol. Optical properties of the smoke particles have also been studied by using the ozone monitoring instrument [32] onboard EOS-Aura satellite. Cloud-aerosol lidar with orthogonal polarization provided the vertical profile of the plume during the smoke episode in 2010 [32]. In addition to comparison with ground-based instruments, the data from the space born instruments has been used in the modeling of smoke plumes (e.g., [10]).

Table 2 Physical properties and chemical components analyzed from the smoke particles in the northern European regions

Property/chemical component	Reference
Levoglucosan	[9, 10, 14, 16–21, 30, 31]
Organic carbon	[10, 16–21, 25, 27, 30, 31]
Water-soluble organic carbon	[18–20, 27, 30]
^{14}C	[31]
BC/BCe^a/EC/LAC^b/BS^c	[9, 10, 14, 16–21, 25–27, 29–31, 33]
Ions	[9, 10, 12–14, 16–23, 27, 30]
Elements	[12]^d; [13, 14]^d; [17]^d; [27]^d; [21–23]
PAHs	[22, 23]
Size-distribution	[10, 14, 27]: [18, 20, 21, 25, 28, 32, 33]
Optical properties	[10, 24, 27, 28, 32, 33]

^{a}Equivalent black carbon
^{b}Light-absorbing carbon
^{c}Black smoke
^{d}Individual particle analysis

4 Properties of LRT Smoke Particles

Particle physical properties typically change under the impact of smoke plume but these changes may not be specific for the wildfire smoke. In addition to biomass burning, particle mass or number concentration can increase due to the biogenic or other anthropogenic sources, e.g., traffic or industrial emissions. Chemical composition of particles is more unique to particle source, however, particles with similar chemistry can have different origin. Physical and chemical properties of the LRT biomass burning particles observed in Northern Europe are discussed below. Physical properties and the chemical components measured from the smoke particles are summarized in Table 2. The measurements of PM mass concentrations are excluded from Table 2 as nearly all the studies had some measurements of particle mass.

4.1 Physical Properties

4.1.1 Particle Mass and Number Concentration

The increase of PM in air is one of the main indicators showing the presence of LRT smoke. The increase in PM due to LRT smoke is more pronounced in PM_1 or $PM_{2.5}$, and most of the mass is usually located in fine particles. However, elevated levels of coarse particles have also been attributed to the LRT smoke [19]. The episodes in April to May and August 2006 can clearly be seen in the PM_1 data from Helsinki for the period of March 2006 to February 2007 (Fig. 2). Also particle number concentration increases with the smoke plume but usually less than the PM mass. In spring

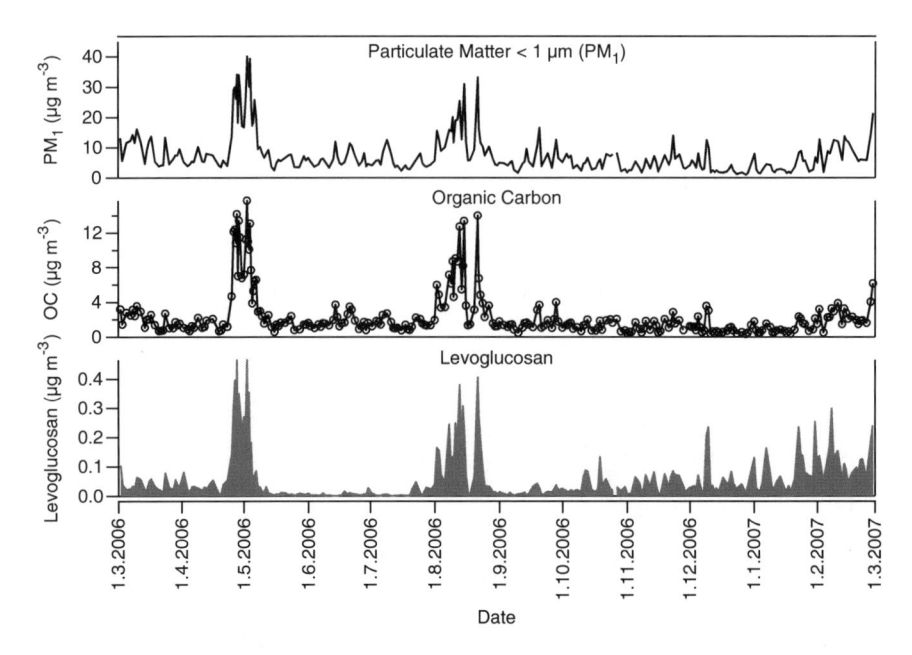

Fig. 2 Daily average concentrations of PM_1, organic carbon, and levoglucosan at an urban background site in Helsinki from March 2006 to February 2007. The elevated concentrations in April to May and in August are due to wildfires. The increase in levoglucosan concentrations during fall and winter is caused by wood burning in domestic heating

2006 at Spitsbergen, the particle number concentrations were about ten times larger during the plume than usually at that time of year [10].

The increase in the number concentration is more pronounced for larger particles. In general, particle number concentration increases for the accumulation mode particles (diameter of 0.1–1 μm) but a decrease of nucleation and Aitken mode particles (<100 nm in diameter) has been observed during the LRT smoke episodes. For example in Hyytiälä in spring 2002, the particle number concentration increased during the smoke episode by a factor of 5.6 in the accumulation mode, decreased slightly in the Aitken mode, and decreased strongly in the nucleation mode [12]. The disappearance of the smallest particles could be due to the condensation of gas-phase components on the larger particles that were abundant in smoke plume. Therefore gas-to-particle conversion could not form new particles during the smoke events. The same behavior was detected in spring 2006 at Spitsbergen where nucleation and small Aitken mode particles were almost completely suppressed [27] during the smokes.

Besides the growth in the number concentration, also the particle size distribution changes during the smoke episodes. In Kuopio 2010, the diameter of the aerosol particles was twice as large during the smoke event as it was in the background conditions [33]. In normal conditions the peak of the size distribution was at 55 nm but during the smoke the peak shifted to 186 nm. The increase in the peak diameter during a LRT smoke episode can affect the amount of potential cloud condensation nuclei. The concentration of particles larger than 100 nm in diameter, which can be considered as a good estimate of the size of the smallest particles contributing to cloud droplet activation, increased from 700 to 2,900 cm^{-3} from background conditions to smoke events, respectively. That increase can have an effect on cloud formation in suitable conditions [33].

4.1.2 Optical Properties

Smoke plumes alter aerosol optical properties. In spring 2006 at Spitsbergen, extreme aerosol optical depth (AOD) values were measured during the two episodes [10]. AOD was approximately a factor of 5 higher than the average values measured at that time of the year suggesting that the smoke strongly perturbed the radiation transmission in the atmosphere. In general, AOD has been found to have a consistent seasonal pattern with peaking values in April and August in Southern Scandinavia [40]. This maximum in April coincides with the snow melt and agricultural burning season in Eastern Europe. In 2010 in Kuopio, aerosol scattering coefficients increased more than 20 times during the smokes [33].

Aerosol optical properties have a strong impact on global warming estimates. It has been suggested that increased pollution events from wildfires in spring will enhance the Arctic warming [41]. However, the radiative forcing of an extreme smoke episode, detected in spring 2006 in the Arctic, showed that instead of warming, smoke aerosols had a strong cooling effect above the ocean [28]. A weak heating effect was found above the area covered by ice and snow like over Spitsbergen due to a much higher surface albedo. Biomass burning aerosols can also cause a considerable reduction in the ultraviolet (UV) radiation reaching the earth surface. Compared to typical aerosol conditions, a reduction up to 35% of midday surface UV irradiance (at 340 nm) was observed in Jokioinen, Finland, during the episode in spring 2006 [24].

4.2 *Chemical Composition*

Chemical composition of aerosol is strongly size-dependent. In general, fine particles consist mostly of organic matter and secondary inorganic ions, sulfate, ammonium, and nitrate, but also from black carbon and metals. Coarse particles are composed of crustal material, their oxides, and large sea salt particles. Particles from different sources, e.g., from biomass burning, traffic, industrial, or biogenic processes, have fairly different chemical composition, but LRT wildfire smokes are often mixed with other emissions, either in the source region or on the way to Northern Europe, making it difficult to separate the impact of biomass burning smokes from the other particle sources. As already mentioned, specific tracer compounds can be used to identify particles originated from biomass burning.

4.2.1 Carbonaceous Matter

Incomplete combustion of biomass is known to produce particulate carbonaceous components, organic matter, and black carbon [42]. The concentration of OM increases significantly during the LRT smoke episodes. During the major episode in spring 2006, the observed OM concentrations were elevated by a factor of 8 in Helsinki (Fig. 2; [19]), factor of 5–11 in Sweden [25], and around tenfold at Spitsbergen [10]. The concentration of OM increased more than that of PM resulting in a larger fraction of organic matter in PM during the smokes. For example in Helsinki, the contribution of OM to $PM_{2.5}$ doubled from the reference period to the episodes [19].

Smoke episodes have no clear effect on the general water solubility of organic matter. In Helsinki in 2006, the water solubility of OM did not change significantly from the reference period to the LRT smoke episodes [19]. Some water-soluble organic compound groups were more pronounced in the smoke samples. Dicarboxylic acids (succinic, malonic, and oxalic acids) were found in the smoke samples collected in August to September 2002 in Helsinki [14]. Dicarboxylic acid concentrations were elevated both in the fine and coarse fraction suggesting that the acids were probably condensed onto existing particles. Especially oxalic acid has been associated with the biomass smoke particles [19] but there are several other sources for it. Also elevated concentrations of PAHs have been attributed to the smoke episodes [22, 23]. Individual particle analysis revealed that during the LRT smoke episode in spring 2004 in Hyytiälä there were particles called as tar balls, a distinct carbonaceous particle type from soot [17]. Tar balls originated from biomass burning, especially during smoldering conditions.

Monosaccharide anhydrides are unique organic tracer compounds for biomass burning. For example in 2006 in Helsinki, levoglucosan concentrations were 18 times higher during the plumes than in the reference period (Fig. 2; [18]). Usually the separation of LRT smoke from the local biomass combustion is challenging. Using the fractions of levoglucosan, mannosan, and galactosan in the total amount of MAs, LRT smoke episodes were distinguished from the local wood burning [38]. The proportions of MAs are different from one type of burnt biomass to the other. The ratio of levoglucosan to galactosan was lower during the LRT smoke plumes in April to May and in August 2006 than in the period with residential wood combustion. That was assumed to be caused partly by the burning of foliar material in wildfires and logs in residential combustion.

Smoke particles contain black carbon but the concentrations are lower than for organic matter. For example in spring 2006 in Sweden, the concentration of OC rose more significantly than that of BC resulting in lower BC/OC ratios during the smoke episodes than in the reference period [25]. In 2010 in Kuopio, black carbon concentrations increased around 12 times during the smoke episodes but the fraction of BC in $PM_{2.5}$ remained nearly the same [33]. In some studies similar material to black carbon has also been measured as elemental carbon, light absorbing carbon, black smoke, or equivalent black carbon depending on the measurement technique but here they are all referred as black carbon in order to keep things simple.

Smoke aerosols are especially interesting in the Arctic because they have a considerable effect on surface albedo when they are deposited on snow [10]. In spring 2006 at Spitsbergen, the concentration of BC doubled during the smoke plumes but a very low fraction of total carbon (TC; OC + BC) consisted of BC during the event. The low fraction for BC was suggested to be due to the condensation of secondary organic material during the transport [10]. The low fraction of BC in smoke can also be related to the burning conditions. In August 2006 in Helsinki, BC concentrations did not increase notably, which indicated that the wildfires were smoldering-dominated, whereas is April to May burning occurred in flaming conditions resulting in higher BC concentrations [18].

4.2.2 Inorganic Fraction

Elevated levels of inorganic secondary ions, sulfate, nitrate, and ammonium, have been associated with the LRT biomass smoke episodes. The presence of a lot of secondary inorganic ions has been speculated to be biomass burning emissions mixed with other emission sources during the transport [14, 16]. However, it is possible that the elevated concentrations of inorganic ions are fire related. The agricultural areas in Eastern Europe receive large nitrogen loads from fertilization and from atmospheric deposition. Nitrogen oxide (NO_x) emissions from the fires can be unusually high in such conditions. Similar to nitrate it is possible that there is some re-emission of deposited sulfur by fires. In case of spring 2006 episodes, it was quite likely that the biomass burning emissions made a significant contribution to the measured values even for sulfate and nitrate [10]. In August 2006, high nitrate concentrations in the plumes were speculated to be caused by high nitrogen content of burnt biomass material rather than the nitrate formation via NO_x emissions [18]. This statement is supported by the fact that relative low burning temperature of wildfire does not favor NO_x formation in flame.

Potassium has been used as an inorganic tracer for biomass burning. In Helsinki during the smoke episodes in August to September 2002 and April to May 2006, the concentration of water-soluble potassium increased clearly [14, 19]. A high portion of K in smoke-related particles has also been observed in individual particle analysis [12, 13], but the percentage of K was different in LRT-episode observed in March 2002 and in August to September 2002. Similar to BC, a reason for the difference was speculated to be dissimilar burning conditions. Smoldering burning conditions are more dominant in boreal forest fires and peat fires (in August to September) than in agricultural waste fires (in March) that mainly burn in flaming conditions. The fraction of K is lower in the emissions from smoldering fires than those from the flaming fires [13].

In August to September 2002, there was calcium (Ca) in the PM_{10} samples indicating that Ca-rich particles in coarse size fraction can originate from the forest and peat fires [13]. Coarse particles generated in biomass combustion are either residual ash, plant tissues released due to incomplete combustion, or soil particles suspended by physical mechanisms. Coarse particles are usually redeposited to the ground close to the emission source but they can also be transported to long distances under favorable conditions. Besides biomass burning, Ca-rich particles are emitted from many other sources, including fossil fuel burning and the cement and metal industries. In August to September 2002, there were also a lot of crustal elements (Fe, Al, Cd, and Si) in both fine and coarse particles collected in Helsinki [14].

Usually urban aerosol in fine fraction is neutral. However, in the smoke plumes observed in August 2006 in Helsinki, there was an excess of cations, mainly ammonium [18]. These particles can be good sites for various heterogeneous reactions involving acidic trace gases. In August 2006 smoke particles had transported only a short distance, and therefore there were probably still plenty of gaseous smoke compounds that had not yet reacted with the particles.

4.3 Transformation During Transport

After release from fires, organic and some inorganic components undergo rapid or more delayed chemical transformation in the atmosphere. The physical properties as well as chemical composition of smoke particles may alter on the way from the source areas (biomass burning areas) to the measurement sites in Northern Europe. There are several reasons why particle properties change. Chemical components can, e.g., become oxidized or substituted in particles, but also the condensation of secondary material onto the LRT particles during the transport changes the particle properties.

It has been suggested that levoglucosan degrades during the transport [10]. In spring 2006 at Spitsbergen, the ratio of levoglucosan to carbon monoxide was more than an order of magnitude lower than the emission factor for levoglucosan from agricultural biomass burning determined earlier. Since the LRT smoke aerosol was assumed not to be removed during the transport, the low emission factor for levoglucosan indicated the degradation during the transport. It has been suggested that levoglucosan may decay in the atmosphere, e.g., by the influence of hydroxyl radicals [43].

Oxidation of organic compounds during the transport changes their water solubility. Two smoke episodes detected in 2006 in Helsinki had significantly different transport times, and therefore also the water solubility of OM. In spring the biomass smokes were carried in the atmosphere several hundreds of kilometers in a time period from 1 day to 4 days whereas in August 2006 the fires were only 200–500 km from Helsinki with a minimum transport time of 5 h. In spring the contribution of water-soluble organic compounds was higher (70%) than in August (56%) indicating that the organic compounds were much less oxidized after the transport of only the few hours in August, and therefore they were still largely water insoluble [18].

In addition to organic compounds also inorganic components may change during the transport. Potassium chloride (KCl) occurs in young smoke, whereas increased amounts of potassium sulfate (K_2SO_4) and potassium nitrate (KNO_3) are present in old smoke. This is due to the rapid substitution of chloride by sulfate and nitrate in a smoke plume. That behavior was seen in Helsinki in March 2002 when sulfur-rich particles collected during the smoke episode contained very little chloride because it had already depleted during the transport [12].

The size distribution of LRT smoke particles may change during the transport [28]. By comparing the data from Spitsbergen to that from the fire area (Minsk in Belarus and Toravere in Estonia), it was found that the average median radii for the fine mode decreased from the source area (0.16 μm) to Spitsbergen (0.12 μm) during the episode in spring 2006. Same behavior was also seen for the coarse mode; however, the main difference between the size distributions was for the mode centered in 5–7 μm that was not found at Spitsbergen at all. In fire areas, the fine and coarse mode concentrations were approximately ten times higher during the smoke events than in the reference days, whereas at Spitsbergen the fine mode concentration increased by a similar factor but the coarse mode was only 1.5–2 times higher during the most intense episode. Thus, even though both fine and coarse mode aerosols were emitted during the fires or were produced by dynamic processes in the fresh smoke, coarse aerosols were deposited during the transport to Spitsbergen whereas smaller aerosols had longer atmospheric residence times and therefore were drifted in the plume into the Arctic region. Deposition of coarse particles during the transport was also seen in optical properties of smoke aerosol. Single scattering albedo (SSA) increased with the distance from the fire source [28]. The increase in SSA indicated that the oxidation and transport of the aerosols from the source area towards the Arctic reduced the absorption of particles, whereas the scattering process became more dominant. Reduced absorption was explained by the deposition of coarse particles that probably had some absorbing aerosols.

5 Modeling

In order to show the dispersion of plumes originated from the fires, LRT smoke plumes have been modeled. FLEXPART particle dispersion model was used to indicate that the smoke was transported from detected source region to Spitsbergen and Iceland in spring 2006 smoke [10]. In addition to the dispersion of smoke, the fire-related PM concentrations in the smoke plumes can be modeled. For example in Helsinki during the episode in April to May 2006, the Finnish emergency and air quality modeling system SILAM [44] and during August 2006 episode the fire assimilation system (FAS; [45]) were used to forecast the $PM_{2.5}$ concentrations generated by biomass burning [18, 19]. In general, SILAM and FAS were able to reproduce PM concentrations, even though gaps were obtained in the modeled time series due to diverse reasons. Receptor modeling (e.g., positive matrix factorization; PMF) can be used to find out the contribution of smoke to the observed PM concentrations as well as to explain the chemical components attributed to biomass burning smokes [22]. PMF was applied to two data sets obtained from Virolahti in 2006. For spring data PMF yielded five sources for PM_{10} of which one was biomass fires, whereas for August 2006 data three factors were resolved of which one was smoke from biomass fires. In total LRT smoke explained 65% of the total variation of PM_{10} in August 2006 in Virolahti.

6 Health Effects

Health effects of residential wood combustion have been investigated broadly (e.g., [46]) but there are fewer studies devoted to the health impacts of LRT biomass smokes. Compared to residential wood combustion, the material burnt as well as the burning conditions are different in wildfires, and therefore also emissions differ. Emissions from residential wood combustion are usually quite fresh at the exposure, whereas LRT smoke particles may have been oxidized during the transport and therefore transformed to more water-soluble form, which may change the toxicity of particles. The exposure to LRT smoke is typically infrequent and lasts only from few hours to several weeks. However, during the smoke episodes ambient PM concentrations can be several times higher than that at any other time. The challenge in assessing the health impacts of LRT biomass smokes is that the emissions are often mixed with other particulate anthropogenic emissions, road dust, ozone, or even pollen. Especially in summertime, the smoke episodes are also often related to high ambient temperatures making the discrimination between the effects of smoke and heat wave difficult.

Smoke particles have acute health effects like the irritation of eyes and airways, but the exposure to the elevated concentrations of smoke-related particles for a long period can also cause morbidity and mortality. In Lithuania in August to September 2002, nearby biomass fires increased respiratory and asthma complaints in the health centers significantly [35]. There were 20 times higher number of respiratory diseases and bad asthma than a year earlier, and the negative health impacts became more abundant when the exposure time to elevated ambient pollutants increased. In case of LRT smoke episode observed in August to September 2002 in Finland, a total of 3.4 million people were exposed to the elevated levels of $PM_{2.5}$ [15]. It was estimated that the episode caused 17 excess deaths in Southern and Eastern Finland where elevated $PM_{2.5}$ levels were measured. That estimation was calculated assuming 1% increase in mortality per 10 $\mu g\ m^{-3}$ of $PM_{2.5}$. In general, LRT air pollution has a significant impact on excess mortality in Finland. Over half of all premature deaths due to anthropogenic primary $PM_{2.5}$ are estimated to be related to $PM_{2.5}$ emissions transported to Finland from other European sources [47].

In toxicological cell or animal studies LRT smoke particles have shown contrasting effects. Aerosol particles collected during the LRT episode in August to September 2002 in Helsinki had significantly lower activity in cytokine production than the corresponding particles in that season on average. Lower activity was suggested to be due to the chemical transformation of the organic fraction during the transport [48]. However, due to the increased PM concentration in the accumulation mode, the smoke episodes can be assumed to be associated with enhanced toxic potential per inhaled cubic meter of air, which may have public health implications. Biomass burning smoke consists of a mixture of different chemical components. In case of Californian wildfires in 2008, PM in coarse size fraction was found to be highly toxic to lung cells (macrophages), the active components being heat-labile organic compounds [49]. In general, there is no reason to assume that the biomass burning-related particles are less dangerous than particles from other sources [46].

7 Conclusions

LRT pollution from the wildfires has been observed in the northern European regions almost every year and recognized as one significant source of atmospheric pollution. As the PM concentrations are typically quite low in Northern Europe, the LRT smoke plumes increase the PM concentrations in several folds even at background sites with no local emissions. During the LRT smoke episodes daily average $PM_{2.5}$ concentrations typically reach 30–50 μg m^{-3}, which is approximately three- to sixfold compared with the annual mean $PM_{2.5}$ concentration and can result in exceedance in the European Union PM_{10} daily limits. The chemical composition of smoke particles differs from that of particles usually measured at the site. That probably has an effect on the toxicity of PM, however, the overall harmfulness of breathing smoky air is probably caused by much increased PM mass in a cubic meter of inhaled air. The most visible effect of the arrival of smoke plume is the smell of especially fresh smoke and decreased visibility but LRT smoke also causes acute health effects, morbidity, and premature deaths in the northern European countries.

Hot and dry meteorological conditions in summer increase the extent of observed biomass fires. For example, the summer of 2010 was exceptionally warm in Eastern Europe and large parts of Russia. With climate change such extreme events are predicted to increase and by the end of the century heat waves of this magnitude are expected to occur every decade [50]. For European Arctic the transport of smokes from more frequent biomass fires may affect the global climate change. Whether the Arctic is getting warmer or colder because of the smoke events is still under evaluation.

Wildfires have the most significant impact on people in the region close to the fires, but as being a transboundary pollutant, smoke also causes troubles thousands of kilometers away from the burning. In order to avoid air quality deterioration both in the areas near the fires as well as in distant countries, people need to be enforced to change the current agricultural burning practices as well as extinguish existing fires more efficiently.

References

1. Giglio L, Csiszar I, Justice CO (2006) Global distribution and seasonality of active fires as observed with the Terra and Aqua MODIS sensors. J Geophys Res. doi:10.1029/2005JG000142
2. Korontzi S, McCarty L, Loboda T, Kumar S, Justice C (2006) Global distribution of agricultural fires in croplanda from 3 years of Moderate Resolution Imaging Spectroradiometer (MODIS) data. Global Biogeochem Cycles. doi:10.1029/2005GB002529
3. Barnaba F, Angelini F, Curci G, Gobbi GP (2011) An important fingerprint of wildfires on the European aerosol load. Atmos Chem Phys 11:10487–10501
4. Anderson JO, Thundiyil JG, Stolbach A (2011) Clearing the air: a review of the effects of particulate matter air pollution on human health. J Med Toxicol. doi:10.1007/s13181-011-0203-1
5. Sillanpää M, Hillamo R, Saarikoski S, Frey A, Pennanen A, Makkonen U, Spolnik Z, Van Grieken R, Braniš M, Brunekreef B, Chalbot M-C, Kuhlbusch T, Sunyer J, Kerminen V-M, Kulmala M, Salonen RO (2006) Chemical composition and mass closure of particulate matter at six urban sites in Europe. Atmos Environ 40S2:212–223
6. Putaud J-P, Van Dingenen R, Alastuey A, Bauer H, Birmili W, Cyrys J, Flentje H, Fuzzi S, Gehrig R, Hansson HC, Harrison RM, Herrmann H, Hitzenberger R, Hüglin C, Jones AM, Kasper-Giebl A, Kiss G, Kousa A, Kuhlbusch TAJ, Löschau G, Maenhaut W, Molnar A, Moreno T, Pekkanen J, Perrino C, Pitz M, Puxbaum H, Querol X, Rodriguez S, Salma I, Schwarz J, Smolik J, Schneider J, Spindler G, ten Brink H, Tursic J, Viana M, Wiedensohler A, Raes F (2010) A European aerosol phenomenology e 3: physical and chemical characteristics of particulate matter from 60 rural, urban, and kerbside sites across Europe. Atmos Environ 44:1308–1320
7. Air Quality in Europe – 2011 Report, EEA Technical Report, No 12, 201, European Environment Agency (2011). doi:10.2800/83213
8. Karppinen A, Härkönen J, Kukkonen J, Aarnio P, Koskentalo T (2004) Statistical model for assessing the portion of fine particulate matter transported regionally and long range to urban air. Scand J Work Environ Health 30(S2):47–53
9. Niemi JV, Saarikoski S, Aurela M, Tervahattu H, Hillamo R, Westphal DL, Aarnio P, Koskentalo T, Makkonen U, Vehkamäki H, Kulmala M (2009) Long-range transport episodes of fine particles in southern Finland during 1999–2007. Atmos Environ 43:1255–1264
10. Stohl A, Berg T, Burkhart JF, Fjæraa AM, Forster C, Herber A, Hov Ø, Lunder C, McMillan WW, Oltmans S, Shiobara M, Simpson D, Solberg S, Stebel K, Ström J, Tørseth K, Treffeisen R, Virkkunen K, Yttri KE (2007) Arctic smoke – record high air pollution levels in the European Arctic due to agricultural fires in Eastern Europe in spring 2006. Atmos Phys Chem 7:511–534
11. Witham C, Manning A (2007) Impacts of Russian biomass burning on UK air quality. Atmos Environ 41:8075–8090
12. Niemi JV, Tervahattu H, Vehkamäki H, Kulmala M, Koskentalo T, Sillanpää M, Rantamäki M (2004) Characterization and source identification of a fine particle episode in Finland. Atmos Environ 38:5003–5012
13. Niemi JV, Tervahattu H, Vehkamäki H, Martikainen J, Laakso L, Kulmala M, Aarnio P, Koskentalo T, Sillanpää M, Makkonen U (2005) Characterization of aerosol particle episodes in Finland caused by wildfires in Eastern Europe. Atmos Chem Phys 5:2299–2310
14. Sillanpää M, Saarikoski S, Hillamo R, Pennanen A, Makkonen U, Spolnik Z, Van Grieken R, Koskentalo T, Salonen RO (2005) Chemical composition, mass size distribution and source analysis of long-range transported wildfire smokes in Helsinki. Sci Total Environ 350:119–135
15. Hänninen OO, Salonen RO, Koistinen K, Lanki T, Barregard L, Jantunen M (2009) Population exposure to fine particles and estimated excess mortality in Finland from an East European wildfire episode. J Expo Sci Environ Epidemiol 19:414–422

16. Saarikoski SK, Sillanpää MK, Saarnio KM, Hillamo RE, Pennanen AS, Salonen RO (2008) Impact of biomass combustion on urban fine particulate matter in central and northern Europe. Water Air Soil Pollut 191:265–277
17. Niemi JV, Saarikoski S, Tervahattu H, Mäkelä T, Hillamo R, Vehkamäki H, Sogacheva L, Kulmala M (2006) Changes in background aerosol composition in Finland during polluted and clean periods studied by TEM/EDX individual particle analysis. Atmos Chem Phys 6:5049–5066
18. Saarnio K, Aurela M, Timonen H, Saarikoski S, Teinilä K, Mäkelä T, Sofiev M, Koskinen J, Aalto PP, Kulmala M, Kukkonen J, Hillamo R (2010) Chemical composition of fine particles in fresh smoke plumes from boreal wild-land fires in Europe. Sci Total Environ 408:2527–2542
19. Saarikoski S, Sillanpää M, Sofiev M, Timonen H, Saarnio K, Teinilä K, Karppinen A, Kukkonen J, Hillamo R (2007) Chemical composition of aerosols during a major biomass burning episode over northern Europe in spring 2006: experimental and modelling assessments. Atmos Environ 41:3577–3589
20. Timonen H, Saarikoski S, Tolonen-Kivimäki O, Aurela M, Saarnio K, Petäjä T, Aalto PP, Kulmala M, Pakkanen T, Hillamo R (2008) Size distributions, sources and source areas of water-soluble organic carbon in urban background air. Atmos Chem Phys 8:5635–5647
21. Aurela M, Sillanpää M, Pennanen A, Mäkelä T, Laakia J, Tolonen-Kivimäki O, Saarnio K, Yli-Tuomi T, Aalto P, Salonen I, Pakkanen T, Salonen RO, Hillamo R (2010) Characterization of urban particulate matter for a health-related study in southern Finland. Boreal Environ Res 15:513–532
22. Anttila P, Makkonen U, Hellén H, Kyllönen K, Leppänen S, Saari H, Hakola H (2008) Impact of the open biomass fires in spring and summer of 2006 on the chemical composition of background air in south-eastern Finland. Atmos Environ 42:6472–6486
23. Makkonen U, Hellén H, Anttila P, Ferm M (2010) Size distribution and chemical composition of airborne particles in south-eastern Finland during different seasons and wildfire episodes in 2006. Sci Total Environ 408:644–651
24. Arola A, Lindfors A, Natunen A, Lehtinen KEJ (2007) A case study on biomass burning aerosols: effects on aerosol optical properties and surface radiation levels. Atmos Chem Phys 7:4257–4266
25. Targino AC, Krecl P, Johansson C, Swietlicki E, Massling A, Coraiola GC, Lihavainen H Deterioration of air quality across Sweden due to transboundary agricultural burning emissions. Boreal Environ Res (in press)
26. Krecl P, Targino AC, Johansson C (2011) Spatiotemporal distribution of light-absorbing carbon and its relationship to other atmospheric pollutants in Stockholm. Atmos Chem Phys 11:11553–11567
27. Treffeisen R, Tunved P, Ström J, Herber A, Bareiss J, Helbig A, Stone RS, Hoyningen-Huene W, Krejci R, Stohl A, Neuber R (2007) Arctic smoke – aerosol characteristics during a record smoke event in the European Arctic and its radiative impact. Atmos Chem Phys 7:3035–3053
28. Lund Myhre C, Toledano C, Myhre G, Stebel K, Yttri KE, Aaltonen V, Johnsrud M, Frioud M, Cachorro V, de Frutos A, Lihavainen H, Campbell JR, Chaikovsky AP, Shiobara M, Welton EJ, Tørseth K (2007) Regional aerosol optical properties and radiative impact of the extreme smoke event in the European Arctic in spring 2006. Atmos Chem Phys 7:5899–5915
29. Hyvärinen A-P, Kolmonen P, Kerminen V-M, Virkkula A, Leskinen A, Komppula M, Hatakka J, Burkhart J, Stohl A, Aalto P, Kulmala M, Lehtinen KEJ, Viisanen Y, Lihavainen H (2011) Aerosol black carbon at five background measurement sites over Finland, a gateway to the Arctic. Atmos Environ 45:4042–4050
30. Timonen H, Aurela M, Carbone S, Saarnio K, Saarikoski S, Mäkelä T, Kulmala M, Kerminen V-M, Worsnop DR, Hillamo R (2010) High time-resolution chemical characterization of the water-soluble fraction of ambient aerosols with PILS-TOC-IC and AMS. Atmos Meas Tech 3:1063–1074
31. Yttri KE, Simpson D, Nøjgaard JK, Kristensen K, Genberg J, Stenström K, Swietlicki E, Hillamo R, Aurela M, Bauer H, Offenberg JH, Jaoui M, Dye C, Eckhardt S, Burkhart JF, Stohl A,

Glasius M (2011) Source apportionment of the summer time carbonaceous aerosol at Nordic rural background sites. Atmos Chem Phys 11:13339–13357
32. Mielonen T, Portin H, Komppula M, Leskinen A, Tamminen J, Ialongo I, Hakkarainen J, Lehtinen KEJ, Arola A (2012) Biomass burning aerosols observed in Eastern Finland during the Russian wildfires in summer 2010 – Part 2: remote sensing. Atmos Environ 47:279–287
33. Portin H, Mielonen T, Leskinen A, Arola A, Pärjälä E, Romakkaniemi S, Laaksonen A, Lehtinen KEJ, Komppula M (2012) Biomass burning aerosols observed in Eastern Finland during the Russian wildfires in summer 2010 – Part 1: in-situ aerosol characterization. Atmos Environ 47:269–278
34. Stohl A, Wotawa G, Seibert P, Kromp-Kolb H (1995) Interpolation errors in wind fields as a function of spatial and temporal resolution and their impact on different types of kinematic trajectories. J Appl Met 34:2149–2165
35. Ovadnevaitė J, Kvietkus K, Maršalka A (2006) 2002 summer fires in Lithuania: impact on the Vilnius city air quality and the inhabitants health. Sci Total Environ 356:11–21
36. Šopauskienė D, Jasinevičienė D (2009) Variations of concentration of aerosol particles (<10 μm) in Vilnius. Lith J Phys 49:323–334
37. Nolte CG, Schauer JJ, Cass GR, Simoneit BRT (2001) Highly polar organic compounds present in wood smoke and in the ambient atmosphere. Environ Sci Technol 35:1912–1919
38. Saarnio K, Niemi JV, Saarikoski S, Aurela M, Timonen H, Teinilä K, Myllynen M, Frey A, Lamberg H, Jokiniemi J, Hillamo R (2012) Using monosaccharide anhydrides to estimate the impact of wood combustion on fine particles in the Helsinki Metropolitan Area. Boreal Environ Res 17:163–183
39. Alfarra MR, Prevot ASH, Szidat S, Sandradewi J, Weimer S, Lanz VA, Schreiber D, Mohr M, Baltensperger U (2007) Identification of the mass spectral signature of organic aerosols from wood burning emissions. Environ Sci Technol 41:5770–5777
40. Toledano C, Cachorro VE, Gausa M, Stebel K, Aaltonen V, Berjón A, Ortiz de Galisteo JP, de Frutos AM, Bennouna Y, Blindheim S, Myhre CL, Zibordi G, Wehrli C, Kratzer S, Hakansson B, Carlund T, de Leeuw G, Herber A, Torres B (2012) Overview of sun photometer measurements of aerosol properties in Scandinavia and Svalbard. Atmos Environ 52:18–28
41. Law KS, Stohl A (2007) Arctic air pollution: origins and impacts. Science 315:1537–1540
42. Frey AK, Tissari J, Saarnio KM, Timonen HJ, Tolonen-Kivimäki O, Aurela MA, Saarikoski SK, Makkonen U, Hytönen K, Jokiniemi J, Salonen RO, Hillamo REJ (2009) Chemical composition and mass size distribution of fine particulate matter emitted by a small masonry heater. Boreal Environ Res 14:255–271
43. Hennigan CJ, Sullivan AP, Collett JL Jr, Robinson AL (2010) Levoglucosan stability in biomass burning particles exposed to hydroxyl radicals. Geophys Res Lett. doi:10.1029/2010GL043088
44. Sofiev M, Siljamo P, Valkama I, Ilvonen M, Kukkonen J (2006) A dispersion modelling system SILAM and its evaluation against ETEX data. Atmos Environ 40:674–685
45. Sofiev M, Vankevich R, Lotjonen M, Prank M, Petukhov V, Ermakova T, Koskinen J, Kukkonen J (2009) An operational system for the assimilation of the satellite information on wild-land fires for the needs of air quality modelling and forecasting. Atmos Chem Phys 9:6833–6847
46. Boman BC, Forsberg AB, Järvholm BG (2003) Adverse health effects from ambient air pollution in relation to residential wood combustion in modern society. Scand J Work Environ Health 29:251–260
47. Tainio M, Tuomisto JT, Pekkanen J, Karvosenoja N, Kupiainen K, Porvari P, Sofiev M, Karppinen A, Kangas L, Kukkonen J (2010) Uncertainty in health risks due to anthropogenic primary fine particulate matter from different source types in Finland. Atmos Environ 44:2125–2132
48. Jalava PI, Salonen RO, Hälinen AI, Penttinen P, Pennanen AS, Sillanpää M, Sandell E, Hillamo R, Hirvonen M-R (2006) In vitro inflammatory and cytotoxic effects of size-segregated particulate samples collected during long-range transport of wildfire smoke to Helsinki. Toxicol Appl Pharmacol 215:341–353

49. Franzi LM, Bratt JM, Williams KM, Last JA (2011) Why is particulate matter produced by wildfires toxic to lung macrophages? Toxicol Appl Pharmacol 257:182–188
50. Barriopedro D, Fisher EM, Luterbacher J, Trigo RT, García-Herrera R (2011) The hot summer of 2010: redrawing the temperature record map of Europe. Science 332:220–224

Residential Wood Burning: A Major Source of Fine Particulate Matter in Alpine Valleys in Central Europe

Hanna Herich and Christoph Hueglin

Abstract Residential wood burning is one of the important sources of ambient particulate matter (PM) in many European regions. Besides total PM, residential wood burning is at many locations an important source of other air pollutants such as polycyclic aromatic hydrocarbons (PAHs), benzene, particulate organic carbon (OC), and black carbon (BC), especially in regions such as the Alpine region, where wood fuel is, on one hand, traditionally used for domestic heating during the cold season in small stoves and, on the other hand, meteorological conditions during winter are often favourable for accumulation of wood smoke in a shallow boundary layer. As a consequence, wood burning in the Alpine region can be the dominating source of PM, OC, and BC during the cold season. This is true for both larger cities and small villages in rural areas. The absolute contribution of wood burning emissions to particulate air pollutants tends in rural environments to be even larger than in urban areas. This chapter gives an overview about the results of studies on ambient particulate pollutants from residential wood burning in the Alpine region.

Keywords Black carbon, Organic carbon, Particulate matter, Residential wood burning

Contents

1 Introduction 124
   1.1 Fine Particulate Matter From Wood Burning 125
   1.2 Volumes of Wood Fuel and Used Appliances 125
2 Methods for Quantification of Source Impacts 126
   2.1 Emission Measurements 126
   2.2 Emission Factors: Examples for Austria, Germany and Switzerland 127
   2.3 PM Source Apportionment Methods 127

H. Herich and C. Hueglin (✉)
Laboratory for Air Pollution and Environmental Technology, EMPA, Swiss Federal Laboratories for Materials Science and Technology, 8600 Duebendorf, Ueberlandstrasse, Switzerland
e-mail: Christoph.Hueglin@empa.ch

3 Overview of Published Studies on the Impact of Wood Burning Emissions on Particulate
   Air Pollutants 129
   3.1 Measurement Sites and Data 129
   3.2 Impact of Wood Burning Emissions on Air Quality 130
4 Conclusion 136
References 137

1 Introduction

Fine particulate matter (PM) is well known to cause serious negative impacts on human health [1–4]. As a consequence, ambient PM concentrations are regulated in many countries worldwide. For example, air quality standards for the mass concentration of particles with aerodynamic diameter less than 10 µm (PM10) are in the European Union set to 40 µg/m^3 (annual mean) and 50 µg/m^3 (daily mean). In the USA, the daily limit value for PM10 is 150 µg/m^3; in addition, the mass concentration of the finer fraction of particulate matter PM2.5 is not allowed to exceed 35 µg/m^3 (annual mean) and 15 µg/m^3 (daily mean), respectively. The World Health Organization (WHO) has set air quality guideline values for the annual mean and daily mean concentrations of ambient PM10 (PM2.5) at 20 µg/m^3 (10 µg/m^3) and 40 µg/m^3 (20 µg/m^3), respectively [5].

During the past two decades, actions have been taken in Europe and in other regions to reduce the emissions of primary PM and precursors of secondary particles that are formed in the atmosphere by gas-to-particle conversion or condensation of gaseous compounds on existing aerosol particles. Air pollution abatement strategies in Europe have been targeted on the reduction of emissions from known anthropogenic sources (e.g. road traffic, power plants, industrial processes) and substantial reductions of the emissions of precursor gases such as SO_2, NO_x and volatile organic compounds (VOCs) have been achieved. Consequently, the PM10 concentrations in Europe showed generally a downward trend during the 1990s, this trend seems to have leveled off since then [60, 61]. Nevertheless, air quality standards are still regularly exceeded in wide parts of Europe.

Today, residential wood burning (WB) is one of the important sources of ambient PM in many European regions as shown in various publications [6–10]. According to Simpson et al. [10] and references therein, wood burning is responsible for 30% of PM at rural and natural sites in Europe. Besides total PM, residential wood burning is at many locations an important or even dominating source of other air pollutants such as polycyclic aromatic hydrocarbons (PAHs), benzene, particulate organic carbon (OC) and black carbon (BC). A review article by Naeher et al. [11] provides an overview about health effects of the entire mixture of gaseous and particulate air pollutants on wood burning emissions, typically denoted as wood smoke.

Especially in regions such as the Alpine region, where wood fuel is, on the one hand, traditionally used for domestic heating during the cold season in small stoves and, on the other hand, meteorological conditions during winter are often favourable for accumulation of wood smoke in a shallow boundary layer. As a consequence, wood burning can in the Alpine region be the dominating source of PM, OC and BC [12–15]. This chapter gives an overview about the results of studies on ambient particulate pollutants from residential wood burning in the Alpine region.

1.1 Fine Particulate Matter From Wood Burning

The chemical composition of fine particulate matter from wood burning depends on many factors such as the type of wood burning appliance, fuel type and the combustion conditions [16]. Incomplete combustion conditions as often prevailing in small residential wood stoves lead to emissions of refractory carbonaceous particles denoted as black carbon, elemental carbon (EC) or soot and also particulate organic carbon including constituents of high toxicity [17, 18]. Under incomplete combustion conditions, VOCs present in wood smoke can act as precursors for secondary organic aerosol (SOA) formation. The contribution of SOA to total wood burning-related PM may even be larger than that of primary particles [19]. Depending on the burning processes wood burning emissions can also be dominated by incombustible inorganic compounds such as potassium and calcium salts and oxides. This is, for example, the case during constant and ideal burning conditions occurring in larger wood-firing appliances (e.g. boilers). Inorganic PM from wood burning is typically considered to be less harmful than the carbonaceous fraction. Particle diameters of both primary and secondary PM from wood burning are predominantly in the lower range of the accumulation mode and therefore have atmospheric lifetimes of several days.

1.2 Volumes of Wood Fuel and Used Appliances

In central Europe the use of wood fuel varies widely between the countries. For some of the Alpine countries statistics about the amount of wood used as fuel is available [20]. According to this publication the use of wood for energy production is currently about 41 Mm^3 in France, 30 Mm^3 in Germany, 14 Mm^3 in Austria, 4.5 Mm^3 in Switzerland and 2 Mm^3 in Slovenia. From this total consumption, private households account for 90% of the wood in France, 44% in Germany, 59% in Austria, 34% in Switzerland and 75% in Slovenia. The remainder is used in power plants, for heat production and other industrial appliances.

The use of wood for heating purposes in the Alpine region has a long tradition and is still very popular because wood is a renewable and locally available source of energy. However, detailed numbers about the temporal development of installed and used wood burning appliances in Alpine countries are difficult to obtain. For Switzerland, detailed wood energy statistics has been published on an annual basis since 1993 by the Swiss Federal Office of Energy. These reports provide an inventory of the wood burning appliances used in the country as retrieved from evaluations of sales volumes and other sources of information. The report for 2011 shows the changes in the stock of different types of wood burning appliances [21]. According to BFE [21], there are currently 630,000 wood burning appliances installed in Switzerland, the stock decreased by 8.9% since 1990. However, the meteorologically adjusted annual use of wood increased in Switzerland by 36.9% since 1990 to the above-mentioned 4.5 Mm^3 in 2011. In Switzerland, there is a clear trend to larger automatic furnaces (heating power > 50 kW) and the stock increased since 1990 by 219%. In contrast, medium-sized wood fired building heating systems (heating power < 50 kW) decreased since 1990 by 53.9%, the number of installed small wood stoves and fireplaces remained between 1990 and 2011 about constant (+2.9%).

2 Methods for Quantification of Source Impacts

2.1 Emission Measurements

Emissions from fireplaces and stoves have been reported in several studies testing furnaces and wood fuel typical for different areas of the world. Wood smoke emissions typical for central Europe were investigated by Schmidl et al. [22, 23]. In these studies, different common European wood types were analysed to derive chemical profiles of wood combustion emissions for various types of wood. An overview of the fuel wood types used in the Alpine countries is given by Kistler et al. [24]. In the above-mentioned emission studies, wood has been burnt in a tiled wood stove [22] and in two automatically and two manually fired appliances [23]. The authors found a high variability for the emissions from small-scale manually fired wood combustion appliances in the performed individual tests.

In addition to the emission studies, information on the chemical signature of PM from wood burning is available. For example, organic aerosol mass spectral signatures from wood burning emissions were described by Alfarra et al. [25] and Weimer et al. [26]. Weimer et al. [26] found a larger influence of the burning conditions (flaming phase and smouldering phase) on the mass spectra than of the wood type used. Heringa et al. [27] have shown in smog chamber studies that aerosol mass spectra can be used to discriminate SOAs from different sources of precursors.

Emission studies are important for understanding the factors determining wood burning emissions (e.g. technology, operation practise, fuel type and quality) and for quantification of emission factors. Moreover, emission studies can provide chemical source profiles that can be used in source apportionment approaches such as chemical mass balance (CMB). However, data derived from emission studies depends strongly on the conditions the measurements have been performed (e.g. sampling temperature, applied dilution factors), and they typically do not account for chemical changes occurring in the atmosphere (e.g. SOA formation) which is of high importance for the determination of the influence of wood burning emissions on ambient air quality.

2.2 Emission Factors: Examples for Austria, Germany and Switzerland

As indicated above, emissions of PM from wood burning (i.e. emission factors) are highly dependent on various factors. In order to provide some insight into the variability of PM emission factors from wood burning typical for the Alpine region, results from studies in Austria, Germany and Switzerland are summarised.

For Austria latest emission factors are only available from the late 1990s [28, 62]. At that time, automatic boilers were hardly considered. The publications give estimates for the total suspended particulates from wood burning which are 148 kg/TJ for small stoves and 90 kg/TJ for central heating systems. Emission data from 2005 is available from studies in Germany [29]. In this report, device-specific wood burning emissions factors are given, the reported emission factors range from 57 kg/TJ for pellet stoves to 142 kg/TJ for and manual loaded boilers.

For Switzerland particulate emissions from the most common wood burning devices are the following: open stoves 100 kg/TJ, tiled stoves 100 kg/TJ and wood chips (<1,000 kW) 90 kg/TJ [63, 30]). Further detailed and device-specific wood emissions are given in the corresponding publications.

2.3 PM Source Apportionment Methods

Several studies have recently focused on determination of the contributions of the main sources of PM and the total carbonaceous fraction (carbonaceous matter, CM) in ambient air at receptor sites. Although the contribution of WB to PM and CM cannot be determined directly, various methods for source apportionment based on statistical approaches and on single or multiple specific tracers are available and have been applied in the past. All of these methods have advantages and limitations, a problem that is common to all of them is – that validation of the results is not possible due to the absence of a "correct" determination method. Therefore, plausibility checks and comparisons of results obtained with different methods are important for quality control.

A widely used approach for estimation of source contributions at receptor sites is receptor modelling [31, 32]. In receptor models, source contributions are estimated based on the measurements of various chemical constituents in a sufficiently large number of ambient PM samples, often filter samples that are collected during 24 h. Depending on the available knowledge about the main sources, CMB or multivariate statistical models can be applied: CMB requires a priori knowledge of the chemical profile of all relevant sources, i.e. the percentage of the chemical constituents measured in the ambient samples in unit emissions of all different sources [32]. The required source profiles can be either obtained from measurements at single emission sources or taken from literature [33, 34]. In contrast to CMB, multivariate statistical models only require qualitative or semi-quantitative a posteriori information about the source emission profiles. Both, source profiles and contributions of the sources are estimated based on the chemical speciation of ambient particulate matter as measured at the receptor site. A widely used receptor model is positive matrix factorization (PMF), [35, 36]. Applications of PMF for source apportionment of ambient particulate patter are, for example, presented by Lee et al. [37] and Pandolfi et al. [38]. It is important to note that CMB and multivariate receptor models rely on the assumptions that characteristic and constant source profiles exist and that the mass of the considered chemical constituents of PM is conserved during transport from the source to the receptor site. Besides determination of source contributions to total PM mass concentration, CMB and multivariate receptor models also allow for quantitative source attribution of the considered chemical constituents such as organic carbon (OC) and elemental carbon (EC).

PMF has successfully been applied to data from aerosol mass spectrometers (AMS, Aerodyne Research Inc., Billerica) for identification of the main sources of particulate organic matter, OM [9, 12, 39]. AMS instruments allow measurement of the mass spectra of the non-refractory fraction of approximately PM1 with high temporal resolution and determination of the concentration of particulate OM, which can be converted to OC by multiplication by a conversion factor [40].

An alternative to the above described approaches is the radiocarbon method that allows a distinction of contemporary carbon (from biogenic emissions and combustion of biomass) and carbon from combustion of fossil fuels in particulate carbonaceous matter [15, 41, 42]. In contrast to fossil fuels where the 14C isotope is completely depleted, CM emitted from WB shows a contemporary radiocarbon level. Radiocarbon measurements are often combined with measurements of complementary source specific tracers (macro-tracer) for additional information of source impacts [14, 43, 44].

The so-called aethalometer model (AE model, see [12, 45–47]) is another method for the distinction between light absorbing carbon (or black carbon, BC) from wood burning and combustion of fossil fuels. The AE model relies on the stronger light absorption of aerosol particles emitted from WB in the ultraviolet (UV) compared to aerosol particles from fossil fuel (FF) combustion. Multi-wavelength aethalometers such as model AE-31 from Magee Scientific Corporation (Berkeley) continuously measure the aerosol light absorption at several wavelengths (λ). With known Angstrom exponents for aerosol particles from FF combustion and WB, assumption of the Lambert-Beer law and the absorption coefficients $b_{abs}(\lambda)$ measured at two different wavelengths, the contribution of WB and FF to $b_{abs}(\lambda)$ can be determined. The BC mass concentration can then be obtained from $b_{abs}(\lambda)$ divided by the mass-specific aerosol light absorption cross section σ_{abs}. BC is consistent with EC if σ_{abs} is calculated from EC analysis (e.g. thermal optical transmission method). In summary, the AE model allows the determination of EC from wood burning.

Finally the so-called mono- and macro-tracer approaches can be applied for determining source contributions. These methods rely on the fact that a number of chemical compounds can be directly linked to biomass combustion emissions. For example, ambient concentrations of water-soluble potassium, certain PAHs, anhydrosugars and many other tracers have been used as indicators for the impact of biomass burning. When the fractions of one of these tracers in PM and carbonaceous aerosols emitted by wood burning are known (emissions ratios), the contribution of wood burning at a receptor site can be calculated based on the concentration of the considered tracer (mono tracer method).

Levoglucosan is the most commonly used mono tracer for wood burning, as it is almost exclusively formed during cellulose pyrolysis. Other tracers for biomass combustion, e.g. water-soluble fraction of potassium, can also originate from other sources than biomass burning such as soil dust, emissions from incinerators and meat cooking [48, 49]. Since levoglucosan is a specific tracer for wood combustion emissions, measurements of the concentration of levoglucosan in ambient air have often been used for estimation of the impact of this source on PM [7, 50, 51].

Instead of single tracers, a set of (macro) tracers such as OC, EC and levoglucosan can be used for source apportionment [43]. In this so-called macro-tracer approach (MTA), the individual tracers are weighted according to published emission ratios. Optionally, the weighting factors can be further tailored according to local conditions [52] or constrained in sensitivity studies to derive an optimal solution [14].

3 Overview of Published Studies on the Impact of Wood Burning Emissions on Particulate Air Pollutants

3.1 Measurement Sites and Data

During the last decade several publications highlighted the influence of wood burning on air quality at various sites in the Alpine region. These studies were focusing on different size fractions, i.e. PM10, PM2.5 or PM1. In the following the results on the contribution of wood burning emissions to EC, OC and PM will be presented.

These results are taken from a total of 18 publications that were found in a literature review. The corresponding measurements were performed at 27 sites in the Alpine region. Figure 1 shows a map of the sampling locations, the sites are additionally listed in Table 1. The measurement sites represent different air pollution situations such as remote, rural, suburban, urban background and urban roadside sites.

The time duration of the measurement campaigns and the sampling intervals are different depending on the considered studies. PM sampling was restricted to wintertime, when emissions from wood burning can be expected to be largest [10]. Exceptions are annual average contributions of wood burning emissions to

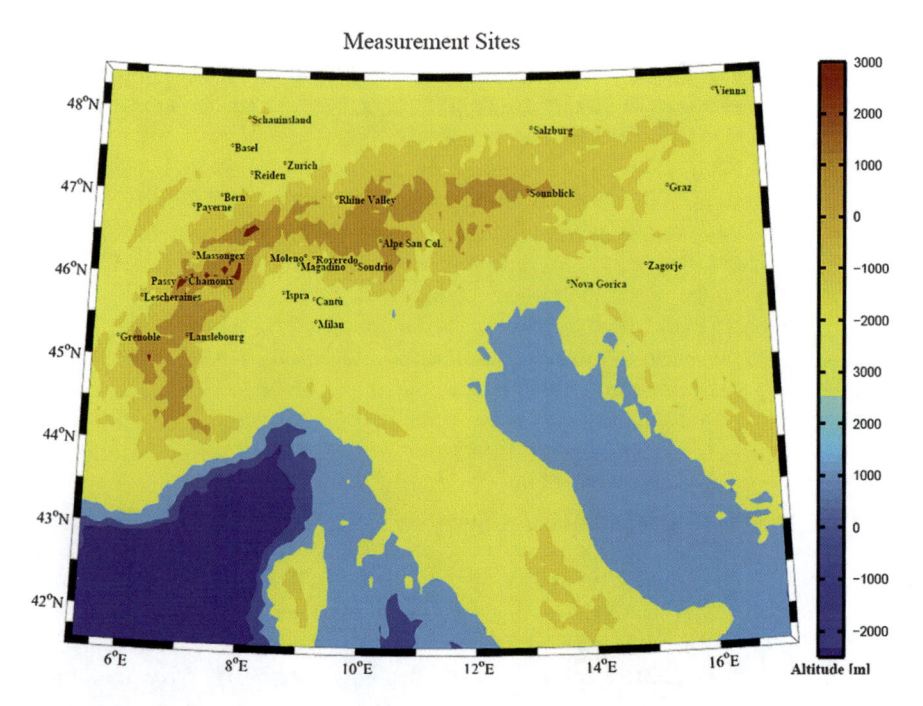

Fig. 1 Map of sites in the Alpine region where results on the contribution of wood burning emissions to PM, OC and EC have been reported

PM10, OC and EC at the Northern Italian sites in Milan, Sondrio and Cantù as reported by Piazzalunga et al. [52].

3.2 Impact of Wood Burning Emissions on Air Quality

Table 1 summarises the results of 18 publications on the impact of wood burning emissions on PM, OC and EC at sites in the Alpine region, denoted here as PM_{wb}, OC_{wb} and EC_{wb}, respectively. The table provides (subject to data availability) absolute contributions from wood burning emissions at 27 measurement sites. In addition relative contributions from wood burning emissions to total EC, OC and PM are presented together with the concentrations of wood combustion tracers levoglucosan and water-soluble potassium as well as the ratio of levoglucosan and mannosan. For all sampling stations, the site type, the PM size-cut, the measurement season and references are also listed. The applied source apportionment approach is given, further details can be found in the corresponding publications as well as in Sect. 2.3. Some sites appeared repeatedly in different studies with different sampling periods. These stations appear in Table 1 repeatedly.

Table 1 Absolute and relative contributions of EC_{wb}, OC_{wb} and PM_{wb} concentrations of levoglucosan, water-soluble potassium, the ratio of two tracers for wood burning emissions (levoglucosan/mannosan) and references

Station	Type	PM	Method	Season	PM_{wb} (μg m^{-3})	OC_{wb} (μg m^{-3})	EC_{wb} (μg m^{-3})	Levo (μg m^{-3})	Potass. (μg m^{-3})	Lev./ Man.	PM_{wb} (%)	OC_{wb} (%)	EC_{wb} (%)	Ref.
Basel	Suburban	10	PMF	Winter 2008/2009	5.3	1.7	0.4		0.40		20	33	30	a
Bern	Urban	10	PMF	Winter 2008/2009	6.9	2.6	0.5	0.31	0.46	7.6	16	37	14	a
Cantù	Urban backgr.	10	Macro-Tracer Appr.	2004–2007	14.2	6.0	0.3	1.38		7.6	18	40	9	e
Cantù	Urban backgr.	10	Macro-Tracer Appr.	Winter 2008	21.0						25			q
Graz	Urban/backgr.	10	Mono-Tracer Appr.	Winter 2004	3.1			0.29		5.6	16	46		g
Lanslebourg	Rural	10	CMB	January 2010	20.2	11.5	0.5	1.51	0.45	5.5	59			d
Lescheraines	Rural	10	CMB	Winter 2009/2010	35.1	20.0	0.9	2.00	0.62	22.1	101			d
Magadino	Rural	10	PMF	Winter 2008/2009	17.3	7.6	1.6	0.99	0.43	10.0	54	75	56	a
Milan	Urban backgr.	10	Macro-Tracer Appr.	2004–2007	9.0	4.7	0.7	0.83		7.2	10	28	16	e
Milan	Urban backgr.	10	Macro-Tracer Appr.	Winter 2008	16.0						18			q
Moleno	Rural, motorway	10	14C	February 2005		12.9	1.0					79	14	h
Nova Gorica		10	AE-Model	January–March 2010		8.3	0.7	0.65	0.40				26	j, jj
Passy	Urban-specific	10	CMB	February 2010	31.5	18.0	0.8	2.80	0.31	13.4	74			d
Payerne	Rural	10	PMF	Winter 2008/2009	6.0	2.0	0.4	0.16	0.32	9.0	38	42	43	a
Roveredo	Rural, motorway	10	14C	January/March 2005		9.8	1.4					91	70	h

(continued)

Table 1 (continued)

Station	Type	PM	Method	Season	PMwb (μg m⁻³)	OCwb (μg m⁻³)	ECwb (μg m⁻³)	Levo (μg m⁻³)	Potass. (μg m⁻³)	Lev./Man.	PMwb (%)	OCwb (%)	ECwb (%)	Ref.
Salzburg	Urban/backgr.	10	Mono-Tracer Appr.	Winter 2004	7.3			0.68		5.4	15	54		g
Sondrio	Urban backgr.	10	Macro-Tracer Appr.	2004–2007	15.9	8.9	1.4	1.51		7.0	21	46	29	e
Sondrio	Urban backgr.	10	Macro-Tracer Appr.	Winter 2008	24.0						28			q
Vienna	Urban/backgr.	10	Mono-Tracer Appr.	Winter 2004	2.4			0.22		5.0	9	35		g
Zagorje		10	AE-Model	November 2009		9.4	1.2	1.41	0.26					j, jj
Zurich	Urban backgr.	10	PMF	Winter 2008/2009	5.5	2.0	0.4	0.23	0.35	7.2	20	42	28	a
Zurich	Urban backgr.	10	14C	February 2003		6.9	1.0	0.62	0.44				29	i

(a) [14], only winter data (the mannosan data, not shown in the paper, was determined together with levoglucosan as described in the manuscript), (b) [46], (c) [13], (d) [54], (e) [53], (f) [15], (g) [8], (h) [16], (i) [43], (j) [65], (jj) [66], (k) [10], (l) [55], (m) [56], (n) [47, 48], (o) [44], (p) [41], (q) [57], (r) [58]

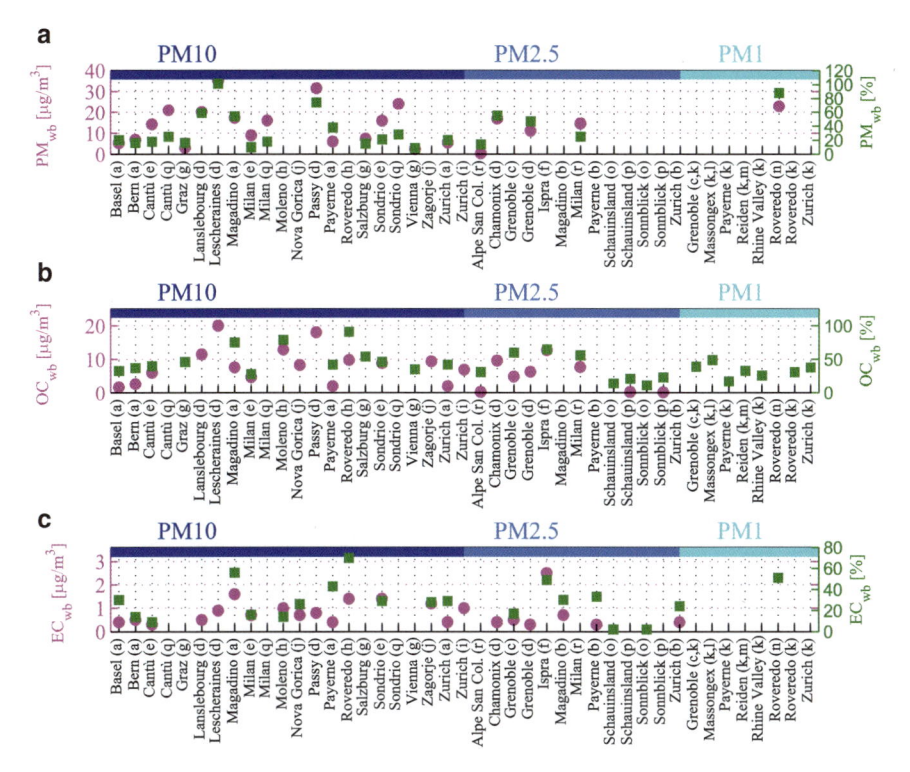

Fig. 2 (**a–c**) Absolute as well as relative contribution of PM, OC and EC emitted from wood burning at various sites in the Alpine region. The letters in brackets refer to the references in Table 1

3.2.1 PM from WB

Figure 2a provides the average contribution of wood burning emissions to PM as listed in Table 1. The absolute contributions of PM_{wb} show a very wide range from 0.4 µg m^{-3} at a remote site in Northern Italy to 35.1 µg m^{-3} at a site in the French Alps that is highly impacted by wood burning emissions. The corresponding relative contributions of PM_{wb} range from 9% to 88%. The relative contribution larger than 100% at the station Lescheraines in France results from a special situation at this site and resulting limitations of the applied source apportionment method (see [53]). Generally lower contributions of PM_{wb} to PM were reported at urban or suburban sites north of the Alps with PM_{wb} during wintertime about or less than 7.3 µg m^{-3} (relative contribution about 12–20%). The highest contributions from wood burning to PM were reported from sites in the French Alps and sites in villages in Alpine valleys, e.g. in Roveredo (Switzerland).

3.2.2 OC from WB

Figure 2b illustrates the contributions of wood burning to OC in the Alpine region as reported in literature and presented in Table 1. Absolute OC_{wb} concentrations range from below 1 µg m^{-3} at remote sites up to 20 µg m^{-3}. The highest concentrations of OCwb (above 10 µg/m^3) were found at the French sites in Lanslebourg, Lescheraines and Passy, at the Northern Italian site Ispra, and at the site in Moleno in Southern Switzerland. These stations are located at the edge of the Alps or in Alpine valleys and are often influenced by periods with strong thermal inversions leading to high concentrations of CM and PM.

Relative contributions of OC_{wb} to total OC are 11–31% at remote sites, 17–91% at rural sites and 28–61% at urban background sites. The contribution of OC_{wb} to OC is comparatively small at the urban sites north of the Alps. For example, at the Swiss sites in Basel, Bern and Zurich, and the Austrian sites in Salzburg and Vienna the contribution of OC_{wb} to total OC was found to be in the range from 33% to 54%.

3.2.3 EC from WB

The contributions of wood burning to elemental carbon at the sites listed in Table 1 are shown in Fig. 2c. The highest average concentrations of EC_{wb} (i.e. concentrations above 1 µg m^3) have been found at rural or urban background sites in Southern Switzerland (Magadino, Moleno and Roveredo), Northern Italy (Sondrio and Ispra) and Zagorje in Slovenia. All these sites are located in Alpine valleys or at the foothills of the Alps. Such locations support the accumulation of pollutants due to reduced air mass exchange, the dominating fraction of EC_{wb} originates very likely from local sources. The relative contribution of wood burning-related elemental carbon to total EC tends to be at these sites also high but depends also on the proximity to road traffic which is typically the most important source of EC.

3.2.4 Comparison Between Site Types

Most of the sites considered in the presented summary of studies on the influence of wood burning emissions on air quality represent rural or urban background situations (see Table 1). Figure 3 shows a box plot of the reported average contributions of wood burning emissions to PM, OC and EC separately for these two site types. Figure 3 includes only results from studies using PM2.5 and PM10 in order to improve the comparability of the concentration values reported by the different studies. For EC and OM most of the mass concentration in PM10 is already found in PM2.5 [58, 59]. Note that PM2.5 to PM10 ratios in Europe range from 0.5 to 0.9 [64]. Figure 3 shows that absolute concentrations of PM_{wb}, OC_{wb} and EC_{wb} tend at rural sites to be higher than in the urban background. The higher concentrations at rural sites might reflect the traditional and until today still

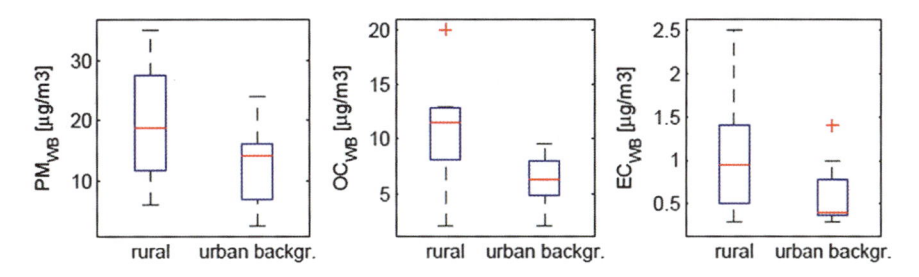

Fig. 3 Box plot of EC_{wb}, OC_{wb} and PM_{wb} from rural and urban background sites measured in PM10 and PM2.5. *Red line*: median, edges of the box are the 25th and 75th percentiles, whiskers extend to the most extreme data points not considered as outliers, outliers are plotted individually

widespread use of wood for residential heating in the Alpine region, often burnt in small stoves or fireplaces at poor operating conditions. It is clear from the absolute contributions of wood burning emissions to the particulate air pollutants as shown in Fig. 3 that wood burning is during wintertime a major and probably at many rural sites dominating source of ambient particulate matter and carbonaceous aerosols.

3.2.5 Diurnal Variations of Elemental Carbon from Wood Burning

Emissions from wood burning vary strongly during the course of the day. The analysis of data from an aerosol mass spectrometer and from an aethalometer is capable of providing source contributions with sufficient temporal resolution to resolve diurnal variation of source contributions. Recent publications show that the highest concentrations of wood burning constituents occur primarily during evening hours, wood burning-related evening peak contributions were found to occur later than the traffic-related evening peaks [9, 45–47].

Here we present diurnal variations of EC_{WB} with exemplary data from the Swiss stations Payerne, Magadino and Zurich. The contribution of WB to EC_{WB} was obtained from continuous optical absorption measurements at several wavelengths during wintertime (December 2010 to February 2011), the data analysis was performed as described in detail by Herich et al. [45]. The resulting diurnal cycles of EC_{WB} in Payerne, Magadino and Zurich are shown in Fig. 4, the results are separately shown for weekdays and weekends. The diurnal variations are similar at all three locations. Apart from the prominent evening peak, concentrations are slightly enhanced prior to noon. Lowest concentrations occur during the afternoon. The diurnal variations are for weekdays and weekends very similar. For all stations daily minimum and maximum concentrations differ by a factor of about 2–3. The enhanced concentrations in the evening may to some extent occur due to meteorological conditions, e.g. a lower mixing layer height can limit dilution processes during the night.

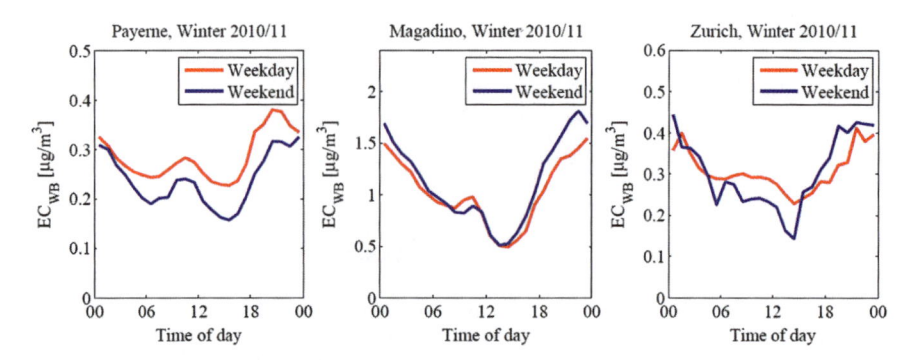

Fig. 4 Diurnal cycles of EC_{WB} at the Swiss measurement sites Payerne, Magadino and Zurich as obtained using the aethalometer model (see Sect. 2.3). The diurnal cycles are shown for weekdays and weekends during winter 2010/2011 (December, January, February)

4 Conclusion

The results of the available studies on the influence of residential wood burning on the concentrations of particulate matter, organic carbon and elemental carbon clearly show that this emission source is an important factor for air quality during wintertime in the entire Alpine region. This is true for both, larger cities and small villages in rural areas. The absolute contribution of wood burning emissions to particulate air pollutants tends in rural environments to be even larger than in urban areas. This is probably, on the one hand, due to the high contribution of wood burning to the energy supply in many villages in the Alpine region and the high fraction of small wood burning appliances that are difficult to operate with low emissions. On the other hand, the meteorological situations often prevailing in loosely populated Alpine or Pre-Alpine valleys during wintertime favour the accumulation of local emissions in a shallow boundary layer, thus leading to poor air quality in these locations.

It is clear from the findings of the studies considered in this overview that strategies for the abatement of particulate air pollution in the Alpine region need to target on reducing emissions from residential wood burning. However, implementation of efficient measures is hampered by the fact that wood is a locally available fuel and its usage has a long tradition in the Alpine region. In addition, there are conflicting interests between air quality and climate issues that need to be resolved because there are good reasons to use wood as a renewable source of energy.

Acknowledgements We are grateful to Grisa Močnik from Aerosol d.o.o. (Slovenia) for providing currently unpublished results from sites in Slovenia that perfectly fit into this overview. Thanks are also due to Christine Piot, Jean-Luc Besombes (both Université de Savoie, Le Bourget-du-Lac, France) and Jean-Luc Jaffrezo (Université Joseph Fourier, Grenoble, France) for helpful discussions about the influence of wood burning on particulate air pollutants at French Alpine sites. Support from the Competence Center Environment and Sustainability of the ETH Domain (CCES) through the research project IMBALANCE is gratefully acknowledged.

References

1. Jerrett M, Burnett RT, Ma R, Pope CA, Krewski D, Newbold KB, Thurston G, Shi Y, Finkelstein N, Calle EE, Thun MJ (2005) Spatial analysis of air pollution and mortality in Los Angeles. Epidemiology 16(6):727–736
2. Kennedy IM (2007) The health effects of combustion-generated aerosols. Proc Combust Inst 31:2757
3. Nel A (2005) Air pollution-related illness: effects of particles. Science 308(5723):804–806
4. Pope CA, Dockery DW (2006) Health effects of fine particulate air pollution: lines that connect. J Air Waste Manag Assoc 56:709–742
5. World Health Organization (WHO) (2005) WHO air quality guidelines global update 2005. Report on a working group meeting, Bonn, 18–20 October 2005
6. Borrego C, Valente J, Carvalho A, Sa E, Lopes M, Miranda AI (2010) Contribution of residential wood combustion to PM10 levels in Portugal. Atmos Environ 44(5):642–651
7. Caseiro A, Bauer H, Schmidl C, Pio C, Puxbaum H (2009) Wood burning impact on PM10 in three Austrian regions. Atmos Environ 43:2186–2195
8. Hellen H, Hakola H, Haaparanta S, Pietarila H, Kauhaniemi M (2008) Influence of residential wood combustion on local air quality. Sci Total Environ 393(2–3):283–290
9. Lanz VA, Prévôt ASH, Alfarra MR, Weimer S, Mohr C, DeCarlo PF, Gianini MFD, Hueglin C, Schneider J, Favez O, D'Anna B, George C, Baltensperger U (2010) Characterization of aerosol chemical composition with aerosol mass spectrometry in central Europe: an overview. Atmos Chem Phys 10:10453–10471. doi:10.5194/acp-10-10453-2010
10. Simpson D, Yttri KE, Klimont Z, Kupiainen K, Caseiro A, Gelencser A, Pio C, Puxbaum H, Legrand M (2007) Modeling carbonaceous aerosol over Europe: analysis of the CARBOSOL and EMEP EC/OC campaigns. J Geophys Res Atmos 112:D23
11. Naeher LP, Brauer M, Lipsett M, Zelikoff JT, Simpson CD, Koenig JQ, Smith KR (2007) Woodsmoke health effects: a review. Inhal Toxicol 19(1):67–106
12. Favez O, El Haddad I, Piot C, Boréave A, Abidi E, Marchand N, Jaffrezo J-L, Besombes J-L, Personnaz M-B, Sciare J, Wortham H, George C, D'Anna B (2010) Inter-comparison of source apportionment models for the estimation of wood burning aerosols during wintertime in an Alpine city (Grenoble, France). Atmos Chem Phys 10:5295–5314
13. Gianini MFD, Fischer A, Gehrig R, Ulrich A, Wichser A, Piot C, Besombes J-L, Hueglin C (2012) Sources of PM10 in Switzerland: an analysis for 2008/2009 and changes since 1998/1999. Atmos Environ 54:149–158
14. Gilardoni S, Vignati E, Cavalli F, Putaud JP, Larsen BR, Karl M, Stenström K, Genberg J, Henne S, Dentener F (2011) Better constraints on sources of carbonaceous aerosols using a combined ^{14}C – macro tracer analysis in a European rural background site. Atmos Chem Phys 11:5685–5700. doi:10.5194/acp-11-5685-2011
15. Szidat S, Prevot ASH, Sandradewi J, Alfarra MR, Synal H-A, Wacker L, Baltensperger U (2007) Dominant impact of residential wood burning on particulate matter in Alpine valleys during winter. Geophys Res Lett 34:L05820. doi:10.1029/2006GL028325
16. Nussbaumer T (2005) Dieselruss und Holzfeinstaub grundverschieden. Holz-Zentralblatt 131 (70):932–933
17. Kelz J, Brunner T, Obernberger I, Jalava P, Hirvonen M-R(2010) PM emissions from old and modern biomass combustion systems and their health effects. In: 18th European biomass conference, 3–7 May 2010, Lyon (F), ETA Florence & WIP Munich
18. Klippel N, Nussbaumer T (2007) Health relevance of particles from wood combustion in comparison to Diesel soot. In: 15th European biomass conference, Berlin 7–11 May 2007
19. Heringa MF, DeCarlo PF, Chirico R, Tritscher T, Dommen J, Weingartner E, Richter R, Wehrle G, Prévôt ASH, Baltensperger U (2011) Investigations of primary and secondary particulate matter of different wood combustion appliances with a high-resolution time-of-flight aerosol mass spectrometer. Atmos Chem Phys 11:5945–5957. doi:10.5194/acp-11-5945-2011

20. Steierer et al (2007) Wood energy in Europe and North America: a new estimate of volumes and flows. UNECE timber committee and the FAO European forestry commission. http://www.unece.org
21. Bundesamt für Energie BFE (2012) Schweizerische Holzenergiestatistik, Erhebung für das Jahr 2011. http://www.bfe.admin.ch/dokumentation/publikationen
22. Schmidl C, Marr IL, Ae C, Kotianová P, Berner A, Bauer H, Kasper-Giebl A, Puxbaum H (2008) Chemical characterisation of fine particle emissions from wood stove combustion of common woods growing in mid-European Alpine regions. Atmos Environ 42:126–141
23. Schmidl C, Luisser M, Padouvas E, Lasselsberger L, Rzaca M, Cruz C, Handler M, Peng G, Bauer H, Puxbaum H (2011) Particulate and gaseous emission from manually and automatically fired small scale combustion systems. Atmos Environ 45:7443–7454
24. Kistler M, Schmidl C, Padouvas E, Giebl H, Lohninger J, Ellinger R, Bauer H, Puxbaum H (2012) 1 Odor, gaseous and PM10 emissions from small scale combustion of wood types indigenous to Central 2 Europe. Atmos Environ 2012(51):86–93
25. Alfarra MR, Prévôt ASH, Szidat S, Sandradewi J, Weimer S, Lanz VA, Schreiber D, Mohr M, Baltensperger U (2007) Identification of the mass spectral signature of organic aerosols from wood burning emissions. Environ Sci Technol 41:5770–5777
26. Weimer S, Alfarra MR, Schreiber D, Mohr M, Prévôt ASH, Baltensperger U (2008) Organic aerosol mass spectral signatures from wood-burning emissions: influence of burning conditions and wood type. J Geophys Res 113:D10304. doi:10.1029/2007JD009309
27. Heringa MF, DeCarlo PF, Chirico R, Tritscher T, Clairotte M, Mohr C, Crippa M, Slowik JG, Pfaffenberger L, Dommen J, Weingartner E, Prévôt ASH, Baltensperger U (2012) A new method to discriminate secondary organic aerosols from different sources using high-resolution aerosol mass spectra. Atmos Chem Phys 12:2189–2203,2205. doi:10.5194/acp-12-2189-2012
28. Spitzer J, Enzinger P, Fankhauser G, Fritz W, Golja F, Stiglbrunner R (1998) Emissionsfaktoren für feste Brennstoffe. Project report. Institut für Energieforschung and Institut für Angewandte Statistik und Systemforschung, Graz. http://www.wien.gv.at/umweltschutz/pool/pdf/brennstoffe.pdf
29. Struschka M, Kilgus D, Springmann M, Baumbach G (2008) Effiziente Bereitstellung aktueller Emissionsdaten für die Luftreinhaltung. Project Report. Forschungsbericht 205 42 322, UBA-FB 001217, Umweltbundesamt, Dessau-Rosslau. http://www.umweltbundesamt.de
30. Nussbaumer T (2010) (Hrsg.): 11. Holzenergie-Symposium: Potenzial und Technik zur Holzenergie-Nutzung. Tagung an der ETH Zürich am 17. September 2010, Verenum Zürich, ISBN 3-908705-21-5
31. Hopke PK (2003) Recent developments in receptor modeling. J Chemom 17:255–265
32. Viana M, Pandolfi M, Minguillon MC, Querol X, Alastuey A, Monfort E, Celades I (2008) Inter-comparison of receptor models for PM source apportionment: case study in an industrial area. Atmos Environ 42:3820–3832
33. El Haddad I, Marchand N, Wortham H et al (2011) Primary sources of PM(2.5) organic aerosol in an industrial Mediterranean city, Marseille. Atmos Chem Phys 11:2039–2058
34. Ke L, Zheng M, Tanner RL, Schauer JJ (2007) Source contributions to carbonaceous aerosols in the Tennessee valley. Atmos Environ 41:8898–8923
35. Paatero P (1997) Least square formulation of robuste non-negative factor analysis. Chemometr Intell Lab Syst 3:23–35
36. Paatero P, Tapper U (1994) Positive matrix factorization: a nonnegative factor model with optimal utilization of error estimates of data values. Environmetrics 5:111–126
37. Lee E, Chan CK, Paatero P (1999) Application of positive matrix factorization in source apportionment of particulate pollutants in Hong Kong. Atmos Environ 33:3201–3212
38. Pandolfi M, Gonzalez-Catanedo Y, Alastuey A, de la Rosa JD, Mantilla E, Querol X, Pey J, Amato F, Moreno T (2011) Source apportionment of PM10 and PM2.5 at multiple sites in the strait of Gibraltar by PMF: impact of shipping emissions. Env Sci Poll Res 18:260–269

39. Lanz VA, Alfarra MR, Baltensperger U, Buchmann B, Hueglin C, Prévôt ASH (2007) Source apportionment of sub- micron organic aerosols at an urban site by factor analytical modelling of aerosol mass spectra. Atmos Chem Phys 7:1503–1522, http://www.atmos-chem-phys.net/7/1503/2007/
40. Puxbaum H, Caseiro A, Sánchez-Ochoa A, Kasper-Giebl A, Claeys M, Gelencsér A, Legrand M, Preunkert S, Pio C (2007) Levoglucosan levels at background sites in Europe for assessing the impact of biomass combustion on the European aerosol background. J Geophys Res 112: D23S05. doi:10.1029/2006JD008114
41. Heal MR, Naysmith P, Cook GT, Xu S, Raventos Duran T, Harrison RM (2011) Application of 14C analyses to source apportionment of carbonaceous PM2.5 in the UK. Atmos Environ 45:2341–2348
42. Szidat S, Jenk TM, Synal H-A, Kalberer M, Wacker L, Hajdas I, Kasper-Giebl A, Baltensperger U (2006) Contributions of fossil fuel, biomass-burning, and biogenic emissions to carbonaceous aerosols in Zurich as traced by 14C. J Geophys Res 111:D07206. doi:10.1029/2005JD006590
43. Gelencsér A, May B, Simpson D, Sánchez-Ochoa A, Kasper-Giebl A, Puxbaum H, Caseiro A, Pio C, Legrand M (2007) Source apportionment of PM2.5 organic aerosol over Europe: primary/secondary, natural/anthropogenic, and fossil/biogenic origin. J Geophys Res 112: D23S04. doi:10.1029/2006JD008094
44. Szidat S, Ruff M, Perron N, Wacker L, Synal H-A, Hallquist M, Shannigrahi AS, Yttri KE, Dye C, Simpson D (2009) Fossil and non-fossil sources of organic carbon (OC) and elemental carbon (EC) in Göteborg, Sweden. Atmos Chem Phys 9:1521–1535
45. Herich H, Hueglin C, Buchmann B (2011) A 2.5 year's source apportionment study of black carbon from wood burning and fossil fuel combustion at urban and rural sites in Switzerland. Atmos Measure Techn 4:1409–1420
46. Sandradewi J, Prevot ASH, Szidat S, Perron N, Alfarra MR, Lanz VA, Weingartner E, Baltensperger U (2008) Using aerosol light absorption measurements for the quantitative determination of wood burning and traffic emission contributions to particulate matter. Environ Sci Technol 42:3316–3323
47. Sandradewi J, Prévôt ASH, Alfarra MR, Szidat S, Wehrli MN, Ruff M, Weimer S, Lanz VA, Weingartner E, Perron N, Caseiro A, Kasper-Giebl A, Puxbaum H, Wacker L, Baltensperger U (2008) Comparison of several wood smoke markers and source apportionment methods for wood burning particulate mass. Atmos Chem Phys Discus 8:8091–8118
48. Schauer JJ, Kleeman MJ, Cass GR, Simoneit BRT (1999) Measurement of emission from air pollution sources. 1. C1 through C29 organic compounds from meat charboiling. Environ Sci Tecnol 33:1566–1577
49. Schauer JJ, Kleeman MJ, Cass GR, Simoneit BRT (2001) Measurement of emission from air pollution sources. 3. C1–C29 organic compounds from fireplace combustion of wood. Environ Sci Tecnol 35:1716–1728
50. Jordan TB, Seen AJ, Jacobsen GE (2006) Levoglucosan as an atmospheric tracer for woodsmoke. Atmos Environ 40:5316–5321
51. Simoneit BRT, Schauer JJ, Nolte CG, Oros DR, Elias VO, Fraser MP, Rogge WF, Cass GR (1999) Levoglucosan, a tracer for cellulose in biomass burning and atmospheric particles. Atmos Environ 33:173–182
52. Piazzalunga A, Bernardoni V, Fermo P, Valli G, Vecchi R (2011) Technical note: on the effect of water-soluble compounds removal on EC quantification by TOT analysis in urban aerosol samples. Atmos Chem Phys 11:10193–10203
53. Piot C (2011) Polluants atmosphériques organiques particulaires en Rhône Alpes - caractérisation chimique et sources d'émission. Ph.D. Thesis, Joseph Fourier University, Grenoble, 294 pp. http://tel.archives-ouvertes.fr/docs/00/66/12/84/PDF/35623_PIOT_2011_archivage.pdf
54. Perron N, Sandradewi J, Alfarra MR, Lienemann P, Gehrig R, Kasper-Giebl A, Lanz VA, Szidat S, Ruff M, Fahrni S, Wacker L, Baltensperger U, Prévôt ASH (2010) Composition and sources of particulate matter in an industrialised Alpine valley. Atmos Chem Phys Discuss 10:9391–9430. doi:10.5194/acpd-10-9391-2010

55. Weimer S, Mohr C, Richter R, Keller J, Mohr M, Prevot ASH, Baltensperger U (2009) Mobile measurements of aerosol number and volume size distributions in an Alpine valley: influence of traffic versus wood burning. Atmos Environ 43:624–630
56. Larsen BR, Gilardoni S, Stenström K, Niedzialek J, Jimenez J, Belis CA (2012) Sources for PM air pollution in the Po plain, Italy: II. Probabilistic uncertainty characterization and sensitivity analysis of secondary and primary sources. Atmos Environ 50:203–213
57. Perrone MG, Larsen BR, Ferrero L, Sangiorgi G, De Gennaro G, Udisti R, Zangrando R, Gambaro A, Bolzacchini E (2012) Sources of high PM2.5 concentrations in Milan, Northern Italy: molecular marker data and CMB modeling. Sci Total Environ 414:343–355
58. Hueglin C, Gehrig R, Baltensperger U, Gysel M, Monn C, Vonmont H (2005) Chemical characterisation of PM2.5, PM10 and coarse particles at urban, near-city and rural sites in Switzerland. Atmos Environ 39:637–651
59. Jaffrezo JL, Aymoz G, Cozic J (2005) Size distribution of EC and OC in the aerosol of Alpine valleys during summer and winter. Atmos Chem Phys 5:2915–2925
60. Harrison RM, Stedman J, Derwent D (2008) New directions: why are PM10 concentrations in Europe not falling? Atmos Environ 42(3):603–606
61. Barmpadimos I, Hueglin C, Keller J, Henne S, Prevot ASH (2011) Influence of meteorology on PM10 trends and variability in Switzerland from 1991 to 2008. Atmos Chem Phys 11 (4):1813–1835
62. Heckmann M, Friedl G, Schwarz M, Rossmann P, Hartmann H, Baumgartner H, Lasselsberger L, Themessl A (2010) Bestimmung von Jahresnutzungsgrad und Emissionsfaktoren von Biomasse-Kleinfeuerungen am Prüfstand. Project report. Bundesministeriums für Verkehr, Innovation und Technologie, Austria. http://www.energiesystemederzukunft.at/edz_pdf/1128a_biomasse_kleinfeuerungen.pdf
63. BAFU (2005) Arbeitsblatt Emissionsfaktoren Feuerungen, Stand September 2005. Bundesamt für Umwelt, Bern
64. Putaud JP, Van Dingenen R, Alastuey A, Bauer H, Birmili W, Cyrys J, Flentje H, Fuzzi S, Gehrig R, Hansson HC, Harrison RM, Herrmann H, Hitzenberger R, Huglin C, Jones AM, Kasper-Giebl A, Kiss G, Kousa A, Kuhlbusch TAJ, Loschau G, Maenhaut W, Molnar A, Moreno T, Pekkanen J, Perrino C, Pitz M, Puxbaum H, Querol X, Rodriguez S, Salma I, Schwarz J, Smolik J, Schneider J, Spindler G, ten Brink H, Tursic J, Viana M, Wiedensohler A, Raes F (2010) A European aerosol phenomenology-3: physical and chemical characteristics of particulate matter from 60 rural, urban, and kerbside sites across Europe. Atmos Environ 44 (10):1308–1320
65. Močnik G, Turšič J, Muri G, Bolte T, Ježek I, Drinovec L, Sciare J (2011) Influence of biomass combustion on air quality in two pre-Alpine towns with different geographical settings. In: 10th international conference on carbonaceous particles in the atmosphere, Vienna (Austria), 26–29 June 2011
66. Močnik G (2012) Private communication

Ammonia Emissions in Europe

Carsten Ambelas Skjøth and Ole Hertel

Abstract Ammonia emissions are mainly related to agricultural activities, and depositions related to these emissions constitute a treat to local ecosystems but possibly also to human health through the contribution for formation of secondary fine fraction particles in ambient air. European ammonia emissions are highly heterogeneously distributed, and the temporal variations in these emissions follow very different pattern as a result of differences in climate but also as a results of significant differences in agricultural practice over Europe. A minor fraction of ammonia emission is related to nonagricultural sources, especially traffic. These sources are mainly found in areas with intense traffic and the use of catalyst converters. Simple and comprehensive models for the spatial and temporal variation in ammonia emissions have been shown useful in modelling of atmospheric nitrogen input to sensitive ecosystems for assessments of critical loads. For the spatial distribution various emission inventories are available at different resolutions. These inventories are derived using different approaches, and as a result they can differ up to a factor of two for certain areas. The overall European ammonia emissions are decreasing as a result of regulation related to the National Emission Ceiling Directive (NEC) and the Integrated Pollution Prevention and Control (IPPC) directive, regulation that has been implemented in national legislation in the single European countries. Some countries have adopted screening methods to be used by local authorities when assessing impact on local ecosystems in relation to applications from farmers to obtain permissions to increase agricultural production. In general Northern European countries have more strict regulation of ammonia emissions compared with Central and Southern European countries.

Keywords Impact of regulation and climate change, Inventories, Models, Spatial and temporal distribution, Trends

C.A. Skjøth and O. Hertel (✉)
Department of Environmental Science, Aarhus University, Frederiksborgvej 399,
P.O. Box 358, 4000 Roskilde, Denmark
e-mail: oh@dmu.dk

Contents

1 Introduction 142
2 Emissions from Different Source Categories 142
   2.1 Animal Houses and Manure Storages 143
   2.2 Application of Manure and Mineral Fertiliser 144
   2.3 Emissions from Grazing Animals 147
   2.4 Emission of NH_3 from Other Agricultural Sources 148
   2.5 Emissions from Nonagricultural Sources, e.g. Urban Areas and Traffic 148
3 Spatial Distributions in Emissions 149
4 Site-Specific Long-Term Trends in Emissions 153
5 Long-Term Trends in Emissions on European Scale 154
6 Regulation of Emissions, Future Scenarios and Impact of Climate Change 155
References 158

1 Introduction

Ammonia (NH_3) emissions lead to significant environmental impacts in Europe not only with respect to formation of ambient air PM with potential health impact but also with respect to nitrogen depositions to natural surfaces with potential impact on biodiversity. NH_3 deposits very fast to most surfaces and may thus through a high atmospheric nitrogen (N) input significantly contribute to stress of ecosystems close to ammonia sources [1], a stress that on the long term may lead to loss of biodiversity [2–4]. Release related to agricultural activities is the dominating source of European NH_3 emissions [5, 6]. In the vicinity of large animal farms, the atmospheric load to ecosystems can be totally dominated by a single farm [7] contributing to N depositions of even 50–80 kg N/ha/year [8, 9], depending on distance to source, source strength and local climatic conditions. Such high local N loadings are a factor of five to ten beyond the critical loads for the sensitive terrestrial ecosystems (e.g. [10], Table 1). In Europe, about 90–95% of NH_3 emissions arise from agricultural sources, with main contributions from animal housings (34–43%), manure handling (22–26%) and mineral fertiliser (17–26%) [11]. It should be noted that organic bound N in the manure is not a direct source of NH_3. The emission strength is therefore mainly related to the manure or fertiliser content of TAN (Total ammonia N (NH_3 and NH_4^+)), the pH in the manure or fertiliser and the ambient air temperature and wind speed. Emission estimates are often derived on the basis of national emission factors for NH_3 relating certain agricultural production methods to specific emission rates [12], e.g. aggregating results from multiplying census data of farm animals for all animal categories with estimated emission factors per animal [13].

2 Emissions from Different Source Categories

Emission of NH_3 to the atmosphere is a physical process taking place from wet surfaces [14]. This process is highly temperature dependent and varies therefore during day and over seasons [15, 16]. The regional variation is a function of production methods and agricultural practice, whereas meteorology but also agricultural practice affects the temporal variations. Furthermore, the emission of NH_3 is related to different anthropogenic activities and sources (especially agriculture) throughout the year. The overall NH_3 emission is therefore highly depending on location and time of the year due to differences in climate and anthropogenic activities between regions. The emissions of NH_3 may be grouped in the following source categories:

Agricultural sources:
- Point sources, i.e. animal houses, manure storages and "slurry lakes" (e.g. [17])
- Application of manure and mineral fertiliser to soil and growing fields (e.g. [18])
- Grazing animals (e.g. [19])

Other sources including plants [20] such as legumes [15] and vegetation during management and senescence [21, 22].

Nonagricultural sources:
- Wild animals [23, 24]
- Catalyst processes, mainly related to road traffic (e.g. [25, 26])
- Manufacturing processes such as production of fertiliser, glass wool, catalysts and cement
- Humans, pets and sewage systems [24, 27]
- Other sources such as landfill and non-anthropogenic sources including natural fires from ecosystems [28, 29]
- Reemission from plants due to compensation point [30, 31]
- Emission from sea surfaces [32, 33]

The last two are not accounted for in this chapter, whereas the other source categories are described in detail in the following. In this description of these categories focus is on parameterisation of spatial and temporal variations and here taking outset in the methodology developed for the Danish area [15, 34] that is to be extended to European coverage within the FP7 project ECLAIRE [6, 11] during 2011–2015. This methodology is currently considered as the best and most advanced of its kind [35], and it has furthermore the advantage that it to some degree is based on generalised physical properties such as volatilisation. It has been derived from European-based studies on agricultural activities in relation to buildings and the interior climatic conditions in these buildings [36, 37]. The parameterisations have been derived for chemistry-transport models (CTMs), and up to now they have been implemented in the long-range transport models, ACDEP [16, 38], DEHM [6] and the EMEP models [39, 40], and the local-scale model, OML-DEP [41].

2.1 Animal Houses and Manure Storages

Significant variations in NH_3 emissions are found in different types of animals and housings [42]. These variations are related to amount of TAN in the manure, stable temperature, ventilation rate and local ambient wind speeds. Highly complex surface models have been developed for describing the NH_3 emissions from agricultural buildings [43–46], but these are generally not applicable for large-scale modelling [15, 47]. A simplified parameterisation based on local wind speed and ambient air temperature has been shown to be a more practical approach in air pollution modelling [6, 10, 15, 16, 41]. Most emission inventories take outset in the total N or TAN content in the manure and applying generalised fixed emission factors (EF). In order to provide proper results, these EFs need to be location specific and account for local conditions. Such data is currently being processed for the entire European area.

On large scale, it has been shown that the variation in emissions from stables and storages (we here term these $E(t)$) may be described by a simple parameterisation [16, 38]:

$$E(t) = C \times T(t)^{0.89} \times V(t)^{0.26} \qquad (1)$$

where C is a constant related to the content of N or TAN in the manure at given time and location, T is the ambient air temperature and V is the ambient air wind speed or the rate of ventilation; the two latter are both functions of time. This formula is applied for distributing a known annual emission into smaller time steps used in the CTMs.

Pig and poultry stables have a relatively high critical temperature for comfort and optimal production of animals compared with cattle stables. This means that in Northern Europe, pigs and poultry stables are heated during winter. The overall annual emission from manure storages varies with the type of storage, whereas the temporal variation follows variations in ambient air temperature [15]. In warm areas and during warm periods of time, the emissions from all buildings will reflect outdoor temperatures [37]. Buildings containing pigs and poultry thus have significant emissions also during cold periods, whereas cattle barns and storages in these situations have low emission rates.

Applying the parameterisation in Eq. (1) together with the simplified functions shown in Fig. 1, the temporal variation in NH_3 emissions may be simulated with good results. Figure 2 shows as an example of calculated temporal variation in NH_3 emission from a pig stable, cattle barn and manure storage. It is easily seen that the variation reflects the differences described in the previous section.

2.2 Application of Manure and Mineral Fertiliser

The emission of NH_3 from field application of manure and mineral fertiliser takes place at distinct times of the year and has a relatively short duration compared with point source emissions. Soil type [49] and the application method are crucial for the magnitude of the emission. Very high emissions are related to the broad spread application methods, whereas soil injection leads to very low emissions. For these emissions, the temporal variations can be described by process-based models on the

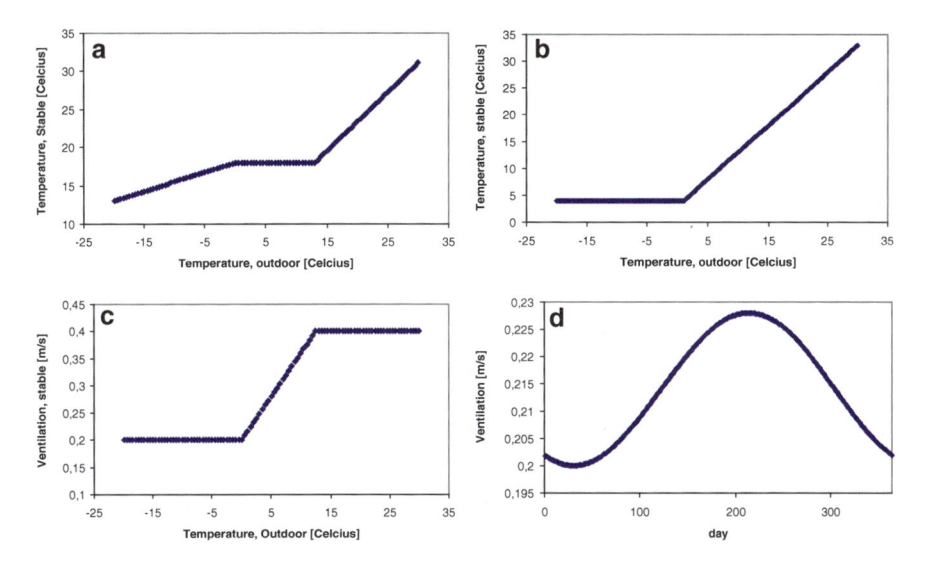

Fig. 1 The *upper figures* show indoor temperature in stables as function of outdoor ambient air temperature for (**a**) isolated stables and (**b**) open barns. The *lower figures* show ventilation rate for (**c**) isolated stables as function of ambient air temperatures and (**d**) open barns as function of day of year

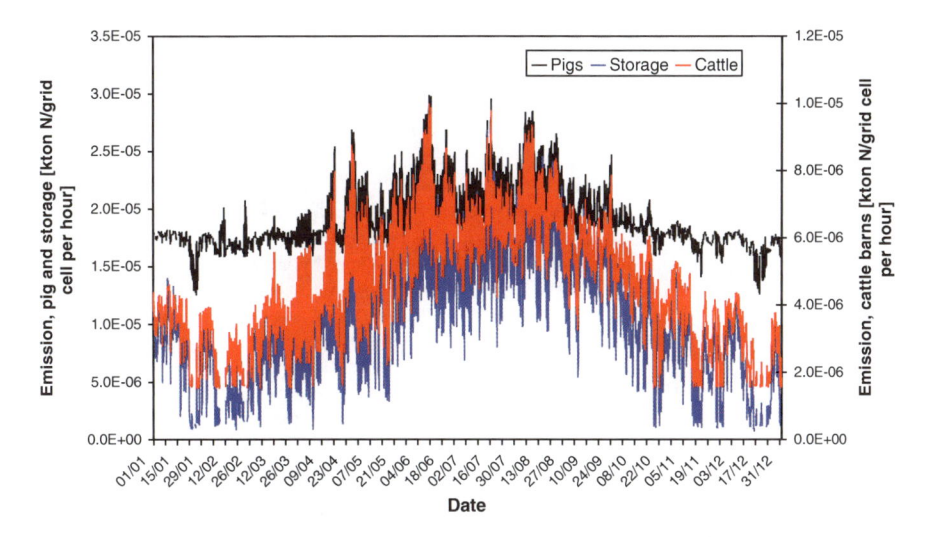

Fig. 2 Simulations of ammonia emission for open and isolated stables and manure storage using hourly meteorological input of temperature and wind speed. All time series were created by using meteorological data for the year 2007 and the emission model available at http://www.atmos-chem-phys.net/11/5221/2011/acp-11-5221-2011.html [48]

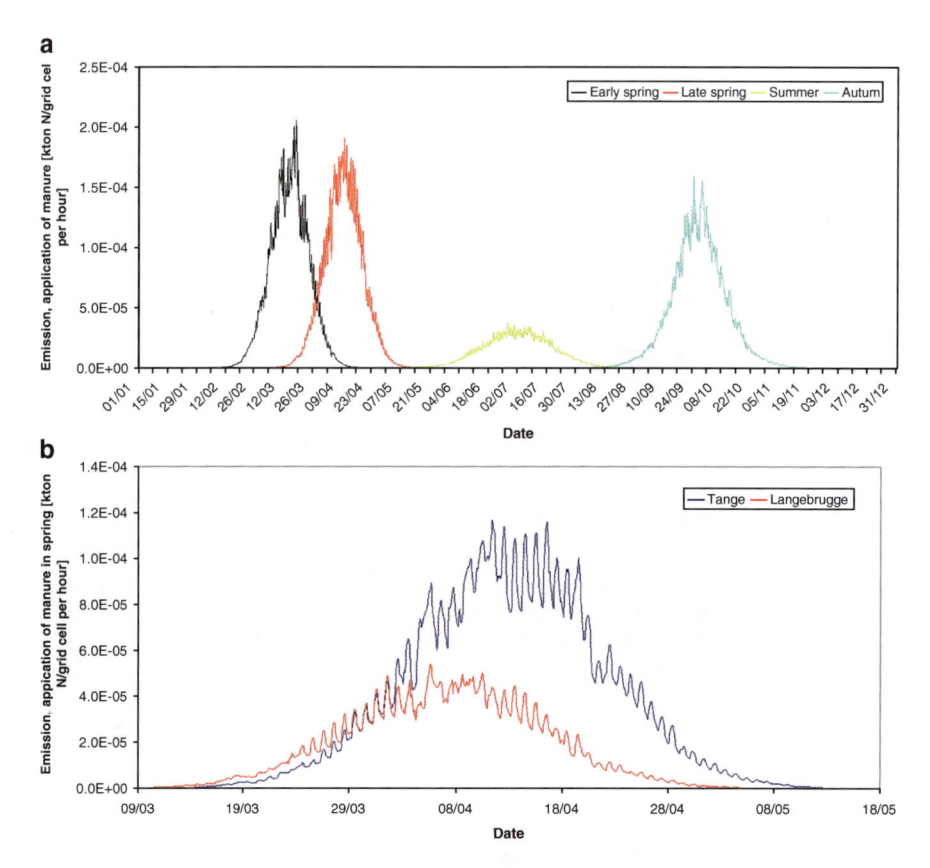

Fig. 3 The emission strength and temporal variation in NH_3 emission in the Lindet area in Denmark related to application of (**a**) manure during spring, summer and autumn and (**b**) emission strength from the Tange area in Denmark and Langebrugge in Germany due to application of manure in growing crops during spring. All time series created by using meteorological data for the year 2007 and the emission model available at http://www.atmos-chem-phys.net/11/5221/2011/acp-11-5221-2011.html

field scale by incorporation of environmental variables, soil characteristics and agricultural management [50]. National regulations can significantly affect the seasonality of the emissions [51], although these emissions have substantial uncertainties [18]. In some countries, manure application is constrained with almost no regulation, whereas in many North European countries, manure application is abandoned during winter time. This constrain is to limit N and phosphorus washout that will be high when there are no crops on the fields. To overcome a shortage in manure storage capacity, farmers often empty their slurry tanks in autumn, which again leads to a late autumn peak in emissions. The timing of mineral fertilisers and the associated release of NH_3 differ also between regions giving different temporal emission patterns, where farmers in the Southern parts of Europe initiate application of fertiliser earlier than those in Northern Europe. Figure 3 shows the temporal variation in NH_3 emission reflecting four typical fertiliser application times during the year 2000 at

Tange in Denmark and emission from two sites in Central Europe (Tange and Langebrugge), where Tange is located about 400 km North of Langebrugge in Germany, and the Tange station hence has a later peak in the emission from spring application directly into growing crops. In general, then the environmental conditions during application time have a large effect on both the temporal pattern and the total amount of emission from the application. A good example is the effect of precipitation: Precipitation before application limits soil infiltration and therefore enhances emission, while precipitation after application can enhance infiltration and reduce the emission [52]. The latter, however, also increases the risk run-off to nearby river systems. This clearly shows how management taking into account environmental variables (e.g. day–night temperatures) can modify emissions. Despite these known effects on environmental variables, the mathematical description accounting for these governing effects still remains to be developed for regional scale application [53, 54].

2.3 Emissions from Grazing Animals

The emission from grazing animals depends on time spend in the field but also on the N content in the grass. As a result, animals feeding from grass with high N contents excrete large amounts of excessive N as TAN compared with animals on grassland with less N-rich grass [19]. Furthermore, urine from grazing animals more easily enters the soil compared with surface-applied slurry [55]. This reduces the emission from grazing animals compared with stable-based systems. The main reason for this difference is that the larger amount of dry matter content of slurries limits infiltration into the soil and thus is allowing for enhanced evaporation of the slurry compared with urine from grazing animals. In Southern Europe, animals are in the field most of the year. Sheep may also stay in the field most of the time, whereas dairy cattle in many countries are inside stables approximately half of the year (see Fig. 2 in Skjøth et al. [6]). The number of grazing animals follows in general the availability of grass, or in other words the season of growth. In principle, the ambient air temperature should increase NH_3 emission related to grazing animals and outdoor yards used by livestock. This was supported by studies by Ryden et al. [56] who found increased NH_3 emissions from the urine fraction during summer compared to autumn. However, clear ambient air temperature responses have not always been seen in practice [57]. This suggests that some of the main processes governing NH_3 emissions from grazing animals have been identified, but several knowledge gabs still exist. The latter include developing a full dynamic description of emission from gazing animals that go beyond the farm scale. Such descriptions would be useful in policy and scenario studies on the international level in relation to air quality. Due to this, there is still significant uncertainty regarding the effect of environmental variables on the NH_3 emission from grazing animals.

2.4 Emission of NH_3 from Other Agricultural Sources

The remaining agricultural sources relate to emission from vegetation. Legumes and plants taking up excess fertiliser are emitting NH_3 [20]. This emission depends on the enrichment of the apoplast with NH_4^+ and the so-called compensation point [30]. The compensation point is a function of the plant status with respect to growth, stress, etc. The emission is still not well described with respect to magnitude, as well as temporal and spatial variation. Emissions from crops are often observed after fertilisation with either manure (e.g. [58]) or mineral fertiliser (e.g. [59]). Grazing (e.g. [60]) as well as cutting (e.g. [61]) of grass is known to release NH_3. The management of the crops can heavily influence the loss of NH_3 to the atmosphere. However, little is known about these processes, and only few studies of N-exchange between atmosphere and vegetation cover the entire season exploring the full cycle of growth and decay [22]. Overall, the emission from vegetation is highly coupled to both crop management and the atmospheric concentrations of NH_3. Several mechanistic descriptions of the compensation point have been derived [31, 62], and these rely highly on detailed information on agricultural production methods which for several decades have been difficult to obtain and generalise (e.g. [6, 63]).

2.5 Emissions from Nonagricultural Sources, e.g. Urban Areas and Traffic

Emissions from nonagricultural sources are in general not well described but include sweat from humans, excreta from pets and wild animals, exhaust from gasoline cars with catalytic converters, stationary combustion sources and industry and evaporation from waste deposits [24]. In Europe, the largest national nonagricultural NH_3 emission has been reported for the UK with a fraction of about 15% [24], a figure that may be compared with about 2.8% from traffic and 0.8% from manufacturing processes and sewage sludge that has been reported for Denmark [64]. A number of European studies have shown elevated NH_3 concentrations near roads and in urban areas (e.g. [65]). Whether these elevated concentrations are due to emission from traffic (e.g. Table 1), humans or sewage or there are other significant sources in the urban area has not been shown. Another possible explanation for elevated concentrations at these locations could be enhanced evaporation of NH_3 from NH_4^+-containing salts, as many of the very high observed values are reported for Southern and Central Europe such as Barcelona [27], Rome [66] or Croatia [67]. Experimental studies have shown that over the sea, the atmospheric fluxes of NH_3 may also be upward or downward [33, 68, 69] depending on the meteorological conditions and the relationship between the pH and the NH_4^+ concentration in the upper surface waters, on the one side, and the NH_3 concentrations in ambient air just above the water surface, on the other side. Similarly studies over forests also indicate bidirectional flux patterns NH_3

Table 1 Annual emission of ammonia (Gg NH$_3$) from different sectors in Europe during the period 1985–2009. Emissions are distributed according to the SNAP categories developed within the EMEP inventories (Webdab extraction, 12th May 2012)

	1985	1990	1995	2000	2005	2009
S1	6	5	4	5	5	6
S2	9	10	9	9	8	7
S3	3	7	6	5	8	6
S4	155	125	91	89	87	37
S5	0	6	5	10	12	11
S6	6	5	5	5	6	6
S7, road	45	16	47	98	83	69
S8	1	1	1	1	1	1
S9	138	136	117	90	85	84
S10	5,068	4,738	4,028	3,840	3,668	3,546

[90, 91, 92, 93]. Bidirectional fluxes over forests have recently been observed during senescence and leaf fall [70], suggesting that a large fraction of N that has been taken up during the growing season is released again in the autumn when litter is decomposed.

3 Spatial Distributions in Emissions

At European level, EMEP (http://www.emep.int) and CORINAIR have compiled inventories of the annual mean emissions on a grid with a spatial resolution of 50 km × 50 km [71]. The EDGAR and GEIA databases are available on 10 × 10° resolutions, and the EUROTRAC GENEMIS project (http://www.gsf.de/eurotrac) compiled inventories with a grid resolution of 16.67 km × 16.67 km. The GENEMIS data was for the year 1994, but this inventory has in some later studies been used to redistribute EMEP emission inventories for subsequent years, assuming unchanged relative distribution over the years [72, 73]. The need for high-resolution inventories has been recognised by the model groups within MACC and MACC-II research programmes (http://www.gmes-atmosphere.eu/). They have therefore adapted a high-resolution inventory on a 7 km resolution of the annual emissions at SNAP code level [54]. This is a combination of the officially reported inventories, information on geographic location of point sources and a correction procedure for inconsistencies. There are therefore different estimates regarding the spatial distribution in NH$_3$ emissions over Europe, a distribution that clearly is non-uniform (Fig. 4).

The distribution of NH$_3$ emissions from nonagricultural sources follows to a large extent the population density but is also associated with the major road network as road traffic contributes an important part of these emissions (Fig. 5). The relative importance of road traffic as a source of NH$_3$ is increasing as other sources are going down and, at the same time, as traffic is still increasing in Europe. In general, nonagricultural sources of NH$_3$ are, however, still poorly explored, and there are thus significant uncertainties associated with both magnitude and distribution of these emissions (Fig. 5).

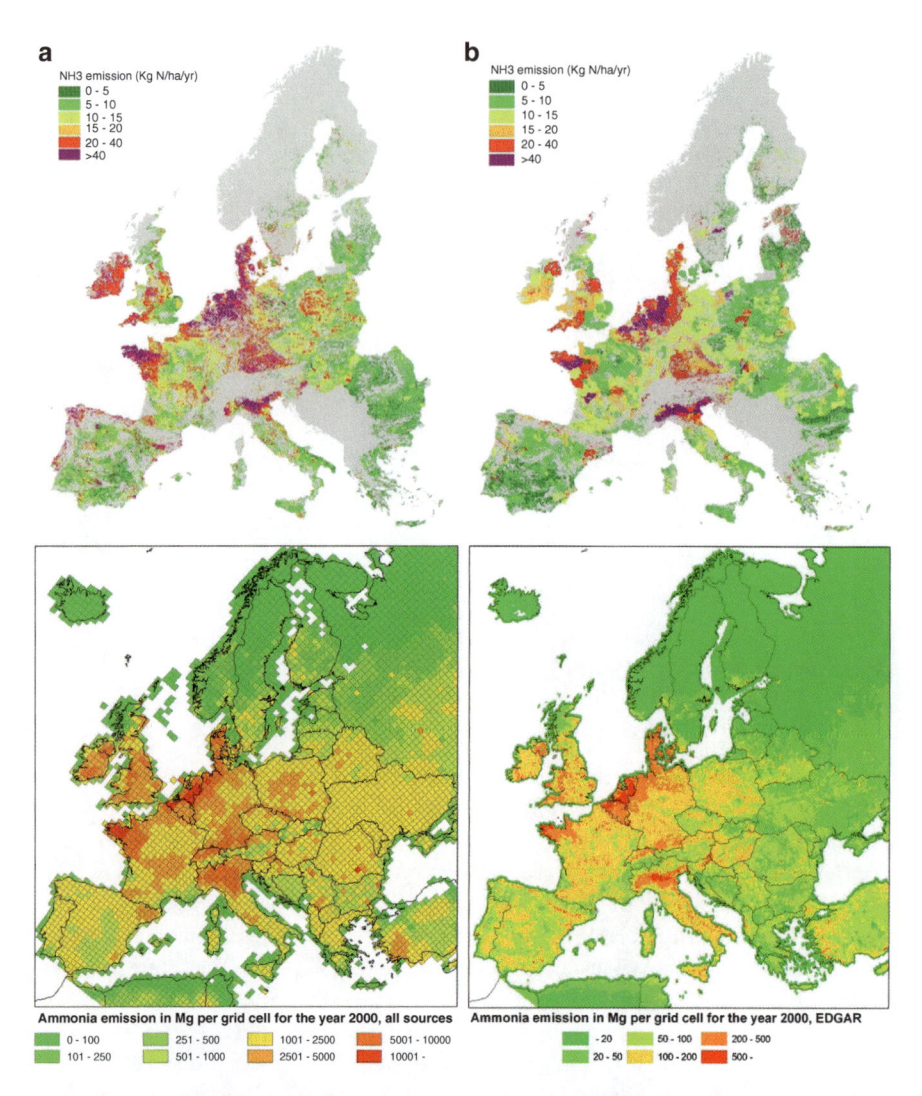

Fig. 4 (**a**) Annual NH$_3$ emissions in Europe for the year 2000 based on (*top left*) IDEAg (published in de Vries et al. [74], (**b**) INTEGRATOR (published in de Vries et al. [74], (**c**) EMEP at 50 km × 50 km and (**d**) EDGAR at '10 × '10

Studies comparing the EMEP inventories (e.g. [75]) with global estimates [76] show that in Europe, the three countries with the highest emission densities (kg NH$_3$-H ha^{-1} year^{-1}) are the Netherlands, Belgium and Denmark. This study, as well as the assessment by de Vries et al. [74], highlights the large uncertainties in national scale emissions, as the different approaches used in, e.g. EMEP, IDEAg, Miterra, IMAGE, GAINS, EDGAR, OECD and Integrator lead to differences of up to almost a factor of two for some countries.

High spatial resolution in NH$_3$ emissions has been shown to be important for a proper mapping of the N deposition to ecosystems for agricultural-intensive areas

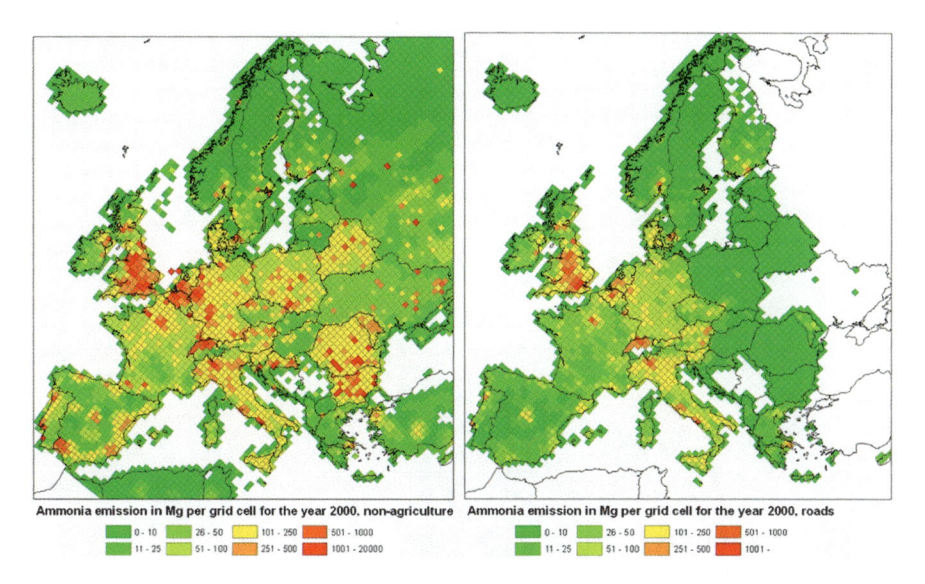

Fig. 5 European NH$_3$ emissions from nonagricultural sources according to the EMEP inventories (webdab system 12 May 2012). Note that the scale in this figure is different from the one that is used in Fig. 4

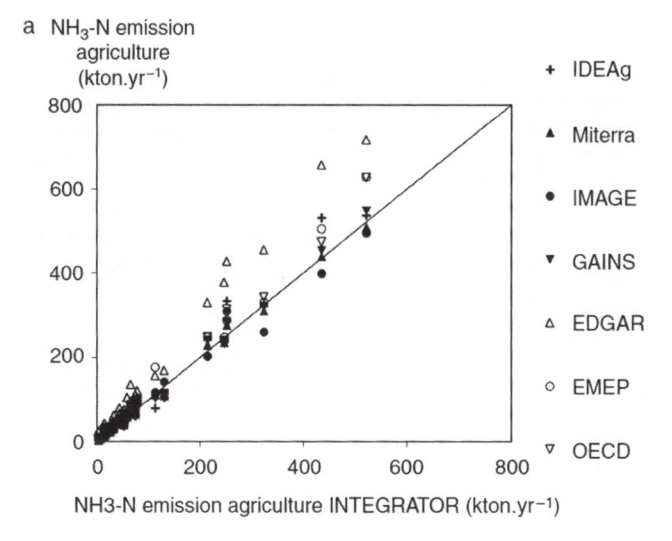

Fig. 6 Comparisons of various emission inventories for the European area performed on national totals for the NH$_3$ emission for each single countries – obtained from de Vries et al. [74]

like Denmark [1, 10], the UK [13], Poland [77] and the Netherlands [78, 79]. In fact, it has been shown that increasing resolution from 5 km × 5 km to 1 km × 1 km and taking into account low-level orographic precipitation and accurate emissions at the same scale provide a more realistic distribution of both concentrations and

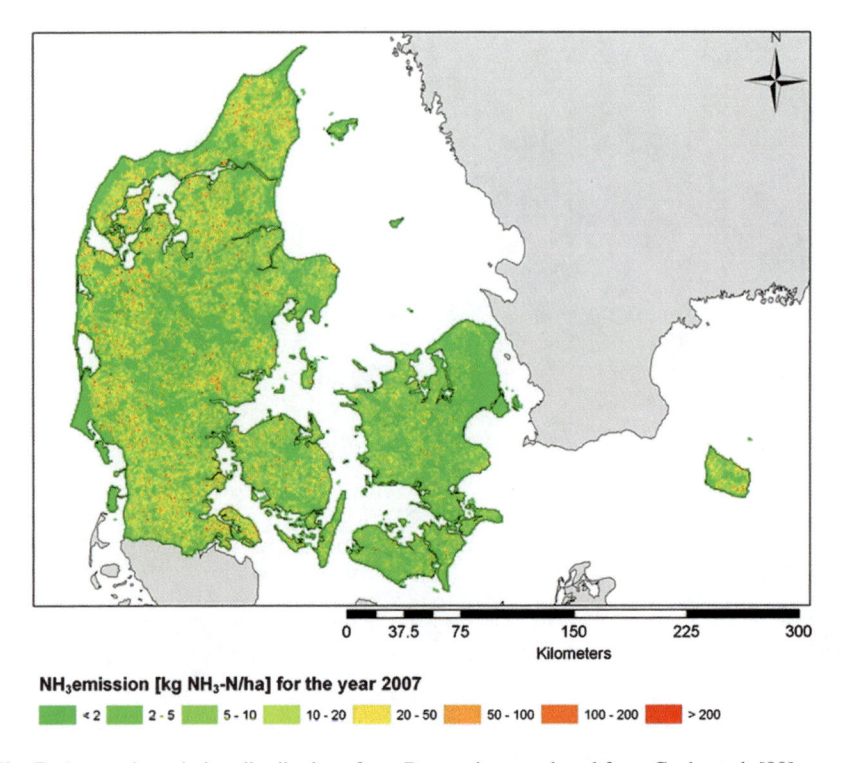

Fig. 7 Ammonia emission distributions from Denmark, reproduced from Geels et al. [80]

depositions of reactive nitrogen [13]. A similar geographical pattern with high diversity and tight connection between emissions and depositions in the high-emission areas is found in many areas of Europe including Denmark [10, 80], the Netherlands [78], Belgium as well as parts of France, Germany and Poland [77].

High-resolution emission inventories are available from the UK [13, 81], Denmark [15, 80], Switzerland [82] and the Netherlands [83]. Emissions are in general among the most important input to all CTMs and have therefore received much attention [5, 48]. Despite this, emission data is considered the largest uncertainty factor in the description of N loadings of sensitive terrestrial ecosystems [5] and similarly an important component among feedback mechanisms in the climate system [11].

Figure 7 shows the high-resolution emission distributions derived for Denmark. The geographical distributions reflect in general the location of the most intensive agricultural areas that constitute the largest NH_3 sources.

The relative national distribution in NH_3 emission between the agricultural source categories for selected European countries is shown in Fig. 8. The figure illustrates the vast differences between countries relating to differences in climate as well as agricultural practice.

Ammonia Emissions in Europe 153

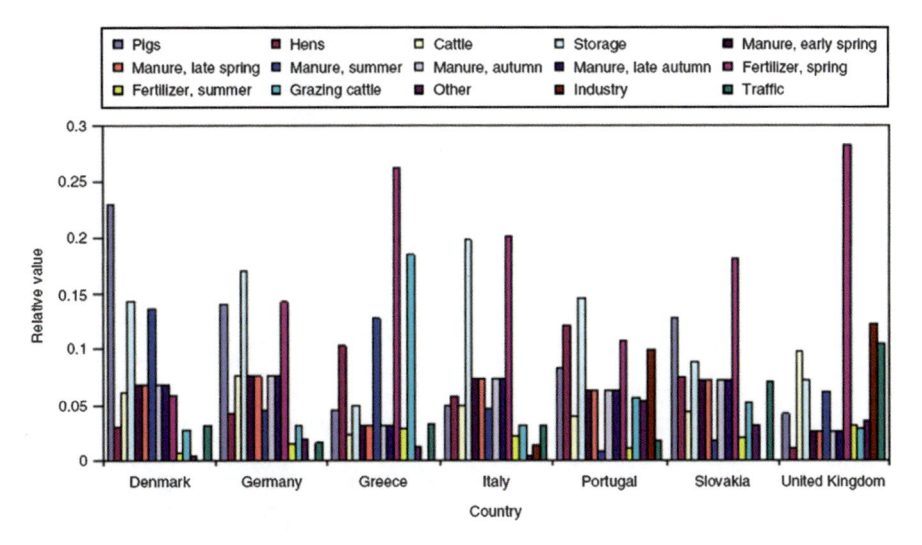

Fig. 8 The relative distribution between different NH_3 sources in national emissions for selected European countries. Data is here derived for the year 2000. Source: [1]

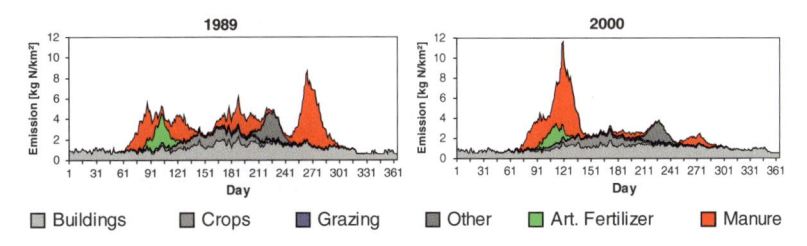

Fig. 9 Temporal variation in daily NH_3 emission [g N/ha/day] from different source categories at the monitoring site Tange, Denmark, for the years 1989 (*left*) and 2000 (*right*). Source: [51]

4 Site-Specific Long-Term Trends in Emissions

Before 1989, Danish NH_3 emissions were relatively low during winter time as a result of low activity and low ambient air temperatures. The accumulated manure during winter was applied not only to crops in the fields during spring but also to grass fields during summer. The manure storages were emptied in autumn, thereby leading to autumn emissions (Fig. 9). This pattern is still typical in Northern Europe with moderate to large agricultural activity and limited legislative control.

In the 1990s, strong regulation was implemented to reduce Danish agricultural emissions to air, soil and water [16, 51, 84]. This regulation had various steps and included improving the entire production chain with respect to reducing NH_3 emissions. The farmers had to increase the fraction of manure applied during growth of the crops in spring and similarly decrease application in summer and autumn. This is seen in the results for 2000 shown to the right of Fig. 9, where the

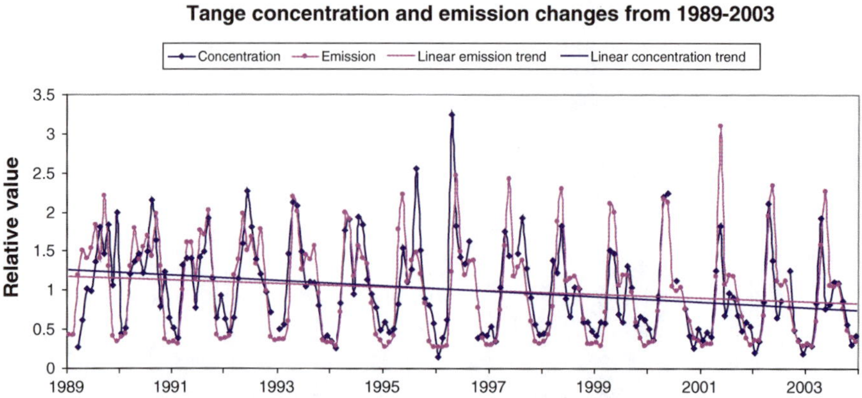

Fig. 10 Comparison of trends in computed monthly mean NH_3 emissions (*pink*) and observed monthly mean ambient air NH_3 concentrations (*blue*) for the Danish monitoring station Tange during the time period 1989–2003. Source: [51]. All values are relative to the annual mean in 1989

emission peak in spring is more pronounced and emissions in summer and autumn reduced when comparing with the situation in 1989 (shown to the left). The overall Danish release of NH_3 was decreased significantly (Table 1) despite for increased animal production [51] (Fig. 10).

5 Long-Term Trends in Emissions on European Scale

The total national NH_3 emissions have been reduced in countries like Denmark, Germany and the Netherlands, whereas for countries like France, Sweden and Norway, only very small changes have occurred over the last 15 years (Table 1). Note the large change in Poland between 1990 and 1995 which is a response in the agricultural emissions related to the large political changes during that period. Over the European area (EU countries), emissions have in overall decreased since 1985 (Fig. 11).

The fact that trends in measured NH_3 concentrations have not reflected reduced animal numbers in Central Europe (e.g. [7]) has been a major topic for debate. Similarly, then the effects of emission abatement policies in the Netherlands [85] and Denmark [51] have not fully corresponded to the expectations. Recent findings (e.g. [86]) suggest that increasing NH_3 concentrations that have been observed in Eastern Europe is a result of both local and regional reduction in SO_2 emissions, which have caused an increased atmospheric lifetime of NH_3. Such relationships are likely to be of general concerns, thus causing uncertainties in direct relation of emission trends with locally observed concentrations (Table 2).

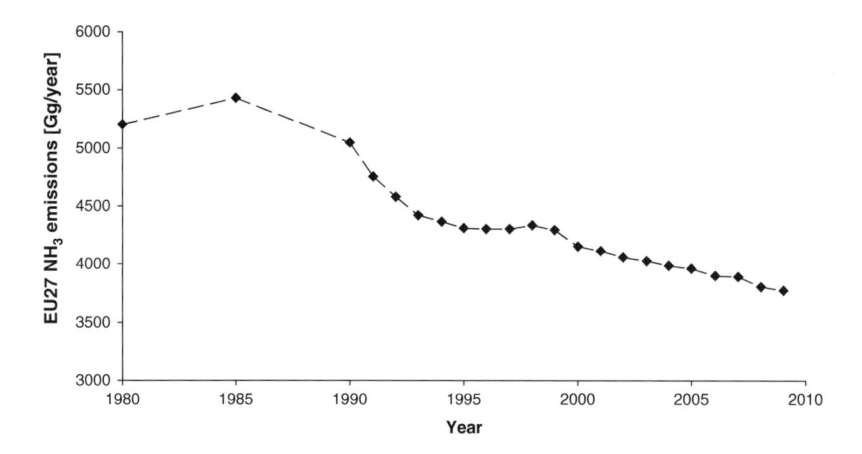

Fig. 11 Trend in total annual NH_3 emissions in Europe over the period 1980–2009. Source: EMEP (http://www.emep.int)

Table 2 Annual emission of ammonia (Gg NH3) from selected European countries during the period 1985–2009 as they are used in the EMEP models

	1985	1990	1995	2000	2005	2009
Denmark	137	134	114	93	84	77
Germany	857	758	642	594	578	597
France	780	787	772	802	751	744
The Netherlands	248	249	193	163	141	125
Norway	23	20	23	24	23	23
Poland	550	511	378	322	326	273
Sweden	54	55	64	56	53	48
United Kingdom	341	382	359	333	311	288

6 Regulation of Emissions, Future Scenarios and Impact of Climate Change

Regulation of NH_3 emissions entered the European legislation by the 1979 signing of the Convention on long-range trans-boundary air pollution (CLRTAP) to abate acidification, eutrophication and ground-level ozone. In 1999, the executive body of the Convention adopted the "Gothenburg Protocol". For NH_3, the objective of this protocol was a 17% reduction in emissions compared with 1990 levels. This was to be obtained through the definition of annual "emission ceiling" not to be exceeded by the signing parties from 2010 onwards. The Gothenburg Protocol came into force in May 2005 and to date signed by 26 ratifying parties. The implementation of the emission reductions has taken place under the National Emissions Ceilings (NEC) directive (2001/81/EC) of the European Union. The decreasing trend in European emissions depicted in Fig. 11 is to a large degree the result of reductions agreed upon in the NEC directive although not European all countries have complied with the emission ceilings for NH_3. Currently the Gothenburg Protocol is under revision with an aim of adopting more stringent

Table 3 Emission factors of total available nitrogen (TAN) in liquid and solid manure from husbandry

	Slurry injection (0 h inc.)	Slurry trailing hose (6 h inc)	Solid broad spread (6 and 12 h inc.)
Cattle	1.9%	10%	27.9%, 39.0%
Pigs	1.3%	6.9%	27.9%, 39.0%

Emissions are given in percentage as function of application method during application on bare soil during winter/spring before sowing. Source: [64]
Note: Trailing hose is also used in growing crops during spring/summer but without incorporation. Here, the emission factors are 19.6% and 13.5% for cattle and pig, respectively

emission ceilings for 2010, accounting for various environmental, socio-economic and political factors. However, NH_3 emissions are not only regulated on national level. The Integrated Pollution Prevention and Control (IPPC) directive aims at preventing or reducing pollutant emissions from industrial installations to air, water and soil through the use of best available techniques (BAT). The IPPC directive was established in 1996 (96/61/EC) and revised in 2008 (2008/1/EC), and it covers new and existing pig and poultry installations that have a capacity >40,000 poultry places, >2,000 production pigs over 30 kg or >750 sows. In order to operate, the farms covered by the IPPC directive need to obtain a permit issued by the relevant member state authorities. The use of BAT is stipulated for NH_3 emissions related to livestock housing, storage of manure and slurry as well as application of manure and slurry to the fields. This BAT can be exemplified by investigating the emission factors that have been used for calculations of emissions related to application of slurry and solid manure by using different application techniques such as broad spread, snake tubes or injection [64].

Clearly the emission factors vary significantly, depending on application technique. However, some of the methods require new investments in production facilities and machinery at the farm level as well as a general increase in efficiency of the N use through the entire production facility [87]. The result is a change in both amount and pattern of the annual NH_3 emissions from agriculture [51]. Some of the techniques to obtain emission reductions can be covering of storage facilities, better utilisation of N in the production chain, change in application techniques of manure (see Table 3), cleaning of the air from the point sources, treatment of slurry before application, etc. [64]. However, the single-member states decide the conditions for detailed regulation of agriculture, and this has resulted in large differences between member states in the implementation of the IPPC directive. Recently, a review of the assessment process in European [88] has stated that the UK, Germany, the Netherlands and Denmark are having the most formalised procedures for protecting sensitive ecosystems against N deposition from anthropogenic activities ([88], Chap. 3, theme 1). According to Theobald [89] and Hertel et al. [10], the today's most advanced permitting processes are the ones applied in England, Wales and Denmark. The applied methodologies are rather similar, as they include identification of Special Areas of Conservation together with a simple screening using either the UK EPAs "Ammonia Screening Tool" (AST) or the

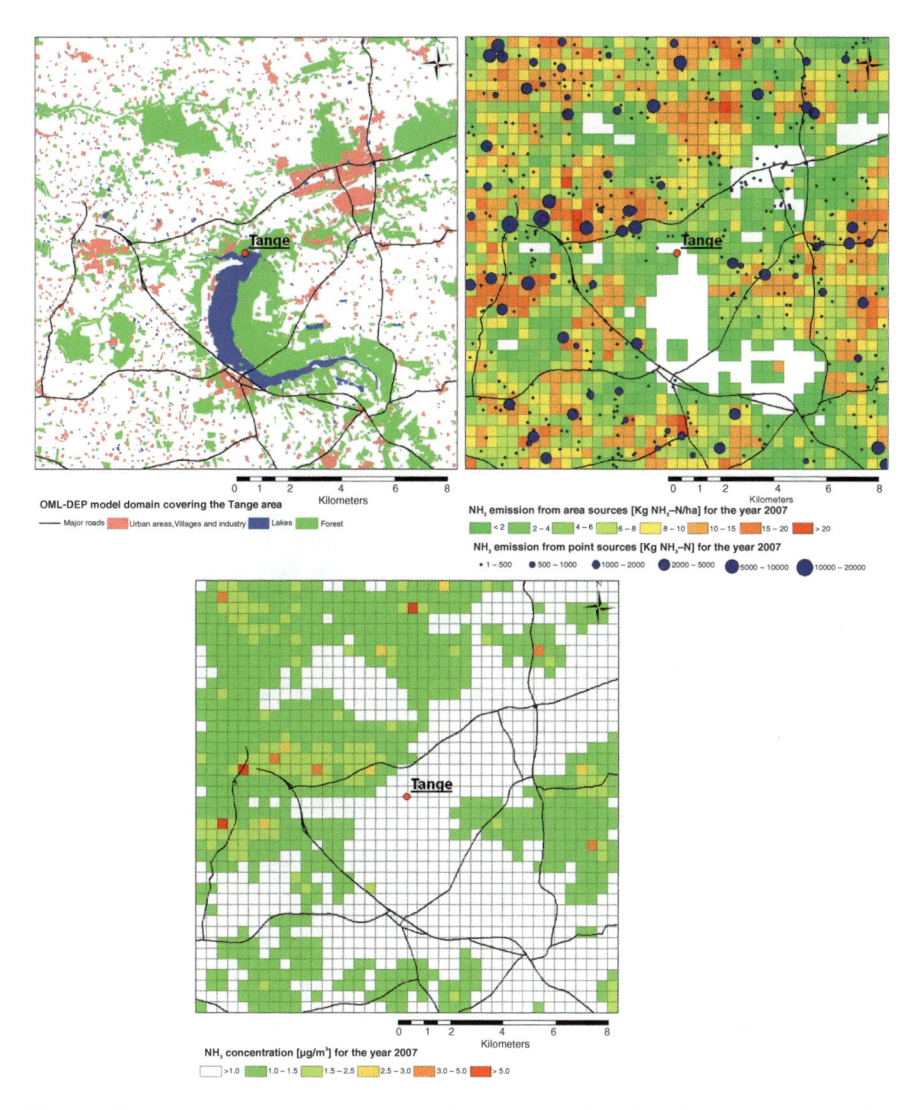

Fig. 12 High-resolution assessment of atmospheric nitrogen deposition in Denmark accounting for emissions from single farms. Figure obtained from Geels et al. [80]

Danish buffer zone approach. In England and Wales, the screening requires detailed modelling in case the AST shows N depositions or NH_3 concentrations that exceed 4% of critical loads or levels for European sites, 20% of Sites of Special Scientific Interest (SSSI) or 50% of threshold values for local conservation sites. In this case, either ADMS or AERMOD is applied for the detailed modelling. A permit is granted only when detailed modelling demonstrates that 20%, 50% or 100% of the critical level or load is not exceeded at the European sites, SSSIs or local conservation sites, respectively. In Denmark, smaller livestock farms are regulated by screening the potential increased N deposition to sensitive ecosystems. Smaller farms are screened within two buffer zones of 300 and 1,000 m, while larger farms are also screened outside these two buffer zones. The screening outcome is that a permission is either granted or declined or that there is a setup requirement for a full model assessment in a similar way as the English methodology by using a local-scale dispersion model and known emissions from the planned livestock farm. The Danish methodology is here based on the OML-DEP model [41, 80] which for regulatory purposes is operated within the DAMOS modelling system [10]. An example of detailed modelling from Denmark is shown in Fig. 12.

References

1. Hertel O, Skjøth CA, Løfstrøm P, Geels C, Frohn LM, Ellermann T, Madsen PV (2006) Modelling nitrogen deposition on a local scale – a review of the current state of the art. Environ Chem 3:317–337
2. Bobbink R, Hicks K, Galloway J, Spranger T, Alkemade R, Ashmore M, Bustamante M, Cinderby S, Davidson E, Dentener F, Emmett B, Erisman JW, Fenn M, Gilliam F, Nordin A, Pardo L, de Vries W (2010) Global assessment of nitrogen deposition effects on terrestrial plant diversity: a synthesis. Ecol Appl 20:30–59
3. Stevens CJ, Dise NB, Mountford JO, Gowing DJ (2004) Impact of nitrogen deposition on the species richness of grasslands. Science 303:1876–1879
4. Stevens CJ, Dupre C, Dorland E, Gaudnik C, Gowing DJG, Bleeker A, Diekmann M, Alard D, Bobbink R, Fowler D, Corcket E, Mountford JO, Vandvik V, Aarrestad PA, Muller S, Dise NB (2010) Nitrogen deposition threatens species richness of grasslands across Europe. Environ Pollut 158:2940–2945
5. Reis S, Pinder RW, Zhang M, Lijie G, Sutton MA (2009) Reactive nitrogen in atmospheric emission inventories. Atmos Chem Phys 9:7657–7677
6. Skjøth CA, Geels C, Berge H, Gyldenkærne S, Fagerli H, Ellermann T, Frohn LM, Christensen J, Hansen KM, Hansen K, Hertel O (2011) Spatial and temporal variations in ammonia emissions – a freely accessible model code for Europe. Atmos Chem Phys 11:5221–5236
7. Sutton M, Asman W, Ellermann T, van Jaarsveld JA, Acker K, Aneja V, Duyzer J, Horvath L, Paramonov S, Mitosinkova M, Tang YS, Achermann B, Gauger T, Bartniki J, Neftel A, Erisman J (2003) Establishing the link between ammonia emission control and measurements of reduced nitrogen concentrations and deposition. Environ Monit Assess 82:149–185
8. Fowler D, Pitcairn CER, Sutton MA, Flechard C, Loubet B, Coyle M, Munro RC (1998) The mass budget of atmospheric ammonia in woodland within 1 km of livestock buildings. Environ Pollut 102:343–348
9. Pitcairn CER, Skiba UM, Sutton MA, Fowler D, Munro R, Kennedy V (2002) Defining the spatial impacts of poultry farm ammonia emissions on species composition of adjacent woodland groundflora using Ellenberg Nitrogen Index, nitrous oxide and nitric oxide emissions and foliar nitrogen as marker variables. Environ Pollut 119:9–21
10. Hertel O, Geels C, Frohn LM, Ellermann T, Skjoth CA, Lofstrom P, Christensen JH, Andersen, HV, Peel RG (2012) Assessing atmospheric nitrogen deposition to natural and semi-natural ecosystems _ experience from Danish studies using the DAMOS system. Atmos Environ, in press, http://dx.doi.org/10.1016/j.atmosenv.2012.02.071
11. Skjøth CA, Geels C (2012) The effect of climate and climate change on ammonia emissions in Europe. Atmos Chem Phys Discuss 12:23403–23431

12. Klimont Z, Brink C (2004) Modelling of emissions of air pollutants and greenhouse gases from agricultural sources in Europe, International Institute for Applied Systems Analysis (IIASA), Laxenburg, Austria
13. Dore AJ, Kryza M, Hall JR, Hallsworth S, Keller VJD, Vieno M, Sutton MA (2012) The influence of model grid resolution on estimation of national scale nitrogen deposition and exceedance of critical loads. Biogeosciences 9:1597–1609
14. Elzing A, Monteny GJ (1997) Ammonia emission in a scale model of a dairy-cow house. Trans ASAE 40:713–720
15. Gyldenkærne S, Ambelas Skjøth C, Hertel O, Ellermann T (2005) A dynamical ammonia emission parameterization for use in air pollution models. J Geophys Res [Atmos] 110, D07108, doi: 10.1029/2004JD005459
16. Skjøth CA, Hertel O, Gyldenkærne S, Ellermann T (2004) Implementing a dynamical ammonia emission parameterization in the large-scale air pollution model ACDEP. J Geophys Res [Atmos] 109, D06306, doi: 10.1029/2003JD003895
17. Sommer SG, Zhang GQ, Bannink A, Chadwick D, Misselbrook T, Harrison R, Hutchings NJ, Menzi H, Monteny GJ, Ni JQ, Oenema O, Webb J (2006) Algorithms determining ammonia emission from buildings housing cattle and pigs and from manure stores. Elsevier Academic, San Diego
18. Sintermann J, Neftel A, Ammann C, Häni C, Hensen A, Loubet B, Flechard CR (2012) Are ammonia emissions from field-applied slurry substantially over-estimated in European emission inventories? Biogeosciences 9:1611–1632
19. Petersen SO, Sommer SG, Aaes O, Søegaard K (1998) Ammonia losses from urine and dung of grazing cattle: effect of N intake. Atmos Environ 32:295–300
20. Larsson L, Ferm M, Kasimir-Klemedtsson A, Klemedtsson L (1998) Ammonia and nitrous oxide emissions from grass and alfalfa mulches. Nutr Cycl Agroecosyst 51:41–46
21. Schjoerring JK, Mattsson M (2001) Quantification of ammonia exchange between agricultural cropland and the atmosphere: measurements over two complete growth cycles of oilseed rape, wheat, barley and pea. Plant Soil 228:105–115
22. Wang L, Schjoerring JK (2012) Seasonal variation in nitrogen pools and 15N/13C natural abundances in different tissues of grassland plants. Biogeosciences 9:1583–1595
23. Anderson N, Strader R, Davidson C (2003) Airborne reduced nitrogen: ammonia emissions from agriculture and other sources. Environ Int 29:277–286
24. Sutton MA, Dragosits U, Tang YS, Fowler D (2000) Ammonia emissions from non-agricultural sources in the UK. Atmos Environ 34:855–869
25. Kean AJ, Littlejohn D, Ban-Weiss GA, Harley RA, Kirchstetter TW, Lunden MM (2009) Trends in on-road vehicle emissions of ammonia. Atmos Environ 43:1565–1570
26. van Vuuren DP, Bouwman LF, Smith SJ, Dentener F (2011) Global projections for anthropogenic reactive nitrogen emissions to the atmosphere: an assessment of scenarios in the scientific literature. Curr Opin Environ Sustain 3:359–369
27. Reche C, Viana M, Pandolfi M, Alastuey A, Moreno T, Amato F, Ripoll A, Querol X (2012) Urban NH3 levels and sources in a Mediterranean environment. Atmos Environ 57:153–164
28. Andreae MO, Merlet P (2001) Emission of trace gases and aerosols from biomass burning. Global Biogeochem Cycles 15:955–966
29. Yokelson RJ, Urbanski SP, Atlas EL, Toohey DW, Alvarado EC, Crounse JD, Wennberg PO, Fisher ME, Wold CE, Campos TL, Adachi K, Buseck PR, Hao WM (2007) Emissions from forest fires near Mexico City. Atmos Chem Phys 7:5569–5584
30. Farquhar GD, Firth PM, Wetselaar R, Weir B (1980) On the gaseous exchange of ammonia between leaves and the environment – determination of the ammonia compensation point. Plant Physiol 66:710–714
31. Massad RS, Nemitz E, Sutton MA (2010) Review and parameterisation of bi-directional ammonia exchange between vegetation and the atmosphere. Atmos Chem Phys 10:10359–10386

32. Barrett K (1998) Oceanic ammonia emissions in Europe and their transboundary fluxes. Atmos Environ 32:381–391
33. Sørensen LL, Hertel O, Skjøth CA, Lund M, Pedersen B (2003) Fluxes of ammonia in the coastal marine boundary layer. Atmos Environ 37:S167–S177
34. Ellermann T, Hertel O, Monies C, Kemp K, Ambelas Skjøth C (2004) Atmospheric deposition 2003 (In Danish: Atmosfærisk deposition 2003) National Environmental Research Institute
35. Pinder RW, Adams PJ, Pandis SN (2007) Ammonia emission controls as a cost-effective strategy for reducing atmospheric particulate matter in the eastern United States. Environ Sci Technol 41:380–386
36. Seedorf J, Hartung J, Schroder M, Linkert KH, Pedersen S, Takai H, Johnsen JO, Metz JHM, Groot Koerkamp PWG, Uenk GH (1998) A survey of ventilation rates in livestock buildings in Northern Europe. J Agric Eng Res 70:39–47
37. Seedorf J, Hartung J, Schroder M, Linkert KH, Pedersen S, Takai H, Johnsen JO, Metz JHM, Groot Koerkamp PWG, Uenk GH (1998) Temperature and moisture conditions in livestock buildings in Northern Europe. J Agric Eng Res 70:49–57
38. Skjøth CA, Hertel O, Ellermann T (2002) Use of the ACDEP trajectory model in the Danish nation-wide Background Monitoring Programme. Phys Chem Earth 27:1469–1477
39. Fagerli H, Simpson D, Svetlana T (2004) Unified EMEP Model: Updates EMEP MSC-W, Norwegian Meteorological Institute. ISSN: 0806-4520
40. Simpson D, Benedictow A, Berge H, Bergström R, Emberson LD, Fagerli H, Flechard, CR., Hayman GD, Gauss M, Jonson JE, Jenkin ME, Nyíri A, Richter C, Semeena VS, Tsyro S, Tuovinen JP, Valdebenito Á, Wind P (2012) The EMEP MSC-W chemical transport model - technical description. Atmos Chem Phys 12:7825–7865
41. Sommer SG, Østergård HS, Løfstrøm P, Andersen HV, Jensen LS (2009) Validation of model calculation of ammonia deposition in the neighbourhood of a poultry farm using measured NH3 concentrations and N deposition. Atmos Environ 43:915–920
42. Koerkamp PWGG, Metz JHM, Uenk GH, Phillips VR, Holden MR, Sneath RW, Short JL., White RP, Hartung J, Seedorf J, Schroder M, Linkert KH, Pedersen S, Takai H, Johnsen JO, Wathes CM (1998) Concentrations and emissions of ammonia in livestock buildings in Northern Europe. J Agric Eng Res 70:79–95
43. Muck RE, Steenhuis TS (1982) Nitrogen losses from manure storages. Agric Wastes 4:41–54
44. Olesen JE, Sommer SG (1993) Modeling effects of wind-speed and surface cover on ammonia volatilization from stored pig slurry. Atmos Environ Part A 27:2567–2574
45. Oudendag DA, Luesink HH (1998) The Manure Model: manure, minerals (N, P and K), ammonia emission, heavy metals and the use of fertiliser in Dutch agriculture. Environ Pollut 102:241–246
46. Zhang RH, Day DL, Christianson LL, Jepson WP (1994) A computer-model for predicting ammonia release rates from swine manure pits. J Agric Eng Res 58:223–229
47. Pinder RW, Strader R, Davidson CI, Adams PJ (2004) A temporally and spatially resolved ammonia emission inventory for dairy cows in the United States. Atmos Environ 38:3747–3756
48. Hertel O, Skjøth CA, Reis S, Bleeker A, Harrison R, Cape JN, Fowler D, Skiba U, Simpson D, Jickells T, Kulmala M, Gyldenkærne S, Sørensen LL, Erisman JW, Sutton MA (2012) Governing processes for reactive nitrogen compounds in the atmosphere in relation to ecosystem, climatic and human health impacts. Biogeosci Discuss 9:9349–9423
49. Loubet B, Génermont S, Ferrara R, Bedos C, Decuq C, Personne E, Fanucci O, Durand B, Rana G, Cellier P (2010) An inverse model to estimate ammonia emissions from fields. Eur J Soil Sci 61:793–805
50. Genermont S, Cellier P (1997) A mechanistic model for estimating ammonia volatilization from slurry applied to bare soil. Agr Forest Meteorol 88:145–167
51. Skjøth CA, Gyldenkærne S, Ellermann T, Hertel O, Mikkelsen MH (2008) Footprints on ammonia concentrations from environmental regulations. J Air Waste Manage 58:1158–1165

52. Smith E, Gordon R, Bourque C, Campbell A, Genermont S, Rochette P, Mkhabela M (2009) Simulated management effects on ammonia emissions from field applied manure. J Environ Manage 90:2531–2536
53. Menut L, Bessagnet B (2010) Atmospheric composition forecasting in Europe. Ann Geophys 28:61–74
54. Pouliot G, Pierce T, Denier van der Gon H, Schaap M, Moran M, Nopmongcol U (2012) Comparing emission inventories and model-ready emission datasets between Europe and North America for the AQMEII project. Atmos Environ 53:4–14
55. Webb J, Anthony SG, Brown L, Lyons-Visser H, Ross C, Cottrill B, Johnson P, Scholefield D (2005) The impact of increasing the length of the cattle grazing season on emissions of ammonia and nitrous oxide and on nitrate leaching in England and Wales. Agric Ecosyst Environ 105:307–321
56. Ryden JC, Whitehead DC, Lockyer DR, Thompson RB, Skinner JH, Garwood EA (1987) Ammonia emission from grassland and livestock production systems in the UK. Environ Pollut 48:173–184
57. Misselbrook TH, Webb J, Chadwick DR, Ellis S, Pain BF (2001) Gaseous emissions from outdoor concrete yards used by livestock. Atmos Environ 35:5331–5338
58. Flechard CR, Spirig C, Neftel A, Ammann C (2010) The annual ammonia budget of fertilised cut grassland – part 2: seasonal variations and compensation point modeling. Biogeosciences 7:537–556
59. Milford C, Theobald MR, Nemitz E, Hargreaves KJ, Horvath L, Raso J, Dämmgen U, Neftel A, Jones SK, Hensen A, Loubet B, Cellier P, Sutton MA (2009) Ammonia fluxes in relation to cutting and fertilization of an intensively managed grassland derived from an inter-comparison of gradient measurements. Biogeosciences 6:819–834
60. Loubet B, Milford C, Hill PW, Tang YS, Cellier P, Sutton MA (2002) Seasonal variability of apoplastic NH_4^+ and pH in an intensively managed grassland. Plant Soil 238:97–110
61. Sutton MA, Milford C, Nemitz E, Theobald MR, Hill PW, Fowler D, Schjoerring JK, Mattsson ME, Nielsen KH, Husted S, Erisman JW, Otjes R, Hensen A, Mosquera J, Cellier P, Loubet B, David M, Genermont S, Neftel A, Blatter A, Herrmann B, Jones SK, Horvath L, Fuhrer EC, Mantzanas K, Koukoura Z, Gallagher M, Williams P, Flynn M, Riedo M (2001) Biosphere-atmosphere interactions of ammonia with grasslands: experimental strategy and results from a new European initiative. Plant Soil 228:131–145
62. Wichink Kruit RJ, Schaap M, Sauter FJ, van Zanten MC, van Pul WAJ (2012) Modeling the distribution of ammonia across Europe including bi-directional surface-atmosphere exchange. Biogeosci Discuss 9:4877–4918
63. Hutchings NJ, Sommer SG, Andersen JM, Asman WAH (2001) A detailed ammonia emission inventory for Denmark. Atmos Environ 35:1959–1968
64. Gyldenkærne S, Mikkelsen MH (2007) Projection of the ammonia emission from Denmark from 2005 until 2025 National Environmental Research Institute, Research Notes from NERI No. 239, ISSN: 1399–9346, available from http://www2.dmu.dk/Pub/AR239.pdf
65. Cape JN, Tang YS, van Dijk N, Love L, Sutton MA, Palmer SCF (2004) Concentrations of ammonia and nitrogen dioxide at roadside verges, and their contribution to nitrogen deposition. Environ Pollut 132:469–478
66. Perrino C, Catrambone M, Di Menno Di Bucchianico A, Allegrini I (2002) Gaseous ammonia in the urban area of Rome, Italy and its relationship with traffic emissions. Atmos Environ 36:5385–5394
67. Alebic-Juretic A (2008) Airborne ammonia and ammonium within the Northern Adriatic area, Croatia. Environ Pollut 154:439–447
68. Lee DS, Halliwell C, Garland JA, Dollard GJ, Kingdon RD (1998) Exchange of ammonia at the sea surface – a preliminary study. Atmos Environ 32:431–439
69. Quinn PK, Charlson RJ, Bates TS (1988) Simultaneous observations of ammonia in the atmosphere and ocean. Nature 335:336–338

70. Zhang Y, Dore AJ, Liu X, Zhang F (2011) Simulation of nitrogen deposition in the North China Plain by the FRAME model. Biogeosciences 8:3319–3329
71. Tørseth K, Aas W, Breivik K, Fjæraa AM, Fiebig M, Hjellbrekke AG, Lund Myhre C, Solberg S, Yttri KE (2012) Introduction to the European Monitoring and Evaluation Programme (EMEP) and observed atmospheric composition change during 1972–2009. Atmos Chem Phys 12:5447–5481
72. Hertel O, Skjøth CA, Frohn LM, Vignati E, Frydendall J, de Leeuw G, Schwarz U, Reis S (2002) Assessment of the atmospheric nitrogen and sulphur inputs into the North Sea using a Lagrangian model. Phys Chem Earth 27:1507–1515
73. Spokes L, Jickells T, Weston K, Gustafsson BG, Johnsson M, Liljebladh B, Conley D, Ambelas-Skjodth C, Brandt J, Carstensen J, Christiansen T, Frohn L, Geernaert G, Hertel O, Jensen B, Lundsgaard C, Markager S, Martinsen W, Moller B, Pedersen B, Sauerberg K, Sorensen LL, Hasager CC, Sempreviva AM, Pryor SC, Lund SW, Larsen S, Tjernstrom M, Svensson G, Zagar M (2006) MEAD: an interdisciplinary study of the marine effects of atmospheric deposition in the Kattegat. Environ Pollut 140:453–462
74. de Vries W, Leip A, Jan Reinds G, Kros J, Lesschen JP, Bouwman AF, Grizzetti B, Bouraoui F, Butterbach-Bahl K, Bergamaschi P, Winiwarter W (2011) Geographical variation in terrestrial nitrogen budgets across Europe. In: The European nitrogen assessment, Cambridge University Press, ISBN-978-1-107-00612-6
75. Vestreng V, Ntziachristos L, Semb A, Reis S, Isaksen ISA, Tarrasón L (2009) Evolution of NOx emissions in Europe with focus on road transport control measures. Atmos Chem Phys 9:1503–1520
76. Beusen AHW, Bouwman AF, Heuberger PSC, Van Drecht G, Van der Hoek KW (2008) Bottom-up uncertainty estimates of global ammonia emissions from global agricultural production systems. Atmos Environ 42:6067–6077
77. Kryza M, Dore AJ, Blas M, Sobik M (2011) Modelling deposition and air concentration of reduced nitrogen in Poland and sensitivity to variability in annual meteorology. J Environ Manage 92:1225–1236
78. Van Pul A, Hertel O, Geels C, Dore AJ, Vieno M, Van Jaarsveld H, Bergstrom R, Schaap M, Fagerli H (2008) Modelling the atmospheric transport and deposition of ammonia at a national and regional scale. In: Sutton M, Reis S, Baker SMH (eds) Atmospheric ammonia - detecting emission changes and environmental impacts. Results of an expert workshop under the convention on long-range transboundary air pollution, 1st edn. Springer Science and Business Media B.V., Berlin, p 464
79. van Pul WAJ, van Jaarsveld JA, Vellinga OS, van den Broek M, Smits MCJ (2008) The VELD experiment: an evaluation of the ammonia emissions and concentrations in an agricultural area. Atmos Environ 42:8086–8095
80. Geels C, Andersen HV, Skjøth CA, Christensen JH, Ellermann T, Løfstrøm P, Gyldenkærne S, Brandt J, Hansen KM, Frohn LM, Hertel O (2012) Improved modelling of atmospheric ammonia over Denmark using the coupled modelling system DAMOS. Biogeosciences 9:2625–2647
81. Dragosits U, Sutton MA, Place CJ, Bayley AA (1998) Modelling the spatial distribution of agricultural ammonia emissions in the UK. Environ Pollut 102:195–203
82. Reidy B, Rhim B, Menzi H (2008) A new Swiss inventory of ammonia emissions from agriculture based on a survey on farm and manure management and farm-specific model calculations. Atmos Environ 42:3266–3276
83. Velthof GL, van Bruggen C, Groenestein CM, de Haan BJ, Hoogeveen MW, Huijsmans JFM (2012) A model for inventory of ammonia emissions from agriculture in the Netherlands. Atmos Environ 46:248–255
84. Grant R, Blicher-Mathiesen G (2004) Danish policy measures to reduce diffuse nitrogen emissions from agriculture to the aquatic environment. Water Sci Technol 49:91–99
85. Erisman JW, Bleeker A, van Jaarsveld JA (1998) Evaluation of ammonia emission abatement on the basis of measurements and model calculations. Environ Pollut 102:269–274

86. Horvath L, Fagerli H, Sutton M (2009) Long-term record (1981–2005) of ammonia and ammonium concentrations at K-puszta Hungary and the effect of SO2 emission change on measured and modelled concentrations. In: Sutton M, Reis S, Baker SMH (eds) Atmospheric ammonia: detecting emission changes and environmental impacts. Springer, Berlin
87. Jarvis S, Hutchings NJ, Brentrup F, Olesen JE, Van der Hoek KW (2011) Nitrogen flows in farming systems across Europe. In: Sutton MA, Howard CM, Erisman JW et al (eds) The European Nitrogen Assessment. Cambridge University Press, Cambridge. ISBN ISBN-978-1-107-00612-6
88. Hicks WK, Whitfield CP, Bealey WJ, Sutton M (2011) Nitrogen Deposition and Natura 2000 - Science & practice in determining environmental impacts: European Science Foundation, COST Office, ISBN 978-91-86125-23-3, http://cost729.ceh.ac.uk/n2kworkshop
89. Theobald MR (2012) An intercomparison of modelling approaches for simulating the atmospheric dispersion of ammonia emitted by agricultural sources E.T.S.I. Agrónomos, Universidad Politecnica de Madrid. PhD Thesis
90. Andersen HV, Hovmand MF, Hummelshøj P, Jensen NO (1999) Measurements of ammonia concentrations, fluxes and dry deposition velocities to a spruce forest 1991–1995. Atmos Environ 33:1367–1383
91. Erisman JW, Wyers GP (1993) Continuous Measurements of Surface Exchange of SO_2 and NH_3-Implications for Their Possible Interaction in the Deposition Process. Atmos Environ Part A-General Topics 27:1937–1949, doi:10.1016/0960-1686(93)90266-2
92. Sutton MA, Perthue E, Fowler D, Storetonwest RL, Cape JN, Arends BG, Mols JJ (1997) Vertical distribution and fluxes of ammonia at Great Dun Fell. Atmos Environ 31:2615–2624, doi:10.1016/S1352-2310(96)00180-X
93. Wyers GP, Erisman JW (1998) Ammonia exchange over coniferous forest. Atmos Environ 32:441–451, doi:10.1016/S1352-2310(97)00275-6

Road Traffic: A Major Source of Particulate Matter in Europe

Fulvio Amato, Martijn Schaap, Cristina Reche, and Xavier Querol

Abstract Gaseous and particulate emissions from vehicles represent a major source of atmospheric pollution in cities. Recent research shows evidence of, along with the primary emissions from motor exhaust, important contributions from secondary (due to traffic-related organic/inorganic gaseous precursors) and primary particles due to wear and resuspension processes. Besides new and more effective (for NO_x emissions) technologies, non-technological measures from local authorities are needed to improve urban air quality in Europe.

Keywords NO_x, PM, Primary, Resuspension, Secondary, Wear

Contents

1 Introduction 166
2 Exhaust Particles 172
   2.1 Black Carbon and Particle Number Measurements in Europe 173

F. Amato
TNO, Built Environment and Geosciences, Department of Climate, Air and Sustainability, P.O. Box 800153508 TA Utrecht, The Netherlands

Institute of Environmental Assessment and Water Research, IDAEA, Spanish Research Council CSIC, c/ Jordi Girona 18-26, Barcelona, Spain

M. Schaap
TNO, Built Environment and Geosciences, Department of Climate, Air and Sustainability, P.O. Box 800153508 TA Utrecht, The Netherlands

C. Reche and X. Querol (✉)
Institute of Environmental Assessment and Water Research, IDAEA, Spanish Research Council CSIC, c/ Jordi Girona 18-26, Barcelona, Spain
e-mail: xavier.querol@idaea.csic.es

3   Non-exhaust Emissions 175
    3.1   Road Dust Resuspension 176
    3.2   Wear Emissions 181
References 183

1   Introduction

Road traffic is widely recognised to be a significant and increasing source of air pollutants in urban and industrial areas but also at a regional scale worldwide. Emissions from road traffic can be divided into three different components: (1) primary and secondary particles from motor exhaust (including fuel evaporation); (2) primary and secondary particles from brake, tyre and road wear; (3) primary particles from resuspension due to wheel-generated turbulence [1–3]. The wear and resuspended part of the emissions can also be grouped together as non-exhaust emissions.

Road traffic emissions consist of particulate (PM) and gaseous emissions, with active carbonaceous products present in both phases. Particles contain potentially toxic components, such as polycyclic aromatic compounds (PAHs) and trace metallic elements [4–6], which are related to acute and chronic cardiovascular and respiratory diseases [7]. Some studies suggest that especially diesel exhaust emissions are responsible for cardiac hospital admissions [8] and for asthma and chronic bronchitis development in children [9] in densely populated cities. Also fine and coarse particles from non-exhaust sources have been associated with short-term mortality and morbidity [10–13].

Although laboratory experiments have shown that organic compounds in both gasoline fuel and diesel engine exhaust can form secondary organic aerosol (SOA), the fractional contribution from gasoline and diesel exhaust emissions to ambient SOA in urban environments is poorly known. Recently Bahreini et al. [14] demonstrated that in Los Angeles the contribution from diesel emissions to SOA formation is very low within their uncertainties and that gasoline emissions dominate over diesel in formation of SOA mass. Chamber studies performed in Europe seem to confirm this hypothesis. These results have important implications for air quality policy.

Verma et al. [15] have shown for Los Angeles in summer that both primary and secondary particles possess high redox activity; however, photochemical transformations of primary emissions with atmospheric aging enhance the toxicological potency of primary particles in terms of generating oxidative stress and leading to subsequent damage in cells.

Studies by Biswas et al. [16] using directly exhaust PM emissions from heavy duty vehicles, with and without emission abatement technologies implemented, suggest that the semivolatile fraction of particles is far more oxidative in nature than refractory particles. It is also possible in our opinion that the SOA formed from the condensation of previously volatilised PM has a highly oxidative character.

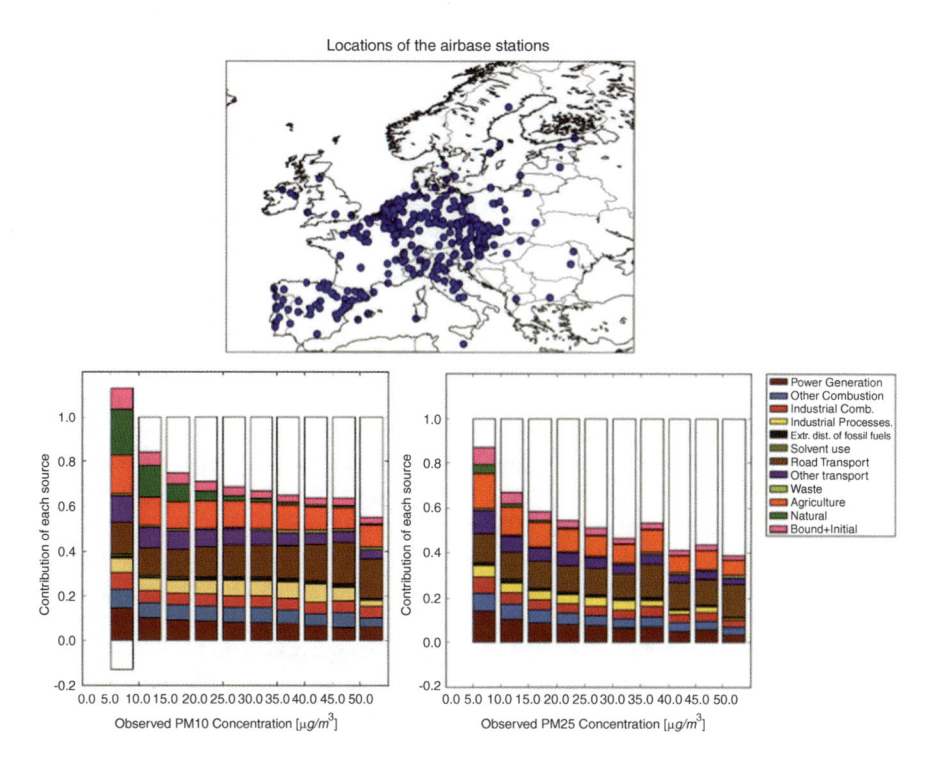

Fig. 1 AIRBASE monitoring sites (*top*) used to calculate average regional contributions to PM$_{10}$ and PM$_{2.5}$ (*bottom*) by means of the LOTOS-EUROS source apportionment tool

Road transport is an important contributor to primary emissions of PM (soot, wear particles and road dust) and also a source of secondary particles formed by condensation of gaseous species (mainly S- and N-compounds and organics) emitted by the tailpipe and partly also by the wear of brakes and tyres. Thus, PM emissions from road traffic are responsible for an important proportion of the exceedances of the PM$_{10}$ and PM$_{2.5}$ Air Quality Limit Values established by the European legislation for the protection of the human health (2008/50/EC; [17]). The daily (50 µg m^{-3}) and annual (40 µg m^{-3}) limit values for PM$_{10}$ (atmospheric particles with mean aerodynamic diameter <10 µm) and the annual limit value for PM$_{2.5}$ (25 µg m^{-3}) (in force from 2015) concentrations in ambient air are indeed exceeded mostly in the urban areas (Fig. 1; [17]).

In spite of the new vehicle emission EUROx regulations, 20% of Europe's urban population is also living in areas where the atmospheric concentrations of nitrogen dioxide (NO$_2$) exceed established air quality standards [18]. This is due to several factors related to the diesel-powered cars [19]: (1) their increasing market penetration across Europe [20]; (2) the NO$_x$ emission factors of diesel cars exceed the emission levels as established during the type approval of these vehicles in the laboratory [19, 21–25] and (3) the fraction of NO$_2$ in the NO$_x$ emissions of diesel cars has been increasing over the past two decades [26, 27], mainly due to the application of oxidative after-treatment systems [28, 29].

Hendriks et al. [30] estimated source contributions from different sectors to PM_{10} and $PM_{2.5}$ at a regional scale for the whole European domain by means of the LOTOS-EUROS labelling source apportionment method. Given the model resolution (25 × 25 km), source contributions within the cities are not captured, but averaged within the larger grid cell. Results show that, averaging all European stations (AIRBASE stations below 700 m altitude), road transport is the largest source of PMx at a European level, responsible for 15–20% and 10–15% of observed PM_{10} and $PM_{2.5}$, respectively. Contributions increase during pollution episodes and are expected to be higher on a local scale and also due to the contribution of secondary organics, currently missing in the modelling approach.

As already mentioned, road traffic emissions comprise a wide variety of particles in terms of their chemical composition (and size). Meteorology plays an important role when determining the impact of emissions on atmospheric concentrations of different components. The impact of dust emissions is in general higher in the large metropolitan areas (London, Paris, Madrid, Barcelona) or in densely populated regions such as the Netherlands and Northern Italy (Fig. 2). Higher contributions are observed in Mediterranean countries due to the infrequent rainfall and arid conditions. Central European countries are more affected by formation of secondary nitrate due to the lower temperatures when compared to South Europe (Fig. 2). Primary emissions from tailpipe (elemental carbon) affect indistinctly all European metropolitan regions (Fig. 2).

From an emission point of view, in spite of the constant technical developments of less pollutant vehicles and the implementation of diverse mitigation strategies for PM [31], atmospheric pollution by road traffic has not diminished for pollutants such as NO_2. Also it has to be highlighted the poor understanding of the so-called non-exhaust emissions as a major source of urban PM. Several studies have shown that the importance of these non-exhaust emissions is comparable (or even higher) to that of emissions from vehicle exhaust systems [3, 32–34] (Fig. 2).

One of the most commonly used tools in the field of atmospheric sciences to quantify the impact of road traffic emissions, both exhaust and non-exhaust, is receptor modeling, which aim to re-construct the source contributions (and profiles) from different sources of atmospheric pollutants, based on the PM chemical characterization data registered at monitoring sites. There are different specific models which differ in the a priori knowledge required about pollution sources. During a review of European source apportionment studies published between 1985 and 2005, Viana et al. [35] identified the almost constant reporting by all studies of a PM source traced by C/Fe/Ba/Zn/Cu, which was interpreted by the different authors as "road traffic" or "exhaust emissions". If interpreted as total traffic, other tracers were crustal elements such as Fe, Ca or Al from road dust. In these cases, this source could be clearly differentiated from the mineral/crustal matter source due to the presence of trace elements linked with brake abrasion (Cu, Ba, Sb) and combustion of lubricating oil (Zn, Ca) [35]. Traditionally, studies accounted for a quantification of the source impact in terms of the particulate mass collected during 24 h for a considerable period of time, including the inorganic fraction. The contribution from such a traffic source

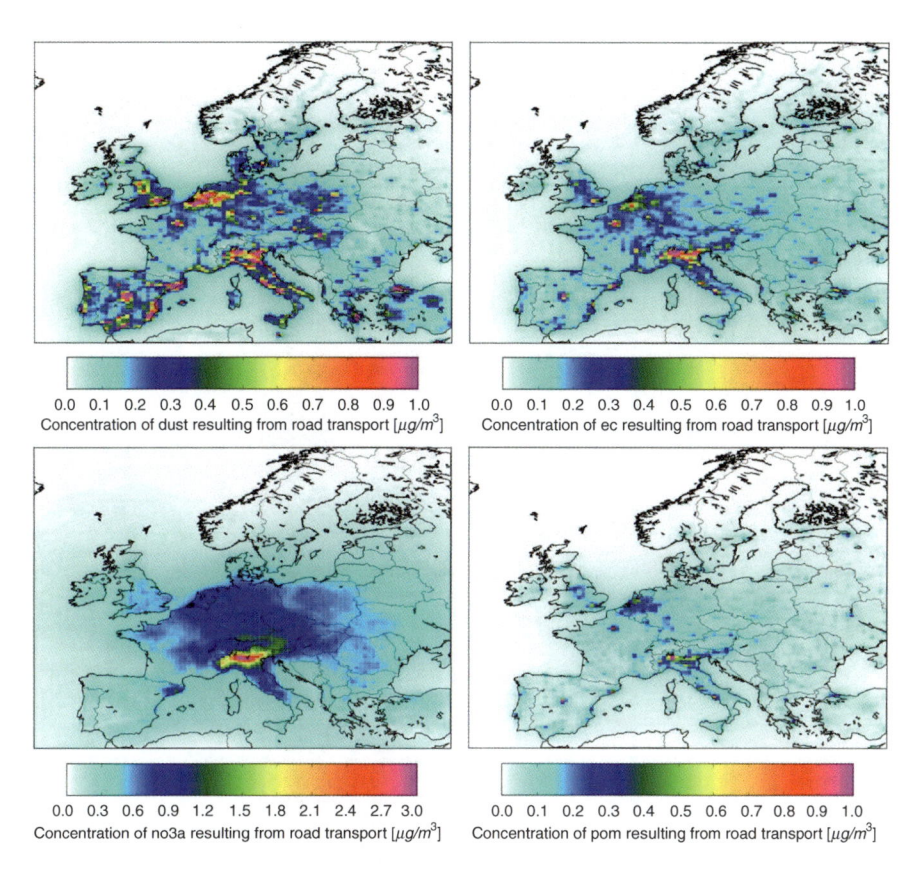

Fig. 2 Absolute contributions from road traffic to dust, EC, nitrate and primary organic matter (pom) modelled by the LOTOS EUROS

ranged from 14% to 48% of PM_{10} and from 9% to 49% of $PM_{2.5}$ in urban areas, and from 1% to 4% of PM_{10} and from 5% to 7% of $PM_{2.5}$ in rural areas (see Table 1 and references therein), a geographic trend has not been observed for the different European regions and the contribution is mostly governed by the traffic volume. Recently, this tool has been applied to particle concentrations in terms of number, obtaining contributions of up to 78% of total particle number in urban areas [41]. Some studies have included organic compounds in particulate or gaseous phase [Volatile Organic Compounds (VOCs) and PAHs] to distinguish between diesel and gasoline vehicles [48, 49].

State-of-the-art methods include the combination of source apportionment models with high time-resolved source apportionment techniques for organic aerosols datasets using the Aerodyne Aerosol Mass Spectrometer (AMS), obtaining contributions from primary organic emissions from traffic [60, 61].

Table 1 Receptor modelled contributions of road traffic reported for different metrics and backgrounds

Country code	Study	City	Background	Metric	Traffic sources	Contribution (%)
CHE	Bukowiecki et al. [33]	Zurich	Traffic	PM_{10}	Exhaust	41
					Non-exhaust	38
CZE	Thimmaiah et al. [36]	Prague	Urban	N	Exhaust and non-exhaust	34.2
DEU	Stölzel [37]	Dresden	Urban	PM_{10}	Exhaust and non-exhaust	36
DEU	Quass et al. [38]	Duisburg	Urban	PM_{10}	Exhaust and non-exhaust	36
DEU	Gerwig [39]	Dresden	Urban	$PM_{2.5}$	Exhaust and non-exhaust	43
DEU	Vallius et al. [40]	Erfurt	Urban	$PM_{2.5}$	Exhaust and non-exhaust	36
DEU	Yue et al. [41]	Erfurt	Urban	N	Exhaust and non-exhaust	78
DEU	Beuck et al. [42]	Mülheim-Styrum	Urban	PM_{10}	Non-exhaust	8
DEU		Eifel Mountains	Rural	PM_{10}	Non-exhaust	2
ESP	Querol et al. [34]	Barcelona	Urban	PM_{10}	Exhaust and non-exhaust	32
ESP	Rodríguez et al. [43]	Barcelona	Urban	$PM_{2.5}$	Exhaust and non-exhaust	49
ESP	Querol et al. [44]	Madrid	Traffic	PM_{10}	Exhaust and non-exhaust	34
ESP	Viana et al. [45]	Albacete	Urban	$PM_{2.5}$	Exhaust and non-exhaust	39
ESP		Barcelona	Urban	$PM_{2.5}$	Exhaust and non-exhaust	53
ESP		Galdakao	Urban	$PM_{2.5}$	Exhaust and non-exhaust	44
ESP		Huelva	Urban	$PM_{2.5}$	Exhaust and non-exhaust	35
ESP		Oviedo	Urban	$PM_{2.5}$	Exhaust and non-exhaust	41
ESP	Nicolás et al. [46]	Elche	Urban	PM_{10}	Exhaust and non-exhaust	13
ESP	Amato et al. [32]	Barcelona	Urban	PM_{10}	Exhaust	30
				$PM_{2.5}$	Exhaust	43
				PM_1	Exhaust	46
				PM_{10}	Non-exhaust	16
				$PM_{2.5}$	Non-exhaust	8
				PM_1	Non-exhaust	2
ESP	Pey et al. [47]	Barcelona	Urban	N	Exhaust and non-exhaust	65
FRA	Gaimoz et al. [48]	Paris	Urban	VOCs	Exhaust and non-exhaust	22
GBR	Yin et al. [49]	Birmingham	Urban	$PM_{2.5}$	Exhaust and non-exhaust	35

GRC	Manoli et al. [50]	Thessaloniki	Urban	PM_{10}	Non-exhaust	28
				$PM_{2.5-10}$	Non-exhaust	57
				PM_{10}	Exhaust	38
				$PM_{2.5-10}$	Exhaust	9
GRC	Karanasiou et al. [51]	Athens	Urban	$PM_{2.5-10}$	Exhaust	8
				PM_{10}	Exhaust	19–27
				$PM_{2.5}$	Exhaust	27–34
				$PM_{2.5-10}$	Non-exhaust	50–53
				PM_{10}	Non-exhaust	25–34
				$PM_{2.5}$	Non-exhaust	18–27
IRL	Yin et al. [52]	Several sites	Rural	PM_{10}	Exhaust and non-exhaust	4
				$PM_{2.5}$	Exhaust and non-exhaust	7
ITA	Contini et al. [53]	Lecce	Urban	PM_{10}	Exhaust and non-exhaust	16.5
ITA	Bernardoni et al. [54]	Milan	Urban	PM_{10}	Exhaust and non-exhaust	14–16
NLD	Vallius et al. [40]	Amsterdam	Urban	$PM_{2.5}$	Exhaust and non-exhaust	30
NLD	Moo:broek et al. [55]	Rotterdam	Urban	$PM_{2.5}$	Exhaust and non-exhaust	9
			Traffic	$PM_{2.5}$	Exhaust and non-exhaust	21
			Rural	$PM_{2.5}$	Exhaust and non-exhaust	6
NOR	Raes et al. [56]	Birkenes	Rural	$PM_{2.5}$	Exhaust and non-exhaust	5
PRT	[57]	Several sites	Rural	PM_{10}	Exhaust and non-exhaust	1
PRT	Oliveira et al. [58]	Oporto	Rural and urban	$PM_{2.5-10}$	Exhaust and non-exhaust	12–55
SWE	Swietlicki et al. [59]	Lund	Urban	PM coarse and fine, SO_2, NO_2, O_3		47
	Viana et al. [35]	Review European cities	Urban, traffic	PM_{10}	Exhaust and non-exhaust	32–55
				$PM_{2.5}$	Exhaust and non-exhaust	30–49

2 Exhaust Particles

Exhaust particles comprise submicron-sized primary particles and micrometric secondary particles (S-N-organics) formed in the atmosphere for condensation of gaseous compounds on existing nuclei.

The airborne particles originated by diesel and gasoline engine exhaust are composed mainly of elemental carbon, adsorbed organic material, inorganic salts and traces of metallic compounds. Soluble organic fractions of the particles contain primarily polycyclic aromatic hydrocarbons, heterocyclic compounds, phenols, nitroarenes and other oxygen- and nitrogen-containing derivatives [62].

The composition and quantity of the emissions from an engine depend mainly on the type and condition of the engine, fuel composition and additives, operating conditions and emission control devices. Thus, exhaust emissions depend mainly on fuel and driving conditions, and any geographical differences across Europe will depend on these parameters (as opposed to climatology and topography). Particles emitted from engines operating with gasoline are different from diesel engine exhaust particles in terms of their size distribution, surface properties and chemical composition. Diesel engines are an especially significant source of fine and ultrafine particles as they exceeded the emissions from gasoline vehicles by a factor of >10 in terms of mass concentration and $>10^4-10^5$ in terms of number concentration [63]. This difference has been significantly reduced in EURO 5 and EURO 6 vehicles, but diesel engines still emit higher particle number concentrations and NO_x than gasoline cars [64].

Important differences between gasoline and diesel relate also to their impact on the formation of secondary organic and inorganic compounds. As already mentioned, recent studies highlight the higher importance of gasoline, compared to diesel, in the formation of secondary organic aerosol [14]. Formation of secondary inorganic aerosols (ammonium sulphate and nitrate) due to NO_x and SO_2 is controlled also by NH_3, another key atmospheric component for urban air quality. NH_3 is emitted mostly by traffic and other fugitive sources such as city waste containers and sewage. This alkaline gas emitted in a high NO_2 scenario may enhance the formation of ammonium nitrate (a major component of $PM_{2.5}$). Furthermore, the levels of ammonium nitrate may also increase due to the marked decrease of SO_2 emissions that yielded a marked decrease of ammonium sulphate levels across the Europe. Since H_2SO_4 is more reactive with NH_3, most of the NH_3 is consumed by sulphuric acid; when sulphuric acid decreases and more NH_3 are available, more NH_4NO_3 can be formed after HNO_3 and NH_3. In regions with high photochemistry the reduction of PM mass pollution and the possible increase in frequency of draughts may yield to the increase of midday nucleation episodes with the consequent increase of levels of secondary nano-particles. Thus, in highly polluted atmospheres the secondary PM mass grows by condensation on pre-existing particles; however, in cleaner conditions, and particularly under high insolation and low relative humidity, new formation of nanoparticles (nucleation) from gaseous precursors may dominate condensation sink in urban areas [65]. The nucleation starts from the oxidation of SO_2 and the interaction with NH_3, and these nanoparticles immediately grow probably by condensation of VOCs on the nucleated particles [66].

VOCs emissions are also due to gasoline vapour emissions [67–69]. Song et al. [70] estimated that 52% of average VOCs concentration in Bejing was due to gasoline exhaust and gas vapour emissions.

2.1 Black Carbon and Particle Number Measurements in Europe

The need for a more specific metric to evaluate the impact of road traffic exhaust emissions on the levels of urban aerosols than the currently regulated ones (PM_{10}, $PM_{2.5}$) is being stated by the scientific community, as a large number of air quality plans, most of them focusing on traffic emission reductions, have been implemented in the last decade, without the subsequent decrease of PM levels in the corresponding cities [31].

Black carbon (BC) is a major contributor to the fine particulate matter in the atmosphere and it is almost exclusively responsible for the short-wave absorption of solar radiation caused by aerosol particles. BC is defined as the carbonaceous material having intense black or dark colour and which absorbs visible light efficiently [71]. By applying a correction fraction, BC and elemental carbon (EC) may be considered equivalent (and reported as equivalent black carbon, EBC). Numerous studies have concluded a strong association between BC and primary road traffic emissions [72–76]. Hamilton and Mansfield [77] already stated in 1990 that in many European cities more than 90% of EBC can originate from traffic. Especially diesel vehicles are known to emit EBC with a large mass fraction in the ultrafine (diameter <0.1 μm) particle size range [78, 79]. In Vienna (Austria), Berner et al. [80] measured the distributions of BC over the particle size spectrum and they found that BC contributed about 50–70% to the total particulate mass in the size range of 0.02–0.1 μm and about 20% in the size range 0.3–2 μm. Harrison and Yin [81] estimated that in Birmingham (UK) about 88% of PM_1 and 6–10% of PM_{10} can be attributed to EC and that this EC mostly originates from traffic. Rodriguez et al. [75] quantified in about 7% of PM_{10} and 13% of $PM_{2.5}$ the contribution of EC in Santa Cruz de Tenerife (Canary Island, Spain).

BC levels have been observed to vary proportionally with those of traffic-related gaseous pollutants, such as CO, NO_2 and NO, regardless of the region of Europe under study [65, 74, 82]. In spite of this high correlation, measurements of BC are highly significant in air quality monitoring sites as the BC/CO, BC/NO_2 and BC/NO ratios can vary widely depending on the distance to traffic emissions, the vehicle fleet composition (percentage of diesel vehicles versus percentage of gasoline vehicles) and the influence of other carbonaceous emission sources such as biomass burning [65].

The combination of PM_{10} and BC in urban background sites potentially constitutes a useful approach for air quality monitoring. While the BC daily cycle is mostly determined by vehicle exhaust emissions, PM_{10} concentrations are also governed by non-exhaust particles resuspended by traffic, by midday atmospheric dilution and by other non-traffic emissions [65]. PM_{10} levels at traffic sites across Europe have been reported to remain high from the morning until the evening traffic peak due to the effects of resuspension processes, whereas in coastal urban background sites, concentrations typically increase at midday when sea breezes transport the re-suspended mineral material from the city towards the monitoring station [34, 83–85].

Particle number concentrations (N) in urban areas are also highly governed by primary vehicle exhaust emissions [47, 74, 86–90]. These emissions show a bimodal size mode, with a nucleation mode below 30 nm and a carbonaceous mode peaking between 50 and 130 nm [91]. Zhu et al. [92] evaluated the number size distribution near a highway and concluded that both atmospheric dispersion and coagulation contributed to the rapid decrease in particle number concentration and change in particle size distribution with increasing distance from the freeway, with the smallest mode (mean diameter 13 nm) disappearing at distances greater than 90 m from the road, while Oliveira et al. [58] estimated that up to 90% of the size spectrum of particle number in a traffic exposed site was dominated by the nucleation mode. Several studies highlighted that exposure to road traffic emissions and the associated health risk may be properly evaluated by combining ambient air measurements of BC with N [90, 93–96].

Peculiarities in the daily cycle of N have been observed for the Mediterranean climate presenting peaks at midday which do not correlate with the levels of BC, indicating that a source other than traffic is influencing this parameter [65, 74, 75, 97]. Similar results were found in Brisbane (Australia) and Los Angeles urban environments, with subtropical climates [98–100]. Conversely, this phenomenon has not been observed for cities in central and northern Europe, where N levels vary proportionally to those of BC during the whole day [101, 102].

Assuming that BC is an accurate tracer of primary traffic emissions, the variability of the ratio N/BC can offer interesting information about the impact of sources other than traffic on N. In Reche et al. [65], it was observed that while this ratio was constant during the whole day in northern and central European cities, maximum values were reported at midday in Mediterranean cities. These N maxima were attributed to midday nucleation episodes favoured by the characteristic factors simultaneously occurring in the southern European cities under study at that time of the day: (1) high solar radiation, (2) high temperature and (3) a contribution of SO_2 from shipping and industry governed by the breeze circulation, all this resulting in enhanced secondary formation processes in these latitudes, giving rise to high particle number concentrations. Therefore, it may be concluded that N is influenced by sources other than traffic in certain European regions (e.g. the Mediterranean region), whereas BC is a more consistent tracer of such influence and it is proposed as a vehicle exhaust emissions tracer to be added to the urban air quality control networks. Nevertheless, N should be measured since ultrafine particles may have large impacts on human health based on the very fine grain size and their potential cardiovascular and cerebrovascular adverse health effects [103].

The current and upcoming EURO regulations are expected to achieve major reductions in ambient concentrations of BC in urban environments and thus, BC is expected to not pose relevant health and air quality problems in the coming future. However, the implementation of technological solutions such as particle filter traps or stripping of exhaust gases has resulted in increased emissions in terms of NO_2, especially primary NO_2. Therefore, the relationships between traffic-related pollutants (e.g., NO_2/BC, NO_2/CO ratios) are also changing as a function of these mitigation strategies.

3 Non-exhaust Emissions

The complexity of the urban environment does not always allow for a clear separation of road traffic sources; consequently, most of source apportionment studies present results only for total contributions from road traffic. It is also common to find studies where the road dust component of traffic emissions is mixed with other mineral/soil sources. Nevertheless, for air quality management and exposure studies it is important to understand the individual source contributions. PM contributions from vehicular traffic should be differentiated between exhaust and non-exhaust. Ideally non-exhaust contributions should be further separated between road dust, brake, tyre and road wear.

Differentiating the contribution of road dust from the exhaust is problematic but a key task since different processes are involved and different control measures/ strategies are necessary. The relative importance of these two categories changes widely in space and in time. Spatially, road dust emissions increase largely not only in Southern Europe (due to drier climate) and Scandinavian countries (due to road sanding and the use of studded tyres) but also within a city environment, e.g. next to construction sites and in heavy traffic roads. Timely, road dust emissions are severely influenced by meteorology (precipitation, insulation, road humidity and droughts). In addition, it is important to monitor the relative increase of non-exhaust emissions (currently uncontrolled) against the motor exhaust emissions, which have been progressively reduced in the last two decades by means of the EUROx legislation. Due to their relatively coarser size distribution (typically between 1 and 10 µm) non-exhaust emissions contribute significantly in terms of mass to the atmospheric concentrations of PM_{10} and $PM_{2.5}$ in large cities causing a high number of the exceedances of Air Quality Limit Values.

Discriminating source contributions within the road traffic sector is also of importance for the health outcomes: low-contributing sources may be more relevant to health. Heavy metals (such as Fe, Cu, Ba, Sb among others) originated by erosion of brake and tyre materials and embedded in road dust induce oxidative stress. Other toxic inorganic (sulphides; [32]) and organic (PAHs, [104]) compounds are enriched in road dust particles. In California, a correlation between atmospheric concentrations of heavy metals (Fe, Cu, Zn, and Ni) and the mortality rate due to ischemic heart disease was recently found [10, 11]. In Stockholm, Meister et al. [12] estimated a 1.68% increase in daily mortality per 10 μg m^{-3} increase in $PM_{2.5-10}$ concentrations, which include road dust and other coarse-size particles. The association with $PM_{2.5-10}$ was stronger for November–May when road dust is most important. Exposure to road dust particles (<below 2.5 microns) was associated with a 7% increment of cardiovascular mortality in Barcelona [13]. Gustafsson et al. [105] found that particles from road wear caused by studded tyres are at least as inflammatory as particles from diesel exhaust.

Given these evidences, the actual pollution scenario is not encouraging. An overview of atmospheric concentrations of heavy metals in Spain (more than 20 monitoring sites) revealed that the highest concentrations of Fe, Cu, Sr, Sb, Ti and Ba (and partly also Zn and Zr) are measured within the cities (rather than at industrial hotspots), where most of the population live and work. Concentrations increase further at roadside locations where also a significant part of population is exposed. Consequently, for these metals, population exposure is much higher than for common industrial tracers such as As and Cd [106]. Under this scenario, investigating the role of non-exhaust emissions in air quality impairment and their impact on health is a non-regret policy and a must for local authorities, mostly considering that such particles are emitted locally, and are therefore easier to control/mitigate, improving public health.

To date, research on non-exhaust emissions has been rather limited due to the difficulties encountered by experimentalists and modellers to characterise and describe the complex phenomenon of road dust resuspension and wear emissions.

3.1 Road Dust Resuspension

Among the non-exhaust sector, road dust resuspension is the largest contributor in terms of mass. The starting point for a better understanding of a pollution process is the characterization of the emission source. For road dust particles, the emission source is the dust reservoir (i.e. the weight of mobilised particles deposited on road pavement and available for resuspension). However not all the sediments are of concern: only the fraction with average diameter below 10 microns is actually measured at the receptors. To our knowledge two devices are currently available for estimating the load of dust reservoir (<10 microns): IDPS (inhalable deposited particles sampler) used to collect only mobilised particles below 10 microns from a delimited area of road pavement after which the sampled filters are gravimetrically and chemically analysed, and PI-SWERL (Portable In-Situ Wind ERosion Laboratory), consisting of an optical counter used to estimate the amount of PM_{10} particles resuspended by a rotating helix within an enclosed chamber.

While the application of PI-SWERL has been limited to US studies [107, 108] the IDPS device has been deployed in several European cities (Barcelona, Zürich, Utrecht, Madrid, Paris and Rotterdam) permitting a first European evaluation of the

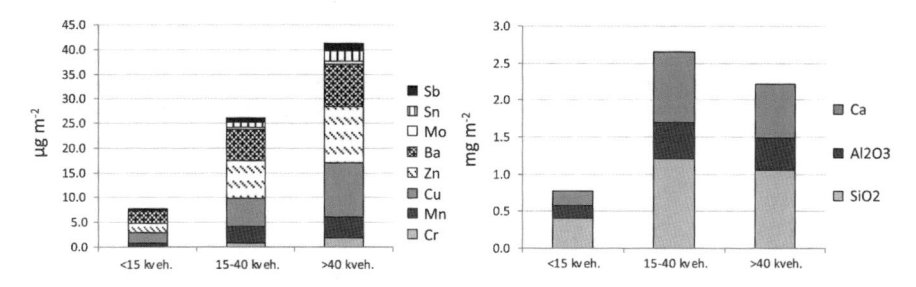

Road Traffic: A Major Source of Particulate Matter in Europe 177

Fig. 3 Variability of road dust components with increasing traffic volumes [109]

road dust emission strength. Road dust loadings were found to increase significantly from Central to Southern Europe [32, 109]. In the city of Utrecht (The Netherlands) road dust loadings were typically around 0.5–2.5 mg m^{-2}, which correspond to higher emissions than the average conditions in Zurich (0.2–1.3 mg m^{-2}) but lower emissions than Madrid or Barcelona (Spain) (3–9 mg m^{-2}). Figure 3 shows the variability of several components with increasing traffic intensity. Brake-related metals show an increase of loading with higher traffic volume (left chart). Current research is investigating whether this is due to higher volumes of vehicles or higher braking frequency. The right chart shows that loadings of typically mineral species (Ca, Al and Si) vary regardless of traffic intensity revealing the existence of additional sources of mineral species rather than only road wear.

The sources involved in the build-up of road dust reservoir are indeed numerous and their contribution changes from one site to another. Determining the concentrations of road dust components allows for the application of multivariate models able to identify the main factors responsible for the build-up of road dust. Typically four main sources are identified: a Mineral source (including road wear and other crustal sources like construction activities), Brake wear, Tyre wear and Motor exhaust. Figure 4 shows the average contributions in streets of Zürich (Switzerland) and Barcelona (Spain). Whilst in Zurich the four source contributions are evenly distributed (about 17–33% each in average), in Barcelona the mineral source is the dominant (72% in average). This information is of primary interest for air quality managers, in order to design targeted mitigation measures to combat road dust build-up and the consequent re-entrainment in the atmosphere. To this aim, further research is currently needed in order to separate contribution of road wear from that of external sources such as construction and spilling from trucks.

Estimating the contribution of road dust emissions on PM levels is problematic. Given that a unique tracer has not been identified yet, several alternative approaches have been followed so far:

- Receptor modelling
- Emission factors/dispersion modelling
- Use of proxies (mineral dust, kerbside increment of coarse PM, etc.)
- Coupling PM mass and size distribution

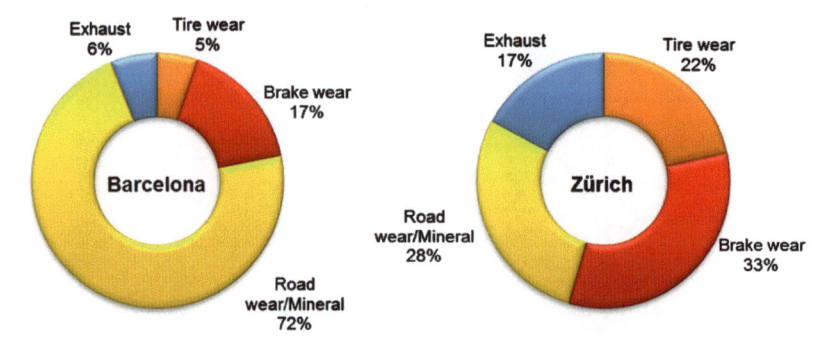

Fig. 4 Average source contributions to road dust loadings measured at different locations in Barcelona and Zurich

Receptor models are widely used tools for apportioning concentrations of pollutants to different sources. They can be factor analytical methods (PMF, PCA, UNMIX, etc.) or chemical mass balance (CMB). On the one hand, these methods revealed to be very valuable to identify the main sources/categories of PM pollution (road traffic, secondary particles, fuel oil combustion, sea salt, etc.) but on the other hand they experienced difficulties in separating the contributions of collinear sources such as mineral dust (natural resuspension) and road dust (anthropogenic) or co-variant sources such as vehicle exhaust and road dust [34, 44, 45, 49, 55, 58, 110–113]). Significant improvements were made with the use of combination of models or constrained models such as the Multilinear Engine (ME-2).

In spite of the low number of studies and the wide range of methods used, it is already possible to observe geographical variability on the road dust emission strength and air quality impact. Central Europe experiences the lowest road dust contributions due to the wet climate and the ban of studded tyres. This low "signal" hampers the task of quantifying road dust contribution. In Germany, Beuck et al. [42] estimated the contribution to PM_{10} from road dust in 2.4 µg/m^3 (8%) and 0.3 (2%) at the urban and regional background sites, respectively. In Stuttgart non-exhaust emissions from road traffic are estimated to be about twice as high as exhaust emissions [114]. Bukowiecki et al. [33] have analysed the traffic-related emissions (through trace elements, BC and nitrogen oxides) at a heavily congested street canyon in Zürich and assigned 21% of the traffic related PM_{10} emissions to brake wear, 38% to road dust and 41% to exhaust emissions. Astel [115] could distinguish road dust contributions from those of soil and vehicular exhaust in Cracow and Vienna by combining several models (CMB, PCA-APCS, PMF and UNMIX).

Thorpe et al. [116] proposed the roadside incremental concentration of coarse particles above the urban background as a first estimate of the sum of source strength road dust resuspension and the coarse fraction of wear emissions. Other studies succeeded in separating different traffic emissions by means of multivariate receptor models applied to PM size distribution data ([84, 117].

In dry climates such as South European countries, the low and infrequent precipitations hamper the wash-out and the moistening of road surface, favouring road dust resuspension by traffic-induced turbulence. Moreover additional inputs of dust come from the urban soil resuspension due to the little vegetal covering and from sporadic intensive deposition of Saharan dust outbreaks or uncontrolled construction/demolition activities.

The experimental evidence is given by the higher suspended PM_{10} mineral matter at the urban areas of Southern Europe as compared to Central Europe [34, 44, 89, 118–124]). In a comparative study between European sites, Querol et al. [44] highlighted that in Central Europe, the mineral contribution increases from 3–5 $\mu g\ m^{-3}$ from urban background sites to 4–7 $\mu g\ m^{-3}$ at kerbside sites. In Spain the increase found induced by traffic resuspension was much higher: from 10 to 16 $\mu g\ m^{-3}$. In Sweden the mineral aerosol accounts for 7–9 $\mu g\ m^{-3}$ in urban background but increases dramatically to 17–36 $\mu g\ m^{-3}$ at the traffic sites, as a result of the sanding and salting of roads during the winter and spring period and the use of studded tyres. Consequently, the local road dust emissions account for up to 9–24 $\mu g\ m^{-3}$ in Sweden, 6 $\mu g\ m^{-3}$ in Spain and for 1–5 $\mu g\ m^{-3}$ for the rest of countries studied: England, Switzerland, UK, Germany and Austria. These differences in levels of crustal components may be attributed largely to the higher dust accumulation and resuspension effect during dry conditions in the southern EU countries and to the high emission during winter-spring times in Scandinavian regions, whereas higher rainfall in the central EU countries may help to clean the road dust from streets.

The application of receptor models permitted to better quantify the contribution of road dust emissions. At the urban background of Barcelona Amato et al. [32] applied a constrained PMF (by means of the ME-2 scripting) revealing that road dust emissions were responsible on average for 16% of PM_{10} concentrations. An interesting outcome of this study was that the contribution did not change over the 5 years of study, contrary to industrial emissions, for example, revealing the existence of a non-controlled sector of transport emissions [32]. The same ME-2 approach was followed by the US EPA that implemented it in EPA PMF v5.0 (of soon release). In Madrid (Spain) similar contributions from vehicular exhaust and road dust emissions (31% and 29% respectively) to kerbside daily PM_{10} measurements were estimated by PMF [125] over 1 month measurements. In Greece, Karanasiou et al. [51] resolved road dust, motor exhaust and a soil factors by coupling PMF2 and PMF3 models on multiple size fractions in Athens, estimating the road contribution to be between 12% and 34% of PM_{10}. Manoli et al. [50] applied in Thessaloniki (Greece) multiple regression on absolute principal component scores estimating that road dust was responsible of 28% and 57% of PM_{10} and coarse PM, respectively.

In Scandinavian countries road dust emissions generate large quantities of coarse particles by enhanced pavement abrasion and mechanical fragmentation of traction sand grains [126–130].

Measurement of road dust emission potentials after road sanding on dry roads indicated a 75% increase in PM_{10} emissions after 2.5 h. This effect was short-lived and emission potentials returned to their pre-sanding levels within 8 h of the sand application [131]. Hussein et al. [132] stated that as compared to friction tyres, studded tyres may increase the road dust resuspension by a factor of 2.0–6.4. Kantamaneni et al. [133] found that the addition of traction material increased emission factor from 1.04 to 1.45 g veh^{-1} km^{-1}. Moreover, when roads were sanded, the correlation found between emission factors and relative humidity was not observed. Concerning road dust contributions Swietlicki et al. [59] estimated that road dust source was explaining 32% and 54% of the variance of PM_{10} and PM coarse levels in the city of Lund (Sweden) during Spring. In Copenhagen Wåhlin et al. [134] estimated by means of COPREM receptor modelling that road dust resuspension accounted for 8 μg m^{-3} of the kerbside PM_{10} mass, while motor exhausts reached 6 μg m^{-3}. Such difference was even greater for particles in the coarse size fraction. In a number of studies from Scandinavia, vehicle exhaust emissions have been found to contribute only around 10% to traffic-related PM_{10} emissions, with much of the remainder accountable for by resuspension [135, 136].

The emission factors (EF) for road dust emissions, expressed in mg per vehicle per km trabelled (mg v kt^{-1}), can be inferred from road dust loadings values (mg m^{-2}) by means of empirical formulas available in literature [137–140]. Beside these methods, there are several experimental approaches followed to estimate road dust EFs. Most used are the so-called upwind–downwind approach [141], the linear relation between NO_x and PM_{10} [142, 143] and the use of SF6 as tracer [144]. The number of studies estimating road dust emission factors is, however, still very limited and observational studies are rather incomplete. In other words, the emission estimates do not fully represent the wide range of conditions that are present within Europe. Emissions vary depending not only on climatic and meteorological conditions (impact of droughts, road moisture, precipitation, etc.) but also on road type (residential, motorway, congested road, etc.), vehicle type (passenger cars, light duty and heavy duty vehicles) and pavement type. In urban paved roads a negative correlation between road humidity and dust emission factors was found by Kantamaneni et al. [133]. Based on the limited literature available, regional differences can be observed, with values ranging from 14–23 mg v kt^{-1} (UK) to 17–92 mg v kt^{-1} (Switzerland, only LDV), 57–109 mg v kt^{-1} and 46–108 mg v kt^{-1} for Germany and Denmark, respectively, 85 mg v kt^{-1} in Spain, 121 mg v kt^{-1} in Finland and 198 mg v kt^{-1} in Sweden [140, 145–147]. On freeways, based on the limited literature available, the reported emission factors lower, ranging within 10–47 mg v kt^{-1} [130, 146, 148, 149].

Once the emission factors and their variability are estimated, dispersion models can be used in order to enable point data to be interpreted in terms of geographical distribution of source contributions, as suggested by the Air Quality Directive (2008/50/EC). This could serve as a basis for calculating the collective exposure of the population living in the area and for assessing air quality with respect to the limit values. Dispersion models are based on the use of meteorological data, modules to account with physico-chemical processes occurring in the atmosphere and EFs.

Schaap et al. [145] proposed a simple methodology to account for traffic re-suspension emissions in a large-scale modelling application by means of the 3D chemistry-transport model LOTOS-EUROS [150]:

$$F_{trs} = C_{clim} C_{rs} \sum_{veh} \sum_{road} EF_{veh,road} D_{veh,road},$$

distinguishing two key factors: car type and road type. The first resembles the aerodynamic properties of the cars and differentiates between heavy duty (HDV) and light duty (LDV) vehicles. The second factor accounts for the dust reservoir as function of traffic characteristics and road surrounding. The method is based upon a vehicle-driven kilometre map ($D_{veh,road}$) over Europe, typified for light duty (LDV) and heavy duty (HDV) traffic and three road type classes (rural roads, urban roads and highways), with corresponding emission factors ($EF_{veh,road}$). The standard emission factor was based on data for central Europe. However, the spatial variation of emission strength is due to the variability in regional climate conditions (aridity) and winter time practices concerning road sanding and studded tyre use. Accounting for these factors is important to translate the emission factors above to a methodology which can also be used in southern Europe and during winter in Scandinavia. Therefore, the methodology applies two factors, C_{clim} and C_{rs}, to account for variability of traffic resuspension due to the variability of climate conditions and road sanding activities scale.

Modelled contributions were validated against mineral dust observations across Europe, revealing a significant improvement in spatial variation when compared to the mineral dust modelling without emission without road dust resuspension. The temporal variation needs improvement as the dependency of resuspension source strength on day-to-day meteorology is crudely parameterized.

With the abovementioned technique Hendriks et al. [30] estimated that road dust resuspension contribute around 10–15% of modelled PM_{10} on a European scale and up to 30% in densely populated area of South Europe (Fig. 5). These estimates are relative to the modelled concentrations, which are lower than observed ones due to uncertainties in primary organic emissions and the lack of secondary organic compounds. However, this overestimate is probably compensated by the fact that in the model, peak contributions (in cities) are not captured due to the model resolution (25 × 25 km).

3.2 Wear Emissions

Wear emissions comprise abraded particles from brake linings, tyres and road pavement. Details on chemical and physical properties of wear particles can be found in Thorpe and Harrison [3]. Kousoulidou et al. [151] showed clear evidence that non-exhaust sources become increasingly important as no emission control strategies are taken by Member states. Among them, road pavement wear is

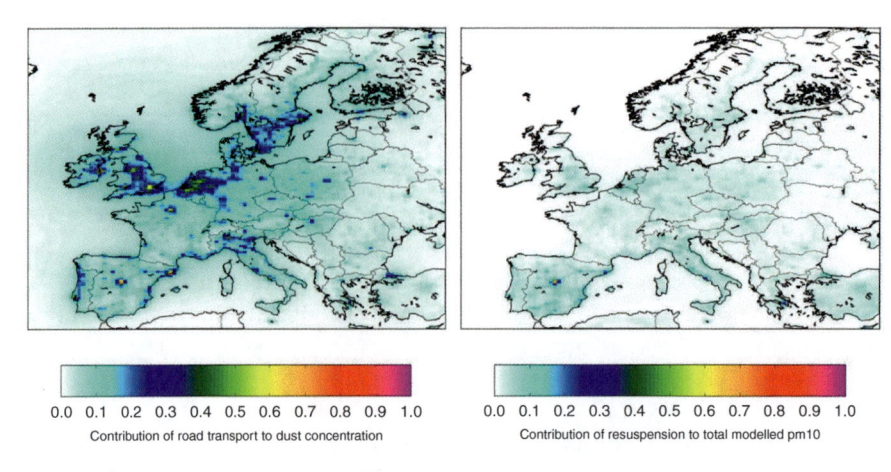

Fig. 5 LOTOS-EUROS relative contributions from traffic-induced resuspension to modelled dust (*left*) and PM_{10} (*right*) concentrations

probably the most difficult source to track given the few existing emission tests [152] and the chemical similarity to other mineral sources (construction materials, soil, etc.). Cu and Sb ambient concentrations and their ratio have already been shown in the literature to be useful chemical tracers for brake wear-related emissions. Brake wear emissions may be responsible for 50–75% of the total copper emissions to air for most of Western Europe [153]. Brake pad compositions normally show values for Cu/Sb in the range of 9–18 [154, 155] although it can be below 2 [156] depending on the manufacturer. When referring to ambient concentrations of Cu and Sb in the PM_{10} fraction the ratio Cu/Sb lies between 4 [157] to more common values of 7–9 in most urban sites [2, 32, 121, 158–162]). In road dust samples [109] found a rather constant Cu/Sb ratio (6.8 ± 0.9) only at one city (Barcelona), while in Zürich and Girona, the ratio was varying considerably, being the average values 13.5 ± 6.1 and 17.0 ± 8.9, respectively.

Contribution estimates of wear emissions in Europe are still few [30, 32, 134, 163, 164]. The few estimates over Europe were found to vary from negligible up to 4.0 $\mu g/m^3$ (10% of daily measured PM_{10} mass. At a regional scale, brake wear emissions contribute up to 2 $\mu g/m^3$ (Fig. 6). Although the contribution of brake particles is not dominant in terms of mass, their health concern might be the most relevant. After the ban of asbestos fibres for brakes manufacturing in the mid-1990s, composition of brake linings has rapidly changed, but there is still a number of possible toxics used and the information on materials employed by each manufacturer is missing. Generally, materials used for brake linings include metallic friction materials (Fe and Fe–Cu oxides), lubricants (graphite and Fe–Sb–Mo–Sn–Mn sulphides) and mineral fibres (Barite, Calcite, Zircon and Al-silicates) used as fillers. However, the composition can change largely from one brand to another and from one country to another. Also, the composition and size of brake particles can change during braking due to the high temperature and friction stress. As an example, at

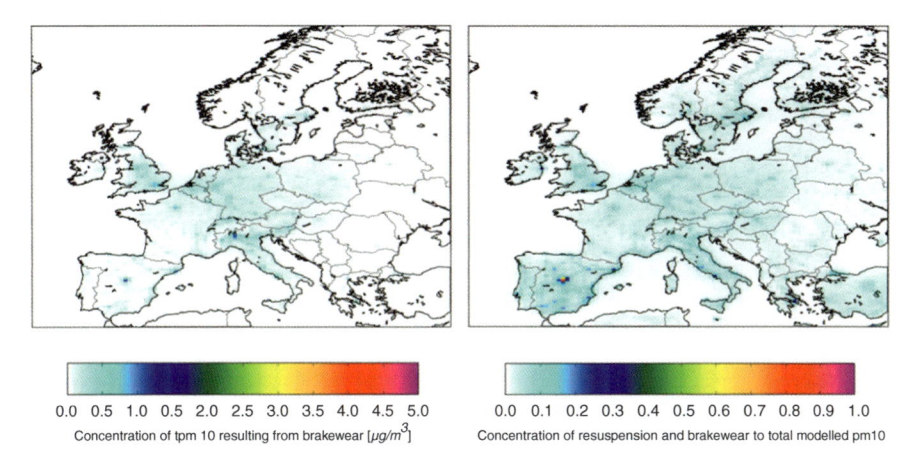

Fig. 6 Modelled contributions from brake wear over Europe (LOTOS-EUROS model; [30])

$T > 850°C$, Sb_2S_3 (Stibnite) is oxidated to Sb_2O_3, which is classified as possible carcinogenic in humans [165]. A detailed inventory, at national level, of manufacturers and brake materials is urgently needed, in combination with field and laboratory tests.

References

1. Raisanen K, Kupiainen K, Tervahattu H (2005) The effect of mineralogy, texture and mechanical properties of anti-skid and asphalt aggregates on urban dust, stages II and III. Bull Eng Geol Environ 64:247–256
2. Schauer JJ, Lough GC, Shafer MM, Christensen WF, Arndt MF, DeMinter JT, Park JS (2006) Characterization of metals emitted from motor vehicles. Health Effects Institute
3. Thorpe AJ, Harrison RM (2008) Sources and properties of non-exhaust particulate matter from road traffic: a review. Sci Total Environ 400:270–282
4. Ntziachristos L, Ning Z, Geller MD, Sheesley RJ, Schauer JJ, Sioutas C (2007) Fine, ultrafine and nanoparticle trace element compositions near a major freeway with a high heavy-duty diesel fraction. Atmos Environ 41:5684–5696
5. Schauer JJ, Kleeman M, Cass GR, Simoneit BT (1999) Measurement of emissions from air pollution sources. 2. C_1 through C_{30} organic compounds from medium duty diesel trucks. Environ Sci Technol 33:1578–1587
6. Schauer JJ, Kleeman MJ, Cass GR, Simoneit BRT (2002) Measurement of emissions from air pollution sources. 5. C 1 C 32 organic compounds from gasoline-powered motor vehicles. Environ Sci Technol 36:1169–1180
7. Künzli N, Kaiser R, Medina S, Studnicka M, Chanel O, Filliger P, Herry M, Horak F Jr, Puybonnieux-Texier V, Quénel P, Schneider J, Seethaler R, Vergnaud JC, Sommer H (2000) Public-health impact of outdoor and traffic-related air pollution: a European assessment. Lancet 356(9232):795–801
8. Le Tertre A, Medina S, Samoli E, Forsberg B, Michelozzi P, Boumghar A, Vonk JM, Bellini A, Atkinson R, Ayres JG, Sunyer J, Schwartz J, Katsouyanni K (2002) Short-term effects of

particulate air pollution on cardiovascular diseases in eight European cities. J Epidemiol Community Health 56:773–779
9. Kagawa J (2002) Health effects of diesel exhaust emissions—a mixture of air pollutants of worldwide concern. Toxicology 181–182:349–353
10. Cahill TA, Barnes DE, Withycombe E, Watnik M (2011) Very fine and ultrafine metals and ischemic heart disease in the California Central Valley 2: 1974–1991. Aerosol Sci Technol 45(9):1135–1142
11. Cahill TA, Barnes DE, Spada NJ, Lawton JA, Cahill TM (2011) Very fine and ultrafine metals and ischemic heart disease in the California central valley 1: 2003–2007. Aerosol Sci Technol 45(9):1123–1134
12. Meister K, Johansson C, Forsberg B (2012) Estimated short-term effects of coarse particles on daily mortality in Stockholm, Sweden. Environ Health Perspect 120(3):431–436
13. Ostro B, Tobias A, Querol X, Alastuey A, Amato F, Pey J, Perez N, Sunyer J (2011) The effects of particulate matter sources on daily mortality: a case-crossover study of Barcelona, Spain. Environ Health Perspect 119(12):1781–1787
14. Bahreini R, Middlebrook AM, De Gouw JA, Warneke C, Trainer M, Brock CA, Stark H, Brown SS, Dube WP, Gilman JB, Hall K, Holloway JS, Kuster WC, Perring AE, Prevot ASH, Schwarz JP, Spackman JR, Szidat S, Wagner NL, Weber RJ, Zotter P, Parrish DD (2012) Gasoline emissions dominate over diesel in formation of secondary organic aerosol mass. Geophys Res Lett 39(6): Art. no. L06805.
15. Verma V, Ning Z, Cho AK, Schauer JJ, Shafer MM, Sioutas C (2009) Redox activity of urban quasi-ultrafine particles from primary and secondary sources. Atmos Environ 43:6360–6368
16. Biswas S, Verma V, Schauer JJ, Cassee F, Cho AK, Sioutas C (2009) Oxidative potential of semi-volatile and non volatile particulate matter (PM) from heavy-duty vehicles retrofitted with emission control technologies. Environ Sci Technol 43:3905–3912
17. EEA (2010) Exceedance of air quality limit values in urban areas (Indicator CSI 004). European Environment Agency. www.eea.europa.eu/data-and-maps/indicators/exceedance-of-air-quality-limit-1/exceedance-of-airquality-limit-2. Accessed 12 Mar 2012
18. EEA (2011) Air quality in Europe – 2011 report. EEA Technical Report No. 12/2011. EEA – European Environmental Agency. Source: http://www.eea.europa.eu/publications/airquality-in-europe-2011. Retrieved 5 Mar 2012
19. Weiss M, Bonnel P, Kühlwein J, Provenza A, Lambrecht U, Alessandrini S, Carriero M, Colombo R, Forni F, Lanappe G, Le Lijour P, Manfredi U, Montigny F, Sculati M (2012) Will Euro 6 reduce the NOX emissions of new diesel cars? – Insights from on-road tests with Portable Emissions Measurement Systems (PEMS). Atmos Environ. doi:10.1016/j.atmosenv.2012.08.056
20. DOE (2012) Vehicles technology program. Fact #716: Diesels are more than half of new cars sold in Western Europe. DOE – US Department of Energy. Source: http://www1.eere.energy.gov/vehiclesandfuels/facts/2012_fotw716.html. Retrieved 9 July 2012
21. Pelkmans L, Debal P (2006) Comparison of on-road emissions with emissions measured on chassis dynamometer test cycles. Transp Res Part D Transp Environ 11:233–241
22. Rubino L, Bonnel P, Hummel R, Krasenbrink A, Manfredi U, De Santi G, Perotti M, Bomba G (2007) PEMS light-duty vehicles application: experiences in downtown Milan. SAE International. Technical Papers 2007-24-0113
23. Rubino L, Bonnel P, Hummel R, Krasenbrink A, Manfredi U, de Santi G (2009) On road emissions and fuel economy of light duty vehicles using PEMS: chase-testing experiment. SAE Int J Fuels Lubr 1:1454–1468
24. Vojtisek-Lom M, Fenkl M, Dufek M, Mareš J (2009) Off-cycle, real-world emissions of modern light duty diesel vehicles. SAE International. Technical Papers 2009-24-0148
25. Weiss M, Bonnel P, Hummel R, Provenza A, Manfredi U (2011) On-road emissions of light-duty vehicles in Europe. Environ Sci Technol 45:8575–8581
26. AQEG (2007) Trends in primary nitrogen dioxide in the UK. AQEG – Air Quality Expert Group. Report published by the Department for Environment, Food and Rural Affairs. London, UK

27. Dünnebeil F, Lambrecht U, Rehberger I (2011) Zukünftige Entwicklung der NO2-Konzentration an Straßen - Szenarien zur Einführung der neuen Grenzwertstufen. Fachgespräch Verkehrsemissionen: Emissionen und Minderungspotenziale im Verkehrsbereich - Was bringt Tempo 30 und wie stark wird Euro 6 die NO2-Emissionen im Realbetrieb senken? 21 July 2011. LUBW – Landesanstalt für Umwelt, Messungen, und Naturschutz Baden Württemberg. Stuttgart, Germany
28. Hausberger S (2011) PHEM - Das Modell der TU Graz zur Berechnung von Kfz-Emissionen und seine Datenbasis bei Euro 5 und Euro 6. Fachgespräch Verkehrsemissionen: Emissionen und Minderungspotenziale im Verkehrsbereich - Was bringt Tempo 30 und wie stark wird Euro 6 die NO2-Emissionen im Realbetrieb senken? 21 July 2011. LUBW – Landesanstalt für Umwelt, Messungen, und Naturschutz Baden Württemberg. Stuttgart, Germany
29. Kleinebrahm M, Steven H (2011) Vermessung des Abgasemissionsverhaltens von zwei Pkw und einem Fahrzeug der Transporterklasse im realen Straßenbetrieb in Stuttgart mittels PEMS Technologie. TÜV Nord. Report Nr. 4500116246/33 for LUBW -Landesanstalt für Umwelt, Messungen und Naturschutz Baden-Württemberg. Karlsruhe, Germany
30. Hendriks C, Kranenburg R, Kuenen J, van Gijlswijk R, Wichink Kruit R, Segers A, Denier van der Gon H, Schaap M (2013) The origin of ambient particulate matter concentrations in the Netherlands. Atmos Environ, 69:289–303
31. Harrison R, Stedman J, Derwent D (2008) New directions; why are PM_{10} concentrations in Europe not falling? Atmos Environ 42:603–606
32. Amato F, Pandolfi M, Escrig A, Querol X, Alastuey A, Pey J, Pérez N, Hopke PK (2009) Quantifying road dust resuspensión in urban environment by Multilinear Engine: a comparison with PMF2. Atmos Environ 43:2770–2780
33. Bukowiecki N, Lienemann P, Hill M, Furger M, Richard A, Amato F, Prevot ASH, Baltensperger U, Buchmann B, Gehrig R (2010) PM10 emission factors for non-exhaust particles generated by road traffic in an urban street canyon and along a freeway in Switzerland (2010). Atmos Environ 44(19):2330–2340
34. Querol X, Alastuey A, Rodríguez S, Plana F, Ruiz CR, Cots N, Massagué G, Puig O (2001) PM_{10} and $PM_{2.5}$ source apportionment in the Barcelona Metropolitan Area, Catalonia, Spain. Atmos Environ 35:6407–6419
35. Viana M, Kuhlbusch TAJ, Querol X, Alastuey A, Harrison RM, Hopke PK, Winiwarter W, Vallius M, Szidat S, Prévôt ASH, Hueglin C, Bloemen H, Wåhlin P, Vecchi R, Miranda AI, Kasper-Giebl A, Maenhaut W, Hitzenberger R (2008) Source apportionment of particulate matter in Europe: a review of methods and results. J Aerosol Sci 39:827–849
36. Thimmaiah D, Hovorka J, Hopke PK (2009) Source apportionment of winter submicron Prague aerosols from combined particle number size distribution and gaseous composition data. Aerosol Air Qual Res 9:209–236
37. Stölzel M (2003) Quellen von Feinstaubpartikeln in Erfurt sowie ihre gesundheitlichen Auswirkungen. Materialien zur Epidemiologie, herausgegeben von HE Wichmann, HE Heilmaier. S. Roderer Verlag Regensburg [in German]
38. Quass U, Kuhlbusch T, Koch M (2004) Identification of source groups for fine dust. Public report to the Environment Ministry of North Rhine Westphalia, Germany. IUTA-Report LP15/2004, p 12
39. Gerwig H (2005) Korngrößendifferenzierte Feinstaubbelastung in Straßennähe in Ballungsgebieten Sachsens, Final report for research project of Dep. 2 of Saxon State Agency for Environment and Geology. http://www.umwelt.sachsen.de/lfug/luft-laerm-klima_5356.html
40. Vallius M, Janssen NAH, Heinrich J, Hoek G, Ruuskanen J, Cyrys J, Van Grieken R, de Hartog JJ, Kreyling WG, Pekkanen J (2005) Sources and elemental composition of ambient $PM_{2.5}$ in three European cities. Sci Total Environ 337:147–162
41. Yue W, Stölzel M, Cyrys J, Pitz M, Heinrich J, Kreyling WG, Wichmann HE, Peters A, Wang S, Hopke PK (2008) Source apportionment of ambient fine particle size distribution using positive matrix factorization in Erfurt, Germany. Sci Total Environ 398(1–3):133–144

42. Beuck H, Quass U, Klemm O, Kuhlbusch TAJ (2011) Assessment of sea salt and mineral dust contributions to PM10 in NW Germany using tracer models and positive matrix factorization. Atmos Environ 45(32):5813–5821
43. Rodríguez S, Querol X, Alastuey A, Plana F (2002) Sources and processes affecting levels and composition of atmospheric aerosol in the Western Mediterranean. J Geophys Res 107 (D24):4777
44. Querol X, Alastuey A, Viana M, Rodríguez S, Artiñano B, Salvador P, Garcia Do Santos S, Fernandez Patier R, Ruiz CR, de la Rosa J, Sanchez de la Campa A, Menendez M, Gil JI (2004) Speciation and origin of PM_{10} and $PM_{2.5}$ in Spain. J Aerosol Sci 35:1151–1172
45. Viana M, Querol X, Götschi T, Alastuey A, Sunyer J, Forsberg B, Heinrich J, Norbäck D, Payo F, Maldonado JA, Künzli N (2007) Source apportionment of ambient $PM_{2.5}$ at five Spanish centres of the European community respiratory health survey (ECRHS II). Atmos Environ 41:1395–1406
46. Nicolás J, Chiari M, Crespo J, Orellana IG, Lucarelli F, Nava S, Pastor C, Yubero E (2008) Quantification of Saharan and local dust impact in an arid Mediterranean area by the positive matrix factorization (PMF) technique. Atmos Environ 42(39):8872–8882
47. Pey J, Querol X, Alastuey A, Rodríguez S, Putaud JP, Van Dingenen R (2009) Source apportionment of urban fine and ultra-fine particle number concentration in a Western Mediterranean city. Atmos Environ 43:4407–4415
48. Gaimoz C, Sauvage S, Gros V, Herrmann F, Williams J, Locoge N, Perrussel O, Bonsang EB, D'Argouges O, Sarda-Estève R, Sciare J (2011) Volatile organic compounds sources in Paris in spring 2007. Part II: Source apportionment using positive matrix factorisation. Environ Chem 8:91–103
49. Yin J, Harrison RM, Chen Q, Rutter A, Schauer JJ (2010) Source apportionment of fine particles at urban background and rural sites in the UK atmosphere. Atmos Environ 44 (6):841–851
50. Manoli E, Voutsa D, Samara C (2002) Chemical characterization and source identification/ apportionment of fine and coarse air particles in Thessaloniki, Greece. Atmos Environ 36 (6):949–961
51. Karanasiou AA, Siskos PA, Eleftheriadis K (2009) Assessment of source apportionment by Positive Matrix Factorization analysis on fine and coarse urban aerosol size fractions. Atmos Environ 43:3385–3395
52. Yin J, Allen AG, Harrison RM, Jennings SG, Wright E, Fitzpatrick M et al (2005) Major component composition of urban PM_{10} and $PM_{2.5}$ in Ireland. Atmos Res 78:149–165
53. Contini D, Genga A, Cesari D, Siciliano M, Donateo A, Bove MC, Guascito MR (2010) Characterisation and source apportionment of PM_{10} in an urban background site in Lecce. Atmos Res 95:40–54
54. Bernardoni V, Vecchi R, Valli G, Piazzalunga A, Fermo P (2011) PM_{10} source apportionment in Milan (Italy) using time-resolved data. Sci Total Environ 409:4788–4795
55. Mooibroek D, Schaap M, Weijers EP, Hoogerbrugge R (2011) Source apportionment and spatial variability of PM2.5 using measurements at five sites in the Netherlands. Atmos Environ 45:4180–4191
56. Raes N, Hubacz R, Hanssen JE, Maenhaut W (2005) Application of three different receptor models to long-term data sets from southern Norway and the Norwegian Arctic. In: Abstracts of the European aerosol conference (EAC2005), Ghent, Belgium, 29/08/05-02/09/05. Abstract no. 288 [ISBN: 9080915939]
57. Pio CA, Castro LM, Cerqueira MA, Santos IM, Belchior F, Salgueiro ML (1996) Source assessment of particulate air pollutants measured at the southwest European coast. Atmos Environ 30(19):3309–3320
58. Oliveira C, Pio C, Caseiro A, Santos P, Nunes T, Mao H, Luahana L, Sokhi R (2010) Road traffic impact on urban atmospheric aerosol loading at Oporto, Portugal. Atmos Environ 44:3147–3158
59. Swietlicki E, Puri S, Hansson H-C, Edner H (1996) Urban air pollution source apportionment using a combination of aerosol and gas monitoring techniques. Atmos Environ 30:2795–2809

60. Lanz VA, Prévôt ASH, Alfarra MR, Weimer S, Mohr C, DeCarlo PF, Gianini MFD, Hueglin C, Schneider J, Favez O, D'Anna B, George C, Baltensperger U (2010) Characterization of aerosol chemical composition with aerosol mass spectrometry in Central Europe: an overview. Atmos Chem Phys 10:10453–10471
61. Mohr C, Richter R, DeCarlo FF, Prévôt ASH, Baltensperger U (2011) Spatial variation of chemical composition and sources of submicron aerosol in Zurich during wintertime using mobile aerosol mass spectrometer data. Atmos Chem Phys 11:7465–7482
62. IARC (1989) IARC monographs on the evaluation of carcinogenic risks to humans, vol 45, Occupational exposures in petroleum refining; Crude Oil and Major Petroleum Fuels, Lyon, 159–201, 219–237
63. Harris SJ, Maricq MM (2001) Signature size distributions for diesel and gasoline engine exhaust particulate matter. J Aerosol Sci 32(6):749–764
64. Ban-Weiss GA, Luden MM, Kirchstetter TW, Harley RA (2009) Size-resolved particle number and volume emission factors for on-road gasoline and diesel motor vehicles. Downloaded at http://escholarship.org/uc/item/7k75r4bf
65. Reche C, Querol X, Alastuey A, Viana M, Pey J, Moreno T, Rodríguez S, Gonzáliz Y, Fernández-Camacho R, de La Rosa J, Dall'Osto M, Pret ASH, Hueglin C, Harrison RM, Quincey P (2011) New considerations for PM, Black Carbon and particle number concentration for air quality monitoring across different European cities. Atmos Chem Phys 11:6207–6227
66. Kulmala M, Kerminen V-M (2008) On the growth of atmospheric nanoparticles. Atmos Res 90:132–150
67. Jo W-K, Song K-B (2001) Exposure to volatile organic compounds for individuals with occupations associated with potential exposure to motor vehicle exhaust and/or gasoline vapor emissions. Sci Total Environ 269(1–3):25–37
68. Hartle RW, Young RJ (1977) Occupational benzene exposure at retail automotive service stations. Draft report, Division of Surveillance, Hazard Evaluations and Field Studies. National Institute for Occupational Safety and Health, Cincinnati, Ohio
69. Kearney CA, Dunham DB (1986) Gasoline vapor exposures at a high volume service station. Am Ind Hyg Assoc J 47:535–539
70. Song Y, Shao M, Liu Y, Lu S, Kuster W, Goldan P, Xie S (2007) Source apportionment of ambient volatile organic compounds in Beijing. Environ Sci Technol 41(12):4348–4353
71. Cachier H (1995) Combustion carbonaceous aerosols in the atmosphere: implications for ice core studies. Ice Core Studies of Global Biogeochemical Cycles, NATO ASI Series, Springer, Berlin, pp 313–346
72. Lyamani H, Olmo FJ, Foyo I, Alados-Arboledas L (2011) Black carbon aerosols over an urban area in south-eastern Spain: changes detected after the 2008 economic crisis. Atmos Environ 45:6423–6432
73. Pakkanen TA, Kerminen VM, Ojanena CH, Hillamo RE, Aarnio P, Koskentalo T (2000) Atmospheric black carbon in Helsinki. Atmos Environ 34:1497–1506
74. Pérez N, Pey J, Cusack M, Reche C, Querol X, Alastuey A, Viana M (2010) Variability of particle number, black carbon, and PM_{10}, $PM_{2.5}$, and PM_1 levels and speciation: influence of road traffic emissions on urban air quality. Aerosol Sci Technol 44:487–499
75. Rodríguez S, Cuevas E, González Y, Ramos R, Romero PM, Pérez N, Querol X, Alastuey A (2008) Influence of sea breeze circulation and road traffic emissions on the relationship between particle number, black carbon, PM_1, $PM_{2.5}$ and $PM_{2.5\text{–}10}$ concentrations in a coastal city. Atmos Environ 42:6523–6534
76. Saha A, Despiau S (2009) Seasonal and diurnal variations of black carbon aerosols over a Mediterranean coastal zone. Atmos Res 92:27–41
77. Hamilton RS, Mansfield TA (1990) Airborne particulate elemental carbon: its sources, transport and contribution to dark smoke and soiling. Atmos Environ 25:715–723

78. Fruin SA, Winer AM, Rodes CE (2004) Black carbon concentrations in California vehicles and estimation of in-vehicle diesel exhaust particulate matter exposures. Atmos Environ 38:4121–4133
79. Kerminen VM, Mäkelä TE, Ojanen CH, Hillamo RM, Vilhunen JK, Rantanen L, Havers N, von Bohlen A, Klockow D (1997) Characterization of particulate phase in the exhaust from a diesel car. Environ Sci Technol 31:1883–1889
80. Berner A, Sidla S, Galambos Z, Kruisz C, Hitzenberger R, ten Brink HM, Kos GAP (1996) Modal character of atmospheric black carbon size distributions. J Geophys Res 101:19559–19565
81. Harrison RM, Yin J (2008) Sources and processes affecting carbonaceous aerosols in central England. Atmos Environ 42:1413–1423
82. Schneider J, Kirchner U, Borrmann S, Vogt R, Scheer V (2008) In situ measurements of particle number concentration, chemically resolved size distributions and black carbon content of traffic-related emissions on German motorways, rural roads and in city traffic. Atmos Environ 42:4257–4268
83. Charron A, Harrison J (2005) Fine ($PM_{2.5}$) and coarse ($PM_{2.5-10}$) particulate matter on a heavily trafficked London highway: sources and processes. Environ Sci Technol 39:7768–7776
84. Harrison J, Yin J, Mark D, Stedman J, Appleby RS, Booker J, Moorcroft S (2001) Studies of the coarse particle (2.5–10um) component in UK urban atmospheres. Atmos Environ 35:3667–3679
85. Querol X, Alastuey A, Viana MM, Rodriguez S, Artiñano B, Salvador P, Garcia Do Santos S, Fernandez Patier R, De La Rosa J, Sanchez De La Campa A, Menendez M (2005) Contaminación atmosférica por partículas en suspensión. In Ministerio de Educación y Ciencia, I. S. d. F. d. P. C. A. d. V. d. L. U. C. d. M. C. d. E., ISBN: 84-369-3924-7., Ed. 2005; pp 133–172
86. Bukowiecki N, Dommen J, Prév t ASH, Weingartner E, Baltensperger U (2003) Fine and ultrafine particles in the Zürich (Switzerland) area measured with a mobile laboratory. An assessment of the seasonal and regional variation throughout a year. Atmos Chem Phys 3:1477–1494
87. Hueglin C, Buchmann B, Weber RO (2006) Long-term observation of real-world road traffic emission factors on a motorway in Switzerland. Atmos Environ 40:3696–3709
88. Morawska L, Jayaratne ER, Mengersen K, Jamriska M, Thomas S (2002) Differences in airborne particle and gaseous concentrations in urban areas between weekdays and weekends. Atmos Environ 36:4375–4383
89. Rodríguez S, Van Dingenen R, Putaud JP, Dell'Acqua A, Pey J, Querol X, Alastuey A, Chenery S, Kin-Fai H, Harrison RM, Tardivo R, Scarnato B, Gianelle V (2007) A study on the relationship between mass concentration, chemistry and number size distribution of urban fine aerosols in Milan, Barcelona and London. Atmos Chem Phys 7:2217–2232
90. Rodríguez S, Cuevas E (2007) The contributions of "minimum primary emissions" and "new particle number formation enhancements" to the particle number concentration in urban air. J Aerosol Sci 38:1207–1219
91. Casati R, Scheer V, Vogt R, Benter T (2007) Measurements of nucleation and soot mode particle emission from a diesel passenger car in real world and laboratory in situ dilution. Atmos Environ 41:2125–2135
92. Zhu Y, Hinds W, Kim S, Shen S, Sioutas C (2002) Study of ultrafine particles near a major highway with heavy-duty diesel traffic. Atmos Environ 36:4323–4335
93. Fischer PH, Hoek G, van Reeuwijk H, Briggs DJ, Lebret E, van Wijnen JH, Kingham S, Elliott PE (2000) Traffic-related differences in outdoor and indoor concentrations of particles and volatile organic compounds in Amsterdam. Atmos Environ 34:3713–3722
94. Harrison RM, Jones AM, Lawrence RG (2004) Major component composition of PM_{10} and $PM_{2.5}$ from roadside and urban background sites. Atmos Environ 38:4531–4538

95. Janssen NAH, Van Mansom DFM, Van Der Jagt K, Harssema H, Hoek G (1997) Mass concentration and elemental composition of airborne particulate matter at street and background locations. Atmos Environ 31:1185–1193
96. Smargiassi A, Baldwin M, Pilger C, Dugandzic R, Brauer M (2005) Small-scale spatial variability of particle concentration and traffic levels in Montreal: a pilot study. Sci Total Environ 338:243–251
97. Fernández-Camacho R, Rodríguez S, de la Rosa J, Sánchez de la Campa AM, Viana M, Alastuey A, Querol X (2011) Ultrafine particle formation in the island sea breeze airflow in Southwest Europe. Atmos Chem Phys 10:9615–9630
98. Cheung HC, Morawska L, Ristovski ZD (2011) Observation of new particle formation in subtropical urban environment. Atmos Chem Phys 11(8):3823–3833
99. Fine PM, Sioutas C, Solomon PA (2008) Secondary particulate matter in the United States: insights from the particulate matter supersites program and related studies. J Air Waste Manage Assoc 58(2):234–253
100. Mejia JF, Morawska L (2009) An investigation of nucleation events in a coastal urban environment in the Southern hemisphere. Atmos Chem Phys 9(20):7877–7888
101. Gomišček B, Hauck H, Stopper S, Preining O (2004) Spatial and temporal variations of PM_1, $PM_{2.5}$, PM_{10} and particle number concentration during the AUPHEP—project. Atmos Environ 38:3917–3934
102. Laakso L, Hussein T, Aarnio P, Komppula M, Hiltunen V, Viisanen Y, Kulmala M (2003) Diurnal and annual characteristics of particle mass and number concentrations in urban, rural and Arctic environments in Finland. Atmos Environ 37:2629–2641
103. Pérez L, Medina-Ramón M, Künzli N, Alastuey A, Pey J, Pérez N, Garcia A, Tobias A, Querol X, Sunyer J (2009) Size fractionated particulate matter, vehicle traffic, and case specific daily mortality in Barcelona (Spain). Environ Sci Technol 43(13):4707–4714. doi:10.1021/es8031488, 2009
104. Pengchai P, Furumai H, Nakajima F (2004) Source apportionment of polycyclic aromatic hydrocarbons in road dust in Tokyo. Polycycl Aromat Compd 24(4–5):773–789
105. Gustafsson M, Blomqvist G, Gudmundsson A, Dahl A, Swietlicki E, Bohgard M, Lindbom J, Ljungman A (2008) Properties and toxicological effects of particles from the interaction between tyres, road pavement and winter traction material. Sci Total Environ 393 (2–3):226–240
106. Querol X, Viana M, Alastuey A, Amato F, Moreno T, Castillo S, Pey J, de la Rosa J, Sánchez de la Campa A, Artíñano B, Salvador P, García Dos Santos S, Fernández-Patier R, Moreno-Grau S, Negral L, Minguillón MC, Monfort E, Gil JI, Inza A, Ortega LA, Santamaría JM, Zabalza J (2007) Source origin of trace elements in PM from regional background, urban and industrial sites of Spain. Atmos Environ 41:7219–7231
107. Etyemezian V, Nikolich G, Ahonen S, Pitchford M, Sweeney M, Purcell R, Gillies J, Kuhns H (2007) The Portable In Situ Wind Erosion Laboratory (PI-SWERL): a new method to measure PM10 windblown dust properties and potential for emissions. Atmos Environ 41 (18):3789–3796
108. Kuhns H, Gillies J, Etyemezian V, Nikolich G, King J, Zhu D, Uppapalli S, Engelbrecht J, Kohl S (2010) Effect of soil type and momentum on unpaved road particulate matter emissions from wheeled and tracked vehicles. Aerosol Sci Technol 44(3):187–196
109. Amato F, Pandolfi M, Moreno T, Furger M, Pey J, Alastuey A, Bukowiecki N, Prévôt ASH, Baltensperger U, Querol X (2011) Sources and variability of inhalable road dust particles in three European cities. Atmos Environ 45(37):6777–6787
110. Richard A, Gianini MFD, Mohr C, Furger M, Bukowiecki N, Minguillón MC, Lienemann P, Flechsig U, Appel K, Decarlo PF, Heringa MF, Chirico R, Baltensperger U, Prévôt ASH (2011) Source apportionment of size and time resolved trace elements and organic aerosols from an urban courtyard site in Switzerland. Atmos Chem Phys 11(17):8945–8963

111. Mijić Z, Stojić A, Perišić M, Rajšić S, Tasić M, Radenković M, Joksić J (2010) Seasonal variability and source apportionment of metals in the atmospheric deposition in Belgrade. Atmos Environ 44(30):3630–3637
112. Masiol M, Rampazzo G, Ceccato D, Squizzato S, Pavoni B (2010) Characterization of PM10 sources in a coastal area near Venice (Italy): an application of factor-cluster analysis. Chemosphere 80(7):771–778
113. Rodríguez S, Querol X, Alastuey A, Viana M, Alarcón M, Mantilla E, Ruiz CR (2004) Comparative PM10-PM2.5 source contribution study at rural, urban and industrial sites during PM episodes in Eastern Spain. Sci Total Environ 328:95–113
114. Ingenieurbüro Lohmeyer, 2004: Maßnahmebetrachtungen zu PM10 im Zusammenhang mit Luftreinhalteplänen. Anhang 2 of Regierungspräsidium Stuttgart (2005)
115. Astel AM (2010) Air contaminants modelling by use of several receptor-oriented models. Int J Environ Pollut 42(1–3):32–57
116. Thorpe A, Harrison RM, Boulter PG, McCrae IS (2007) Estimation of particle resuspension source strength on a major London Road. Atmos Environ 41:8007–8020
117. Gu J, Pitz M, Schnelle-Kreis J, Diemer J, Reller A, Zimmermann R, Soentgen J, Stoelzel M, Wichmann H-E, Peters A, Cyrys J (2011) Source apportionment of ambient particles: comparison of positive matrix factorization analysis applied to particle size distribution and chemical composition data. Atmos Environ 45(10):1849–1857
118. Amato F, Querol X, Johansson C, Nagl C, Alastuey A (2010) A review on the effectiveness of street sweeping, washing and dust suppressants as urban PM control methods. Sci Total Environ 408(16):3070–3084
119. Putaud JP, Van Dingenen R, Alastuey A, Bauer H, Birmili W, Cyrys J, Flentje H, Fuzzi S, Gehrig R, Hansson HC, Harrison RM, Herrmann H, Hitzenberger R, Hüglin C, Jones AM, Kasper-Giebl A, Kiss G, Kousa A, Kuhlbusch TAJ, Löschau G, Maenhaut W, Molnar A, Moreno T, Pekkanen J, Perrino C, Pitz M, Puxbaum H, Querol X, Rodriguez S, Salma I, Schwarz J, Smolik J, Schneider J, Spindler G, ten Brink H, Tursic J, Viana M, Wiedensohler A, Raes F (2010) A European aerosol phenomenology – 3: physical and chemical characteristics of particulate matter from 60 rural, urban, and kerbside sites across Europe. Atmos Environ 44(10):1308–1320
120. Putaud JP, Raes F, Van Dingenen R, Bruggemann E, Facchini MC, Decesari S, Fuzzi S, Gehrig R, Hüglin C, Laj P, Lorbeer G, Maenhaut W, Mihalopoulos N, Müller K, Querol X, Rodriguez S, Schneider J, Spindler G, ten Brink H, Tørseth K, Wiedensohler A, Wiedensohler A (2004) A European aerosol phenomenology-2: chemical characteristics of particulate matter at kerbside, urban, rural and background sites in Europe. Atmos Environ 38 (16):2579–2595
121. Perez N, Pey J, Querol X, Alastuey A, Lopez JM, Viana M (2008) Partitioning of major and trace components in PM10, PM2.5 and PM1 at an urban site in Southern Europe. Atmos Environ 42:1677–1691
122. Ariola V, D'Alessandro A, Lucarelli F, Marcazzan G, Mazzei F, Nava S, Garcia-Orellana I, Prati P, Valli G, Vecchi R, Zucchiatti A (2006) Elemental characterization of PM10, PM2.5 and PM1 in the town of Genoa (Italy). Chemosphere 62(2):226–232
123. Marelli L, Lagler F, Borowiak A, Drossinos Y, Gerboles M, Buzica D, Szafraniec K, Niedzialek J, Jimenez J, De Santi G (2006) PM measurements in Krakow during a winter campaign. In: JRC enlargement and integration workshop, "Outcome of the Krakow Integrated Project": particulate matter: from emissions to health effects, Krakow Municipal Office, 15–16 May 2006
124. Querol X, Alastuey A, Puicercus JA, Mantilla E, Miró JV, López-Soler A, Plana F, Artíñano B (1998) Seasonal evolution of suspended particles around a large coal-fired power station: particle levels and sources. Atmos Environ 32:1963–1978
125. Karanasiou A, Moreno T, Amato F, Lumbreras J, Narros A, Borge R, Tobías A, Boldo E, Linares C, Pey J, Reche C, Alastuey A, Querol X (2011) Road dust contribution to PM levels –

evaluation of the effectiveness of street washing activities by means of positive matrix factorization. Atmos Environ 45(13):2193–2201
126. Norman M, Johansson C (2006) Studies of some measures to reduce road dust emissions from paved roads in Scandinavia. Atmos Environ 40:6154–6164
127. Tervahattu H, Kupiainen KJ, Räisänen M, Mäkelä T, Hillamo R (2006) Generation of urban road dust from anti-skid and asphalt concrete aggregates. J Hazard Mater 132:39–46
128. Areskoug H, Johansson C, Alesand T, Hedberg E, Ekengren T, Vesely V, Wideqvist U, Hansson HC (2004) Concentrations and sources of PM10 and PM2.5 in Sweden. ITM Report no. 110, Stockholm
129. Kupiainen K, Tervahattu H, Raisanen M, Makela T, Aurela M, Hillamo R (2005) Size and composition of airborne particles from pavement wear, tires, and traction sanding. Environ Sci Technol 39:699–706
130. Kupiainen K, Tervahattu H, Raisanen M (2003) Experimental studies about the impact of traction sand on urban road dust composition. Sci Total Environ 308:175–184
131. Kuhns H, Etyemezian V, Green M, Hendrickson K, McGown M, Barton K, Pitchford M (2003) Vehicle-based road dust emission measurement – part II: effect of precipitation, wintertime road sanding and street sweepers on inferred PM10 emission potentials from paved and unpaved roads. Atmos Environ 37:4573–4582
132. Hussein T, Johansson C, Karlsson H, Hansson HC (2008) Factors affecting non tailpipe aerosol particle emissions from paved roads: on-road measurements in Stockholm, Sweden. Atmos Environ 42(4):688–702
133. Kantamaneni R, Adams G, Bamesberger L, Allwine E, Westberg H, Lamb B, Claiborn C (1996) The measurement of roadway PM10 emission rates using atmospheric tracer ratio techniques. Atmos Environ 30(24):4209–4223
134. Wåhlin P, Berkowicz R, Palmgren F (2006) Characterization of traffic-generated particulate matter in Copenhagen. Atmos Environ 40:2151–2159
135. Forsberg B, Hansson HC, Johansson C, Areskoug H, Persson K, Jarvholm B (2005) Comparative health impact assessment of local and regional particulate air pollutants in Scandinavia. Ambio 34:11–19
136. Omstedt G, Bringfelt B, Johansson C (2005) A model for vehicle-induced non-tailpipe emissions of particles along Swedish roads. Atmos Environ 39(33):6088–6097
137. EPA (2006) AP-42. Compilation of air pollutant emission factors, 5th edn. U.S. Environmental Protection Agency, Research Triangle Park. http://www.epa.gov/otaq/ap42.htm (Chapter 13.2) fugitive dust sources
138. Gámez AJ, Düring I, Bösinger R, Rabl P, Lohmeyer A (2001) Determination of the 99.8-percentile of NO_2 concentrations and PM10 emissions for EIA studies. In: Third international conference on urban air quality (measurement, modelling, management), Loutraki, 19–23 March 2001
139. Düring I, Jacob J, Lohmeyer A, Lutz M, Reichenbacher W (2002) Estimation of the 'nonexhaust pipe' PM10 emissions of streets for practical traffic air pollution modelling. In: 11th international conference 'transport and air pollution', Graz, June 2002
140. Amato F, Nava S, Lucarelli F, Querol X, Alastuey A, Baldasano JM, Pandolfi M (2010) A comprehensive assessment of PM emissions from paved roads: real-world emission factors and intense street cleaning trials. Sci Total Environ 408(20):4309–4318
141. Gertler A, Kuhns H, Abu-Allaban M, Damm CR, Gillies J, Etyemezian V, Clayton R, Proffitt D (2006) A case study of the impact of winter road sand/salt and street sweeping on road dust re-entrainment. Atmos Environ 40:5976–5985
142. Johansson C, Norman M, Burman L (2009) Road traffic emission factors for heavy metals. Atmos Environ 43(31):4681–4688
143. Bukowiecki N, Lienemann P, Hill M, Figi R, Richard A, Furger M, Rickers K, Falkenberg G, Zhao Y, Cliff SS, Prevot ASH, Baltensperger U, Buchmann B, Gehrig R (2009) Real-world emission factors for antimony and other brake wear related trace elements: size-segregated values for light and heavy duty vehicles. Environ Sci Technol 43(21):8072–8078

144. Claiborn C, Mitra A, Adams G, Bamesberger L, Allwine G, Kantamaneni R, Lamb B, Westberg H (1995) Evaluation of PM10 emission rates from paved and unpaved roads using tracer techniques. Atmos Environ 29(10):1075–1089
145. Schaap M, Manders AMM, Hendriks ECJ, Cnossen JM, Segers AJS, Denier van der Gon HAC, Jozwicka M, Sauter FJ, Velders GJM, Matthijsen J, Builtjes PJH (2009) Regional modelling of particulate matter for the Netherlands. www.pbl.nl
146. Gehrig R, Hill M, Buchmann B, Imhof D, Weingartner E, Baltensperger U (2004) Separate determination of PM10 emission factors of road traffic for tailpipe emissions and emissions from abrasion and resuspension processes. Int J Environ Pollut 22(3):312–325
147. Ketzel M, Omstedt G, Johansson C, During I, Pohjola M, Oettl D, Gidhagen L, Gidhagen L, Wåhlin P, Lohmeyer A, Haakana M, Berkowicz R (2007) Estimation and validation of PM2.5/PM10 exhaust and nonexhaust emission factors for practical street pollution modeling. Atmos Environ 41(40):9370–9385
148. Amato F, Karanasiou A, Moreno T, Alastuey A, Orza JAG, Lumbreras J, Borge R, Boldo E, Linares C, Querol X (2012) Emission factors from road dust resuspension in a Mediterranean freeway. Atmos Environ 61:580–587
149. INFRAS, HBEFA Handbuch Emissionsfaktoren des Strassenverkehrs. Version 2.1. INFRAS, UBA Berlin, UBA Wien, BUWAL
150. Schaap M, Timmermans RMA, Roemer M, Boersen GAC, Builtjes PJH, Sauter FJ, Velders GJM, Beck JP (2008) The LOTOS-EUROS model: description, validation and latest developments. Int J Environ Pollut 32(2):270–290
151. Kousoulidou M, Ntziachristos L, Mellios G, Samaras Z (2008) Road-transport emission projections to 2020 in European urban environments. Atmos Environ 42(32):7465–7475
152. Gehrig R, Zeyer K, Bukowiecki N, Lienemann P, Poulikakos LD, Furger M, Buchmann B (2010) Mobile load simulators – a tool to distinguish between the emissions due to abrasion and resuspension of PM10 from road surfaces. Atmos Environ 44(38):4937–4943
153. Denier van der Gon HAC, Hulskotte JHJ, Visschedijk AJH, Schaap M (2007) A revised estimate of copper emissions from road transport in UNECE-Europe and its impact on predicted copper concentrations. Atmos Environ 41(38):8697–8710
154. Adachi K, Tainosho Y (2004) Characterization of heavy metal particles embedded in tire dust. Environ Int 30:1009–1017
155. Iijima A, Sato K, Yano K, Tago H, Kato M, Kimura H, Furuta N (2007) Particle size and composition distribution analysis of automotive brake abrasion dusts for the evaluation of antimony sources of airborne particulate matter. Atmos Environ 41(23):4908–4919
156. Von Uexküll O, Skerfving S, Doyle R, Braungart M (2005) Antimony in brake pads – a carcinogenic component? J Clean Prod 13(1):19–31
157. Zereini F, Alt F, Messerschmidt J, Wiseman C, Feldmann I, Von Bohlen A, Müller J, Liebl K, Püttmann W (2005) Concentration distribution of heavy metals in urban airborne particulate matter in Frankfurt am Main, Germany. Environ Sci Technol 39(9):2983–2989
158. Weckwerth G (2001) Verification of traffic emitted aerosol components in the ambient air of Cologne (Germany). Atmos Environ 35:5525–5536
159. Sternbeck J, Sjödin A, Andreasson K (2002) Metal emissions from road traffic and the influence of resuspension—results from two tunnel studies. Atmos Environ 36:4735–4744
160. Gietl JK, Lawrence R, Thorpe AJ, Harrison RM (2010) Identification of brake wear particles and derivation of a quantitative tracer for brake dust at a major road. Atmos Environ 44 (2):141–146
161. Pey J, Querol X, Alastuey A (2010) Discriminating the regional and urban contributions in the North-Western Mediterranean: PM levels and composition. Atmos Environ 44 (13):1587–1596
162. Fujiwara FG, Gómez DR, Dawidowski L, Perelman P, Faggi A (2011) Metals associated with airborne particulate matter in road dust and tree bark collected in a megacity (Buenos Aires, Argentina). Ecol Indic 11(2):240–247.

163. Arditsoglou A, Samara C (2005) Levels of total suspended particulate matter and major trace elements in Kosovo: a source identification and apportionment study. Chemosphere 59 (5):669–678 (Original Research Article)
164. Glaser B, Dreyer A, Bock M, Fiedler S, Mehring M, Heitmann T (2005) Source apportionment of organic pollutants of a highway-traffic-influenced urban area in Bayreuth (Germany) using biomarker and stable carbon isotope signatures. Environ Sci Technol 39(11):3911–3917
165. IARC (1989) International Agency for Research on Cancer Antimony trioxide and antimony trisulfide. In: IARC (ed) IARC monographs. IARC, Lyon, p 291

Source Apportionment of Airborne Dust in Germany: Methods and Results

U. Quass, A.C. John, and T.A.J. Kuhlbusch

Abstract Methodologies and results of approaches used for the source apportionment of particulate matter in Germany are reviewed. Due to the relatively large number of interested parties and stakeholders, in particular the 16 German Federal States and the Federal Environment Agency, the information was found to be quite dispersed.

Based on the PM levels measured in the state monitoring networks the incremental increase of PM from rural to hot-spot conditions is one of the most widely investigated aspects. As a general conclusion a large-scale PM10 background contribution of ca. 50% appears to be typical, with the other 50% originating from urban and local (traffic, industrial) influences. Combination of this spatial information with emission registers reveal detailed information on the shares of the various sources; however, PM formation processes not included in the emission inventories as well as trans-boundary impacts are neglected in such analyses. Complementary information thus is provided by receptor models and chemical dispersion models, both showing significant importance of secondary aerosol formation and, especially in the eastern part of the country, transboundary intrusion.

Many source categories have been investigated in more detail and are presented in separate sections, as e.g. exhaust and non-exhaust traffic emissions and domestic wood combustion.

Keywords Back-trajectories, Chemical composition, Chemical transport models, Hot-spots, Lenschow, PM10, PMF, Urban air

---

Dedicated to Prof. Peter Bruckmann, Head of the Department "Air Quality" at the North Rhine-Westphalia State Agency for Nature, Environment and Consumer Protection (LANUV), on occasion of his 65th birthday.

U. Quass (✉), A.C. John, and T.A.J. Kuhlbusch
Institute of Energy and Environmental Technology (IUTA e.V.), Air Quality & Sustainable Nanotechnology Unit, Bliersheimer Str. 60, 47229 Duisburg, Germany
e-mail: quass@iuta.de

Contents

1 Introduction 196
2 Source Apportionment Approaches 197
   2.1 Spatial Increment Approach 197
   2.2 Lenschow Approach 198
   2.3 Mass Closure/Tracer Based Approaches 199
   2.4 Statistical Receptor Models 199
   2.5 Dispersion and Chemical Transport Models 200
   2.6 Back-Trajectory Modelling 201
3 Source Apportionment Studies Carried Out in Germany 202
   3.1 Overview 202
   3.2 General Source Apportionment Results for Germany 203
   3.3 Main Sources and Processes in Detail 207
4 Resume and Outlook 213
References 214

1 Introduction

Source apportionment of airborne dust is a technique allowing us to identify and quantify sources contributing significantly to ambient air pollution. Airborne particles have various environmental and health effects. They are, by Aeolian transport of particles, important nutrient sources but also cause negative health effects for humans especially after inhalation. Sources and meteorological conditions significantly influence ambient particle concentrations and hence human and environmental exposure. Exposure also varies significantly throughout Europe, and within each country. In case of Germany, a large fraction of the population lives in urban agglomerations which historically have been centred along trading routes, rivers and industrial (coal mining and steel production) areas. Thirteen of the EU's 50 largest cities by population are located in Germany. Hence identification and quantification of source contributions to ambient particles is of interest especially for urban areas with their high population density and proximity to significant sources of air pollution.

Although the air quality situation has been improving throughout the recent decades (cf. Fig. 1), these agglomeration areas still face air quality problems. Dust, with high levels of coarse particles, and sulphur dioxide led to frequent (winter) smog episodes in the middle of the twentieth century. At present, fine dust (PM10, PM2.5, particles with aerodynamic diameters <10 and <2.5 µm, respectively) and nitrogen dioxide pose the major issues. In cities, the latter pollutant is largely influenced by automotive traffic; a multitude of sources, however, is contributing to the PM levels. This complicates the development of mitigation actions and calls for deeper insights into the source–receptor relationships.

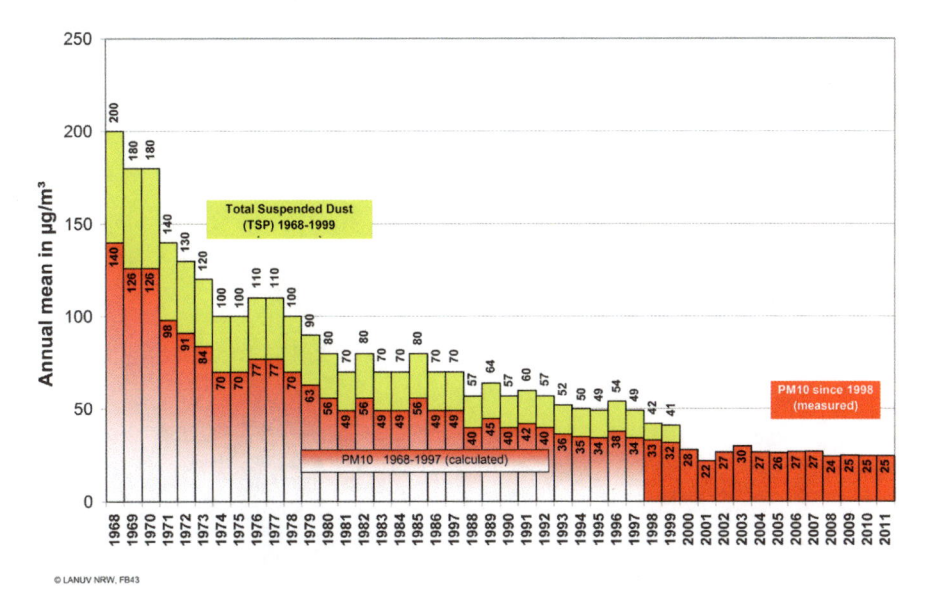

Fig. 1 Time trend of dust concentrations in the Rhine-Ruhr agglomeration (reprinted by courtesy of LANUV, 2012, Diagram provided by Ulrich Pfeffer, personal communication)

2 Source Apportionment Approaches

A number of different methods and models exists to apportion particle contributions to a specific source or source group. Unfortunately there is not a one perfect method to be employed. Different approaches are currently used depending on the type of data available and the preferences of the researchers. The various approaches vary with respect to the computational effort and to the degree of knowledge needed on the source-related emission rates and emission composition. Also, the outcomes vary between the different approaches. Nevertheless some comparison studies showed that major sources are identified by most methods and agree in their trends. Reviews can be found in, e.g. [1, 2].

2.1 Spatial Increment Approach

The most simple and widely used "spatial increment" approach compares concentration levels measured in different environments, assuming that the actual level at a given site is the sum of emissions released on regional, urban, and local scales (cf. Fig. 2). Hence, by calculation of the spatial increments (e.g. traffic-urban background, urban background-rural background) basic assessments of the shares of emissions from the different "source areas" can be obtained. This approach constitutes the first step within a source apportionment method first proposed by Lenschow et al. [3].

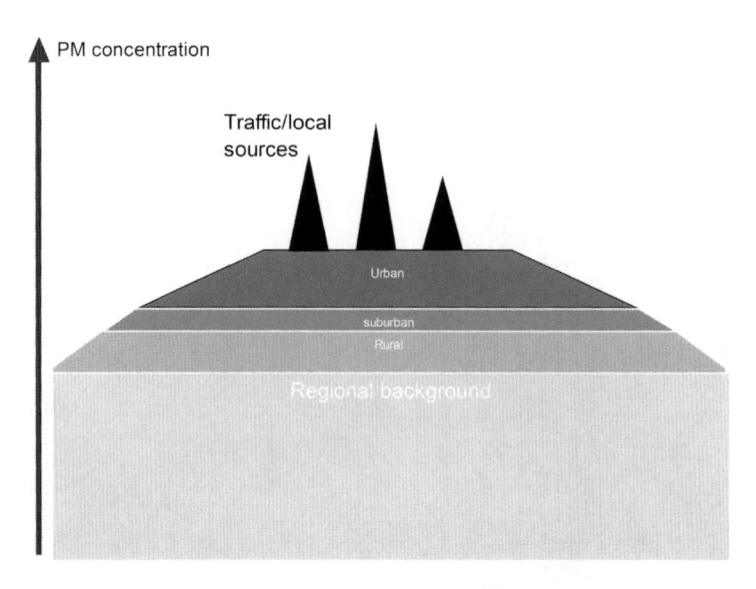

Fig. 2 Illustration of the "spatial increment" approach according to Lenschow et al. [3]

2.2 Lenschow Approach

The basic increment approach described before has further been extended [3] to include chemical composition data (for the major compounds) as well as emission inventories for the corresponding areas. Using this data, the measured compound concentrations are apportioned according to the relative shares of the corresponding emissions for each environment.

The Lenschow method has become particularly attractive for Germany's air quality management authorities (e.g. [4]) since it directly relates measured PM levels to source categories listed in official emission inventories. For example, measured regional levels for elemental carbon (EC) are distributed according to the shares of various EC emission sources registered in the national emission inventory. Similarly, the urban increments are related to the urban emission register. The local increment, e.g. measured at a traffic site, is considered to be exclusively caused by the local traffic emissions, if chemical composition does not indicate otherwise. Similar computation is performed for all other measured compounds. Clearly, the validity of the result depends on the accuracy and completeness of the emission registers. Moreover, for some source types the distribution depends on assumptions that may be considerably uncertain (e.g. tailpipe/non-tailpipe traffic emissions). Natural sources as well as contributions by trans-boundary long-range transport are not always included, which may cause overestimation of the relative contributions by national emission sources [5].

2.3 Mass Closure/Tracer Based Approaches

A further, from computational point of view, rather simple method is based on general information about emission source profiles which is used to assort the measured PM compounds into different source categories [6]. This method can be applied, for instance, easily and rather accurately in case of marine aerosols which mainly comprise sodium and chloride. Based on the measurements of one or both compounds the fraction of PM10 generated by sea spray emissions can be computed from the well-known sea water composition. Sodium can be considered as the more suitable tracer for inland sites [7] as the marine aerosol is frequently subject to an exchange of chloride for nitrate (reaction with gaseous nitric acid) during the transport. Similarly, contributions by mineral dust from soil erosion may be assessed measuring silicon and/or aluminium concentrations (e.g. [8]). Elemental carbon is the most widely used indicator compound for traffic exhaust emissions, while elements like Cu, Ba and Sb are suitable tracers for non-exhaust traffic emissions from brake and clutch wear [9]. Organic carbonaceous matter in many cases reflects both natural (secondary aerosols from natural VOC emissions) and anthropogenic (e.g. wood burning) sources. A major difficulty with tracer-based approaches is to find the correct relationships between the tracer compound and the total particle mass assigned to the source or process. The multiplication factors to be used may have significant spatial and temporal variability.

In addition, most chemical analyses used for the mass closure and tracer approaches are made from bulk samples collected on filters. Consequently, particles with different origin but similar chemical composition can hardly be distinguished. To cope with such overlaps methods based on electron microscopy coupled with X-ray spectroscopy have been developed [10, 11].

2.4 Statistical Receptor Models

The problem mentioned before has led to the development of statistical receptor models (Fig. 3) which nowadays are the most widely used tools for PM source apportionment. They can be applied even to a single site and need a time series of PM mass concentrations and corresponding chemical composition data. Depending on the method, these analyses are based only on the receptor data or additionally use information on the chemical composition from the relevant emission sources (emission profiles).

In Germany, as well as generally in Europe, multivariate methods (PCA and PMF) are the predominantly applied tools (see [1]). Their main advantage is that no information about emission sources is needed, and sources or processes so far not registered in an emission inventory may be detected. On the other hand, the analytical effort to be invested is considerably higher than for the previous methods since these models need "enough" data to disentangle the sources and source

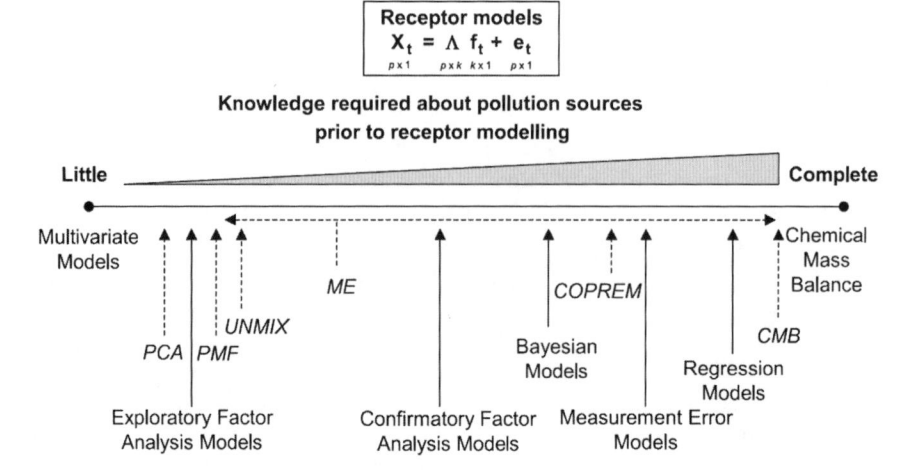

Fig. 3 Statistical receptor model categories and specific models (*italics/dashed arrows*) (reprinted from [1] as modified from [12] with permission from Elsevier)

contributions based on data intrinsic correlations between the different compounds. Trace compounds (heavy metals, polycyclic aromatic hydrocarbons, organic tracers like levoglucosan) in the obtained source profiles, that is a source group-specific chemical composition of PM, are important auxiliary indictors for source identification. Temporal variation of factors and tracer compounds (seasonal, weekday/week-end and diurnal) may provide additional information to verify the source assignments (e.g. [13]).

A further complication is the basic assumption of the statistical methods that source profiles neither change during air transport nor with time. Therefore they cannot be applied strictly to secondary aerosol constituents formed in the atmosphere by gas-to-particle conversion processes. Still, the secondary aerosol constituents tend to be grouped into one source group since they have a common "source", i.e. formation in air triggered by solar irradiation.

2.5 Dispersion and Chemical Transport Models

Small scale dispersion models are traditionally used in environmental assessment studies for new industrial plants to evaluate the add-on impact on air pollution levels. State-of-the-art Lagrange models calculate the trajectories of particles released by the source within three-dimensional wind fields which are generated from typical meteorological data including information on orography, surface roughness and obstacles (buildings). Although the computational effort is significant such models are a valuable complementary approach in a PM source apportionment, particularly addressing large point sources. The add-on concentrations delivered by dispersion models have to be completed by an assessment of regional background levels and evaluations/modelling of small scale local processes (e.g. [4]). Dispersion modelling on the urban scale is possible using special Gaussian plume models which may include canyon-plume-box approaches for street canyons (as implemented, e.g. in the IMMIS – air quality models [14, 15], a widely applied tool for air quality assessment in Germany)

Even higher computation capacities are needed for chemical transport models (CTMs) which extend the pollution transport by modules simulating the atmospheric multiphase chemistry and aerosol physics. CTMs are usually operated in a nested mode, with large grid cell sizes (e.g. 125 km × 125 km) on hemispheric or European scale and small grids for the region of interest (down to 1 km × 1 km). Several CTMs have been developed in Europe, in Germany mainly REM-Calgrid [16] and EURAD [17] have been used. Still, CTM models appear to have problems with accurate modelling of PM concentrations, particularly at high PM levels [18]. Nevertheless they may be used for assessing contribution of sources, in the most simple case by comparing model runs with the emissions of a target source category either switched on or off. Considerable computational resources are needed since these runs have to be done sequentially (or on parallel systems). To overcome this situation methods have been recently developed [19] to "tag" or "label" emitted compounds in such models and hence to detect the various contributions in the modelled PM concentrations. The latter approach may also be viewed as more realistic since it also includes all non-linear effects.

2.6 Back-Trajectory Modelling

Source apportionment studies are frequently complemented by air mass origin analyses to identify where high pollution levels or certain compounds of interest come from. For small to meso scales this can be done by, e.g. wind rose analysis, whereas long-range transport is better assessed by using back trajectories calculated by a suitable dispersion model [20].

Most of the CTM models mentioned in the previous section allow back-trajectory calculations; however, access is possible only through the research groups operating these models. Therefore, the most widely used tool is the open access NOAA Hysplit model [21].

In a basic approach for each measurement day several trajectories are calculated varying in the time of arrival and height above ground level for backward periods of usually 3 or 4 days. Such single-day calculations have been proven to be particularly useful in case of short-term dust events caused, e.g. by long-range dust intrusions from arid regions or wildfires. It should be noted that due to spatial resolution limits (typically grid size is 1° × 1°) back trajectories cannot be used for the identification of sources located close to the receptor site.

To get a temporally more representative picture of the distant regions associated with, e.g. episodes with high PM levels computation of trajectory data is needed for longer periods (up to several years) and multiple sites. The obtained trajectory cloud can be further processed by means of statistical methods like clustering [22] to identify the most relevant types of air mass transport to the sites or areas under consideration.

Moreover, advanced evaluation methods exist to apportion PM levels measured at the receptor site to the trajectory segments (e.g. the residence time weighted concentration method [23]) or to regions (grid cells) hit by trajectory ensembles (e.g. the potential source contribution function, PSCF [24, 25]). Software tools are available which facilitate such calculations and visualisations of the results [26].

3 Source Apportionment Studies Carried Out in Germany

3.1 Overview

Source apportionment studies have been carried out since the late 1990s in a number of the German Federal States (see Table 1 and Fig. 4). Source attributions were done by chemical analysis and tracer assignments in basic approaches, often extended by basic Lenschow analyses (spatial increment calculations); the full Lenschow approach including emission register assignments has been used less frequently. Also receptor modelling using multivariate factor analyses (either PCA or PMF) were applied several times, sometimes supported or complemented by results of dispersion or chemical transport models and back-trajectory analyses.

In Fig. 4 an overview of the "chemical composition", already grouped according to source processes, of PM10 in Germany is shown. The unaccounted PM mass in these mass closures is frequently around 10–20% which can reasonably be assigned to particle bound water and analytical uncertainties. Higher unaccounted mass fractions in most cases also comprise the mineral dust fraction since analyses of silicon and aluminium are scarcely done in routine. In some studies only the main ionic PM constituents were analysed, and the unaccounted mass then became the major fraction and comprises carbonaceous particles as well.

Apparently there is only a limited spatial variability in the overall chemical composition. This indicates that a large fraction of PM, in particular the secondary inorganic and organic aerosol, is formed and dispersed on a rather large spatial scale and can be found everywhere. The increase of PM concentrations in urban background compared to rural conditions thus can be (at least partly) explained by less favourable atmospheric dispersion conditions within the cities leading to accumulation of the particles advected from the regional surroundings. However, in case of influences by strong local sources e.g. high-trafficked street canyons, a clear shift in the chemical profile towards higher shares of related compounds (carbonaceous aerosols and metals) can be seen.

Table 1 Overview on PM source apportionment (SoAp) studies carried out in Germany

No.	Sites	Year(s)	SoAp-method	Reference
	tr, ub, rb	2008	Spatial increments	[27]
	Various	2011	Back trajectory clustering	[28]
	Not specified	2002	CTM modelling	[29]
1	tr, ub, surb, rb	2006	Increments	[30]
2	tr, ub, rb	1992; 1999	Lenschow	[3]
3	Various	1990–2000	Increments	[31]
4	tr, surb rb	2001–2002	Lenschow	[32]
5	tr, rb	2002–2003	Increments	[85]
6	tr, ub, rb	2001–2002	Tracer assignments, trajectories	[33]
7	tr, ub, surb	2003–2004	PCA, Lenschow	[45]
8	ub, surb, rb	1999; 2000	Mass closure, tracer	[6]
9	tr	1995–1998 1998–1999	Tracer assignments PMF	[34, 35]
10	tr, ub, rb	2006–2007	Mass closure, tracers	[36]
11	ub	2002/2003	PMF	[13]
12	ind	2005–2006	Single particle analysis, mass closure, tracer assignments	[86]
13	ub	2006/2007	PMF, cascade impactor data	[37]
14	ub, rb	2008/2009	PMF, mass closure	[7]
15	tr, ind, ub, rb	2001–2002	Lenschow	[38]
16	ub		Single particle analysis, mass closure, tracer assignments	[87]
17	tr, ub, rb	2007–2008	Lenschow, PMF	[39, 40]
18	ub	2006/2007, winter	PMF	[41]
19	tr	2006	Increments	[88]
20	Various	2004	Dispersion and resuspension models	[4]
21	tr, ub, rb	2006–2007	Mass closure, tracer assignment	[42]
22	tr, ub, rb	2007–2010	Mass closure, tracer assignment	[43]

Site indicators: *tr* traffic, *ub* urban background, *surb* sub-urban background, *rb* rural/regional background

3.2 General Source Apportionment Results for Germany

3.2.1 Spatial Analyses/Source Regions

The first three studies listed in Table 1 aimed at deriving source-related information for the whole of Germany. PM10 levels measured around and within nine urban agglomerations in Germany were evaluated by the increment approach for the years 2003–2005 [27, 44]. Typically 40–50% of the PM10 concentration measured at an urban traffic site can already be observed in the rural vicinity. Thus this fraction is likely caused by advection of particles stemming from regional sources including atmospheric transformation processes. The remaining fraction of 40–60% can be attributed to general urban sources and the traffic emissions close to the measurement location. The actual levels and shares at a given site depend on many factors,

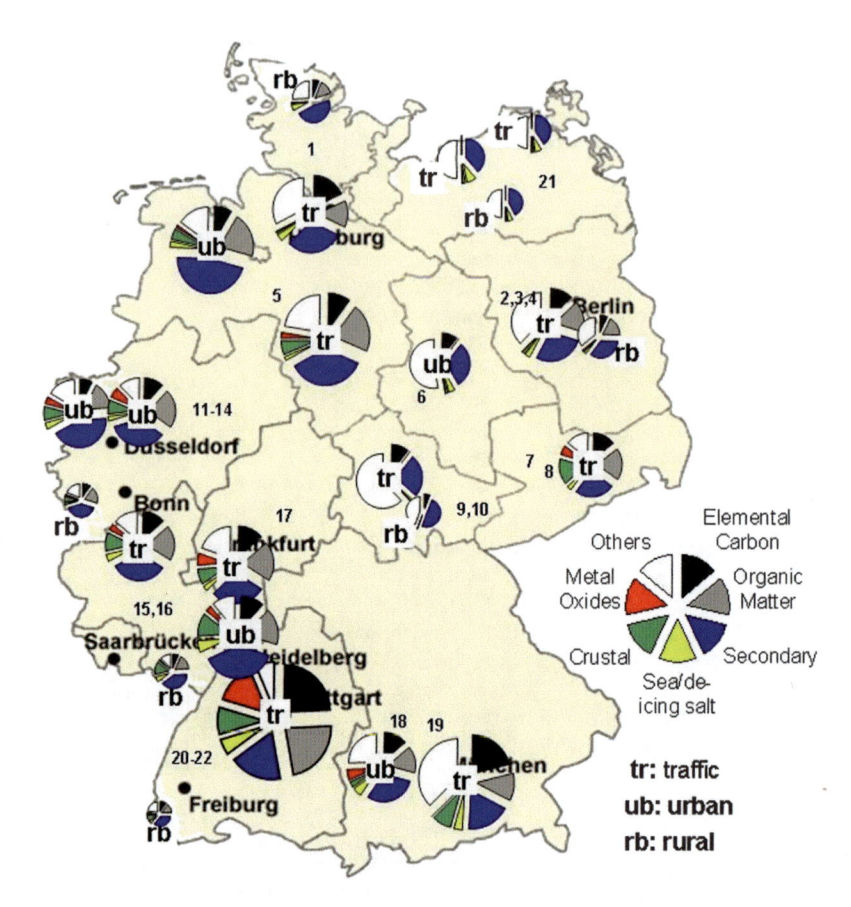

Fig. 4 Average chemical composition of PM10 at various sites; reference numbers in the map relate to numbers in first column of Table 1. Size of circles is proportional to annual averages of PM10 mass concentrations (maximum: 55 μg/m)

as indicated by the variance shown in Fig. 5. A look at weekly profiles (Fig. 6) shows the impact of anthropogenic activities by the increase of PM concentrations over the week and decrease during the weekend for nearly all site types. Interestingly, in regional and often in rural background sites minimum PM concentrations are observed on Mondays, probably due to lower secondary aerosol concentrations which reflect lowered emissions of gaseous precursor compounds (particularly NOx) over the weekend.

More recently, a statistical analysis of back trajectories [28] was carried out for the years 2005–2009 and five German regions. For each region, 12 trajectory clusters were identified (example shown in Fig. 7a) and evaluated with respect to their seasonality and correlation with high PM10 levels. While regional to local sources appear to be more important in the two westerly regions, a pronounced

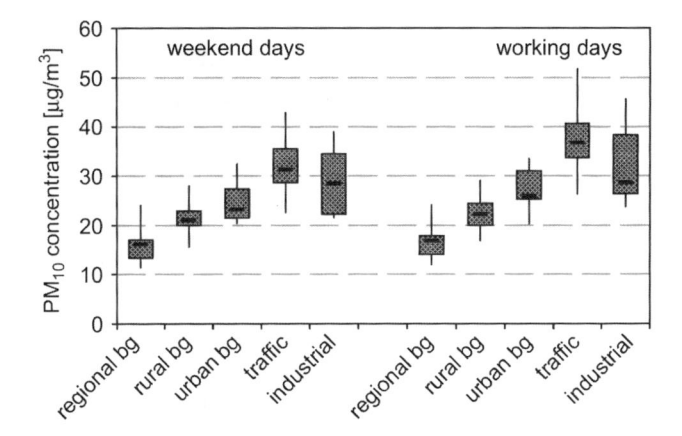

Fig. 5 Box–whisker plot with minimum/maximum (*vertical line*), 25th and 75th percentiles (*box*), and median (*horizontal line*) of PM10 concentrations (averaged for similar site types in the considered agglomerations) (from [44] adapted with permission from Informa Healthcare)

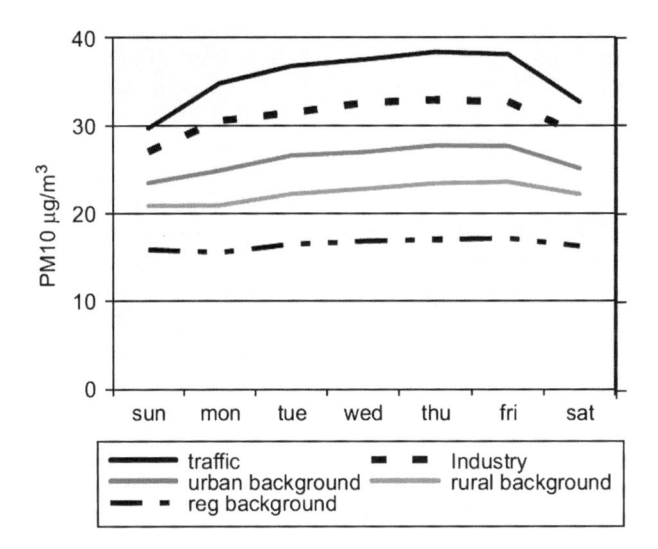

Fig. 6 Averaged PM10 weekly variation of different site types and comparison for weekend and work days (from [44] adapted with permission from Informa Healthcare)

influence of trans-boundary transport on the PM10 levels was evident for the three easterly regions.

The general incremental structure described before was confirmed as shown in Fig. 7b. For the eastern German region shown, episodes with high PM10 concentrations are associated with weather periods characterised by low wind speeds and predominantly south-easterly directions.

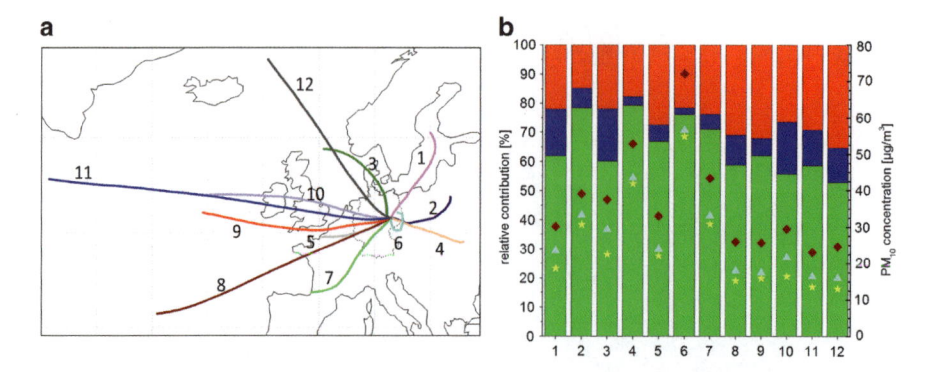

Fig. 7 (**a**) Trajectories representing the clusters for an eastern German region (**b**) relative contributions to PM10 levels (*bars*) and mean PM10 concentrations (*markers*) for rural (*green, stars*), urban (*blue, triangles*) and traffic (*red, diamonds*) sites per trajectory cluster (from [28] with permission from authors and publisher)

This result is in line with CTM model calculations of the contribution of inland and foreign emissions to PM levels in Germany [29]. According to that study, regional PM10 levels would decrease to less than 10 μg/m if only the German emission sources are considered, and below of ca. 20 μg/m in agglomeration areas (Fig. 8a). This corresponds (Fig. 8c) to ca. 40% and ca. 60%, respectively, of the PM10 levels modelled in a base run including all emissions. The complementary impact of non-German emissions on the PM level and their relative share to the base run levels are shown in Fig. 8b, d, respectively.

3.2.2 "Lenschow" Studies

The source apportionment approach according to the full approach proposed by Lenschow et al. in 2001 has not very often been carried out in Germany despite being a rather straightforward method. Following the first original analysis a more extended study based on comparisons of several traffic influenced, urban background and rural background sites in and around Berlin was carried out for the years 2001/2002 [32]. For a traffic site in Dresden a 1 year study on the chemical composition of various PM size fractions was made which included short comparative measurements with urban and rural sites. A Lenschow analysis was carried out based on the results of these campaigns [45]. A third study used chemical composition data obtained during the cold season in the region of Frankfurt [40]. The source apportionment results obtained for the hot-spot (traffic) sites are summarised in Table 2. A similar overall contribution of 40–50% by traffic emissions is apparent. For the other source categories more variable shares were obtained. This, on the one hand, reflects the different measurement periods (higher shares for domestic heating in the cold season campaign in Frankfurt a. M.), and, on the other hand, the different scope of regional and local emission registers.

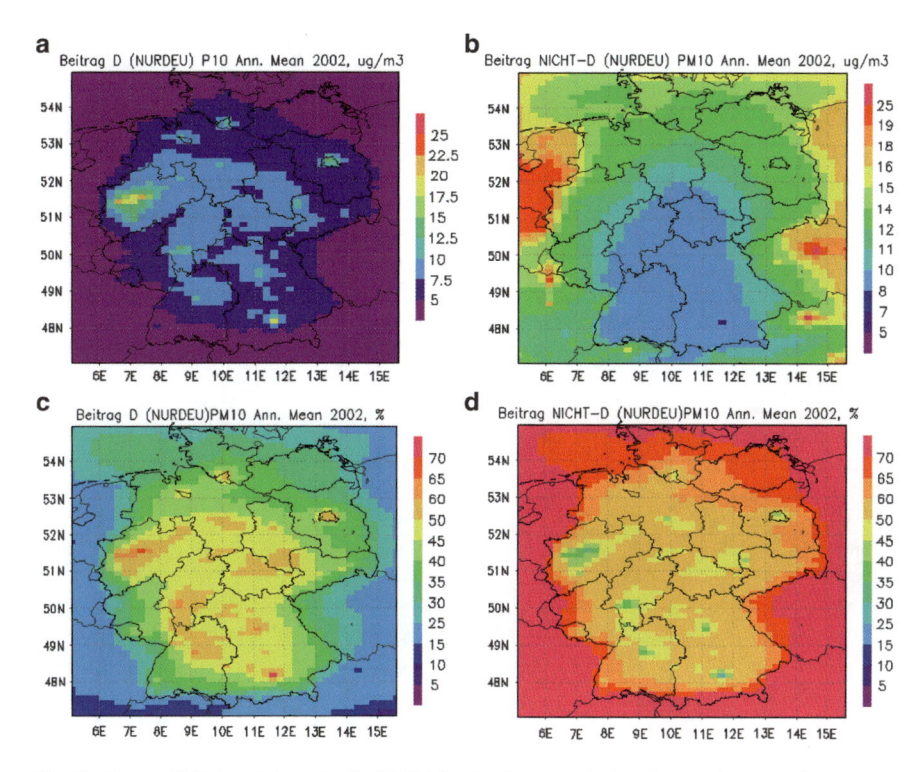

Fig. 8 *Upper*: CTM model results for PM10 levels. *Lower*: relative shares of sources in percentage compared to base run. (**a, c**) German sources on, foreign sources off; (**b, d**) German sources off, foreign sources on; from [29] with permission from publisher

3.3 Main Sources and Processes in Detail

3.3.1 Natural Sources

On the global scale, natural emissions of particulate matter from wildfires, volcanoes, sea spray, wind erosion and the biosphere are estimated to exceed by far the emissions by anthropogenic activities (Hainsch [31] quoting Warneck [47]). In densely populated and industrialised countries like Germany this relation changes and on average natural aerosols are thought to play a minor role compared to anthropogenic particle emissions. However, short-term events like volcano eruptions [48] and trans-boundary dust intrusions from arid areas (Fig. 9) have sporadically led to elevated PM concentrations in parts of Germany [49].

Besides such high PM events, more frequent natural contributions to PM mass concentrations may occur from biogenic particles [50] and from sea salt.

Biogenic particles which comprise primary (fungal spores, bacteria, viruses, plant debris) and secondary organic aerosol (SOA) from biogenic non-methane VOCs are part of the commonly measured organic carbon fraction. Model results [51] indicate a

Table 2 Results from source apportionment studies applying the "Lenschow approach"

Region	Berlin		Frankfurt	Dresden
Period of measurements	1 year 1998	1 year 2001–2002	Winter October 2006–April 2007	2 weeks winter/3 weeks summer 2004
Study	[3]	[46]	[40]	[45]
Mean PM10 concentration ($\mu g/m^3$)	52	33	33	27
	Source contributions (%)			
Traffic (exhaust)	33	28	27	
Traffic (non-exhaust)	17	21	23	44
Other mobile sources	3	3		
Aviation industry	21	15	17	25
Small enterprises (combustion)	1	9	2	
Solvent production and use	0	4		
Domestic heating	5	8	11	4
Private consumption	1		4	
Waste treatment	0	1		
Agriculture	3	4		28
Biogenic	11		13	
Other	5	7	4	

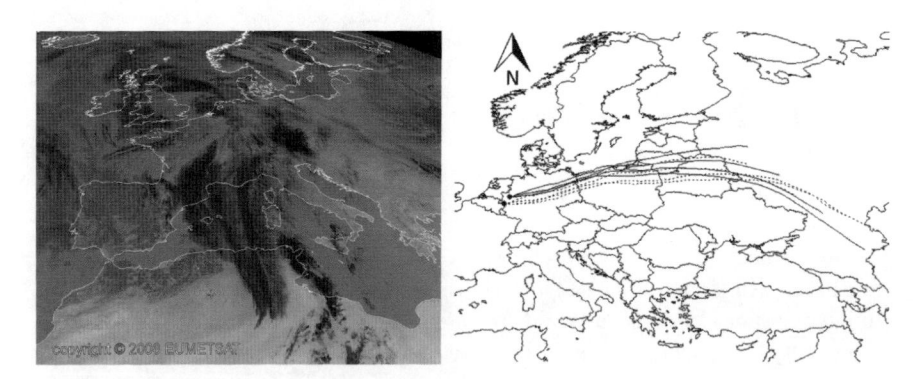

Fig. 9 *Left*: satellite photo (infrared, 10.8 μm) of the Sahara dust plume on 28 May 2008, 12 UTC. *Source*: EUMETSAT Meteosat [49]. *Right*: 72 h backward trajectories for 26 September 2008 showing air mass origin from arid regions north of the Caspian sea (NOAA Hysplit 4.9, data processed by TrajStat; reprinted from [7] with permission from Elsevier)

contribution of about 1–2 $\mu g/m^3$ from SOA, and less than 1 $\mu g/m^3$ from primary particles, mainly fungal spores [52]. However, quantification of the natural portion of SOA in urban environments by measurements remains a challenge and is a topic of current research [53, 54].

Table 3 Estimates for mineral dust concentrations for regional and urban background in northwest Germany, based on different calculation equations (reprinted from [7] with permission from Elsevier)

Model	Included compounds	Mean regional ($\mu g/m^3$)	Mean urban ($\mu g/m^3$)
PMF MinDust	All measured	1.7	4.3
Crustal sum	Mg, Ca, Si, Ti, K, nss-Na	1.2	3.1
Crustal sum + Fe	Mg, Ca, Si, Ti, K, bss-Na, Fe	1.4	4.7
IMPROVE	Al, Ca, Si, Ti, Fe (K)	1.5	5.3
RIVM	Al, Si	1.0	2.5
nss-Ca^{2+} Tracer	Ca, Na	0.7	2.1
Ca^{2+} Tracer	Ca	0.8	2.2
Al Tracer	Al	1.5	3.8

3.3.2 Sea Salt

The impact of sea salt, based on sodium measurements, on the PM concentrations has been evaluated in numerous studies throughout Europe and compared to model results [55]. Sea salt sodium concentrations clearly depend on the distance to coastal areas, with a concentration range from <0.1 $\mu g/m^3$ at continental sites and >3 $\mu g/m^3$ close to the shores. In Germany, higher sea salt fractions in PM are only observed during westerly wind directions and high wind speeds. High turbulence interconnected with the high wind speed also lead to fast dilution of emissions. Hence, the sea salt fraction is inversely correlated with PM mass concentration and only has minor effects on the exceedance of PM limit values [7].

3.3.3 Mineral Dust

Mineral dust (MD) is a ubiquitous PM component generated by multiple sources including natural wind erosion, agriculture and other anthropogenic activities (e.g. construction works, re-suspension by traffic). Its natural elemental composition consists of Si, Al, Ca, K, Ti, Fe, Na (non-sea-salt) and oxygen, as most of the compounds are found in form of their oxidic minerals.

A number of different calculation schemes have been developed for the assessment of the mineral dust PM fraction (Table 3). Some of these schemes are valid only for sites with negligible anthropogenic influence, since, e.g. the use of Fe as tracer may lead to an overestimation in environments influenced by traffic or steel industry. Si and Al are known as the most suitable tracer components [8] as they represent most of the MD mass and have hardly any anthropogenic source except fly ash from coal combustion [56]; however, both components are not commonly measured in German air quality networks. For regional and remote background situations also water soluble Ca was shown to be a suitable tracer [7, 57]. In an eastern German study the MD contribution was calculated from Ti values, a method which for other, more industrialised regions with titanium production or processing may lead to overestimation.

Contrary to southern Europe where Sahara dust intrusions occur more frequently particle size-resolved analyses show almost no mineral dust in the submicron particle fraction in central and northern Europe ($d_{ae} < 1$ µm) [7]. As mentioned before, the mineral dust fraction occasionally can become dominating in case of long-range dust intrusion events.

Table 3 also shows that an MD factor resolved by receptor modelling (PMF) reveals higher concentrations than tracer-based calculations. This may be due to the fraction of biogenic organic matter being present in soils unaccounted by tracer approaches.

3.3.4 Secondary Aerosols

Secondary aerosols comprise the inorganic ions sulphate, nitrate and ammonium, as well as a fraction of the organic matter (OM). The inorganic ions have their origin in emissions of the gaseous pollutants SO_2, NOx and NH_3, and as such are indicators for emissions from combustion processes and agricultural activities, respectively. In Germany only negligible contributions to these PM compounds are expected to stem from natural sources as, e.g. volcanic activities, wildfires and thunderstorm flashes. As indicated in Fig. 4, the inorganic secondary aerosol almost everywhere makes up a major fraction of the PM10 mass concentration. Their share typically rises disproportionately high with increasing PM levels [46], particularly in case of "PM episodes", which are driven by special meteorological conditions with a low atmospheric mixing [58].

Size-resolved analyses show that these compounds, being products of gas-to-particle conversion processes, are predominantly found in the fine particle fraction. This makes them highly susceptible to long-range transport.

A receptor model analysis in western Germany separated nitrate-rich from sulphate-rich secondary aerosols, with the latter being accompanied with vanadium and nickel [7]. Such factor composition pinpoints to heavy oil combustion sources which can be found, e.g. in oil refineries, off-shore platforms and overseas ships. In addition, trans-boundary pollution from eastern European countries is a significant source.

Nitrate-rich secondary aerosol, mainly composed of ammonium nitrate, on the other hand, is predominantly formed on medium spatial scale and hence may have its major origin within Germany. Large sources for NH_3, one of the aerosol precursor compounds, are located in the agricultural areas of north-western Germany ("Swine belt"). The other precursors, oxides of nitrogen, stem mainly from industrial and traffic-related combustion sources.

3.3.5 Traffic

In most urban regions in Germany traffic emissions are considered the major local PM source. In street canyons PM concentrations caused by traffic emissions may become equal to or higher than the urban background concentrations, hence leading

Table 4 Emission factors reported for non-exhaust traffic emissions

Country	Traffic situation	Brake wear (mg/km/veh.)	Resuspension (mg/km/veh.)	Reference
Germany	Motorway, free flow, 20% heavy duty	9^a	5^a	[44, 64]
Switzerland	Motorway, free flow, 15% heavy duty	3^a ($1.6^b/9^c$)	48^a ($28^b/160^c$)	[65]
	Urban street canyon	15^a ($8^b/81^c$)	27^a ($1^b/262^c$)	

[a]Total fleet
[b]Low duty vehicles
[c]Heavy duty vehicles

to frequent exceedances of the daily PM limit value. Even the annual PM limit cannot be attained at a few locations (e.g. Stuttgart, Fig. 4).

PM from traffic comprises compounds which originate from the combustion exhaust (primary and secondary carbonaceous aerosol, secondary nitrate) and from non-exhaust processes (brake, clutch and tyre wear and road dust re-suspension). As a "rule-of-thumb", the ratio of exhaust to non-exhaust contributions for the urban environment was found to be 60:40 in Germany [59]; however, the ratio is highly variable for different traffic situations, fleet compositions and road conditions. Based on the various studies carried out in Germany and neighbouring countries during the last decade generalised procedures to calculate non-exhaust PM emission factors for various traffic situations have been derived [60]. The range of emission factors, which will be imbedded into the handbook on traffic emission factors (HBEFA [61]), is 26–45 mg/km for passenger cars/light duty vehicles and 100–1,200 mg/km for heavy duty vehicles.

Still, considerable uncertainties exist with regard to the different kinds of non-exhaust PM emissions. In recent studies tyre abrasion, initially estimated to make up to 10% of total traffic PM10 contribution, has been shown with a new method for analysing specific rubber components to have much lower impact [62]. Accordingly, most tyre particles have aerodynamic diameters above 10 μm and their contribution to PM10 is 0.5 mass% at maximum.

Similarly, different results have been found for the impact of brake wear particles. While unequivocally a strong enrichment of brake-related chemical trace elements (Sb, Ba, Cu) is found at trafficked sites [63], the quantification of overall PM contribution from brake wear is associated with uncertainties. An even wider range of emission factors was found for re-suspension of road dust (cf. Table 4). It should be noted that re-suspension may be a strong source of PM during wintertime when de-icing salt is spread out. For a traffic site in southern Germany exceedance of the daily PM limit value could be tracked back to road salting in 12 of 43 cases [66].

3.3.6 Domestic Heating

Traditionally, hard coal and lignite were used in wide parts of Germany for house heating purposes. This practice has almost ceased today and natural gas or light fuel oil is used predominantly. This development has considerably contributed to better air quality in cities. However, wood burning has been steadily increasing in recent years, due to ever-rising costs of fossil fuels and policies favouring renewable energy supply as a countermeasure against global warming. Total PM emissions from domestic heating appliances have been investigated [67] and are estimated to meanwhile exceed those from traffic exhaust [68]. Consequently, urban areas are facing a new problem with particulate emissions originating from small single stoves and open fireplaces.

Dispersion model calculations for the city of Augsburg [69] revealed that the average additional PM10 load during the heating period is highest close to the sources and can go up to 3.5 $\mu g/m^3$. From a study based on measurements of tracer components (K, Levoglucosan, organic carbon) it was concluded that domestic heating may contribute up to 30% to the PM10 concentrations observed at traffic sites [66].

3.3.7 Industry

Industrial production is a major economic sector in Germany and so industry emissions were and still are important regarding air quality. However, since the 1960s mitigation actions based on national and EU regulations have led to considerable emission reductions, most prominently for (coarse) dust, soot and sulphur dioxide [89], toxic compounds like heavy metals [70] and dioxins [71].

The German Federal Environment Agency and a number of German Federal States have conducted dedicated emission measurement programs to evaluate the stack emissions of airborne particles and the shares of various PM fractions [72]. For about 70% of the more than 100 investigated facilities total dust emissions were mainly characterised by PM10 particles. Substantial coarse dust (>PM10) emissions were found in the cement and mineral processing industry, as well as agricultural installations (e.g. poultry stables). High shares of PM2.5 and PM1.0 were particularly found for thermal installations. These results contributed to the improvement of emission inventories and are essential as input for chemical transport models.

Industrial emissions may contribute significantly to the PM10 burden at selected receptor sites as was also shown using the Lenschow approach. To a large part this assessment is due to the assignment of measured secondary aerosol components to industrial emissions of sulphur dioxide and nitrogen oxides which are not produced locally. Statistical receptor models like PMF, on the other hand, are able to identify local industrial impacts which can be seen in elevated levels of trace compounds, but hardly attribute secondary aerosol compounds to any specific industrial source. For example, a PMF study was carried out for a receptor site located a few kilometres east of the Rhine river where one of the largest European steel plants and other metallurgical industries can be found. By wind direction analyses, a factor with high shares of iron and zinc could be attributed to these sources, contributing on average 7% to PM10 [13]. Further investigations close to the steel plant [73–75] clearly showed local influences (surplus burden 8.7 µg/m^3) and provided insights into the contribution of various production processes.

3.3.8 Agriculture

Particularly in the northern part of Germany, agricultural activities have a significant impact on the chemical composition of airborne particles. Agriculture is by far the major emission source for ammonia [76] which to a large part ends up in ammonium salts formed by reaction with sulphuric and nitrous acid. While ammonium sulphates are stable solids, ammonium nitrate is a semi-volatile compound exhibiting a pronounced seasonal variation with lowest values in summer [77]. Besides ammonia emissions agriculture also contributes to Aeolian soil dust suspension [78], carbonaceous particles from off-road vehicles and direct release of biogenic particles from stables [79].

4 Resume and Outlook

The source apportionment approaches described and exemplary results presented give insight into sources, source areas and processes influencing the PM levels in Germany. From the array of methods presented it is obvious that, depending on the source to be tackled and data available for interpretation, different source apportionment techniques have to be used. They do not always agree, but mostly show similar trends, as it was shown for the regional background contribution to hot-spot locations.

It is difficult to set up a general picture on the relevance of the various PM sources for Germany due to the high regional and local diversity. Almost all anthropogenic emission sources contribute to the background levels that can be found at rural sites far from roads, industry and urban areas. A significant part of this background originates from gaseous precursor compounds being emitted from combustion processes (industry, traffic, heating).

Clearly, in highly populated conurbations traffic emissions play an important role with average contributions of up to 60% of the total measured PM10 concentrations. Next to this, emissions from domestic heating, after steep decrease due to replacement of coal and lignite burning in the early 1990s, has again become relevant with the increasing number of biofuel (wood) burning facilities. In wintertime episodes this source may easily provide 30% of PM10 and trigger limit value exceedance. Last but not least, industrial activities may significantly affect the local to regional scale situation, especially in case of processes leading to fugitive emissions, as e.g. frequently observed in metallurgic production sites, stone and earth industries and harbour operations.

The results obtained form source apportionment provide valuable information to set up cost-effective and efficient mitigation options which are especially of relevance for agglomeration areas. Most prominently, low emission zones have been introduced in many cities and conurbations. The effectiveness of this measure has been proven by significant decrease of black carbon (soot) concentrations; however, it does not affect non-exhaust traffic emissions, thus being less impressing when looking at its effect on PM10 levels [80].

Since regulations and limit values target mass concentrations of PM10 and PM2.5 these metrics were in the focus of most source apportionment studies and mitigation actions. However, epidemiological and toxicological results indicate that ultra-fine aerosol particles (UFP) with low contribution to PM mass but high number concentrations might cause particular adverse health effects [81, 82]. Similarly, elemental carbon/black carbon concentration has been proposed as a target metric that exhibits stronger health effects than total PM mass [83]. Therefore, reduction of UFP and EC/BC levels might offer a high potential to improve the air quality in European cities with regard to negative health effects, probably being more efficient than only pure PM mass-related actions. In Germany, as in some other European countries, a (research oriented) monitoring network for ultrafine particles [84] has already been established.

References

1. Viana M, Kuhlbusch TAJ, Querol X, Alastuey A, Harrison RM, Hopke PK, Winiwarter W, Vallius M, Szidat S, Prévôt ASH, Hueglin C, Bloemen H, Wåhlin P, Vecchi R, Miranda AI, Kasper-Giebl A, Maenhaut W, Hitzenberger R (2008) Source apportionment of particulate matter in Europe: a review of methods and results. Aerosol Sci 39:827–849
2. Moussiopoulos N, Douros J, Tsegas G (eds) (2010) Evaluation of source apportionment methods. Deliverable D4.6 MEGAPOLI scientific report 10-22 MEGAPOLI-25-REP-2010-12, 53 pp
3. Lenschow P, Abraham H-J, Kutzner K, Lutz M, Preuß J-D, Reichenbächer W (2001) Some ideas about the sources of PM10. Atmos Environ 35(Suppl 1):23–33
4. UMEG (2005) Ursachenanalyse für PM10 im Rahmen der Erarbeitung von Luftreinhalte- und Aktionsplänen in Baden-Württemberg nach § 47 BImschG für das Jahr 2004. UMEG-Bericht Bericht-Nr.: 4-04/2005
5. Winiwarter W, Kuhlbusch TAJ, Viana M et al (2009) Quality considerations on European PM emission inventories. Atmos Environ 43:3819–3828
6. Herrmann H, Brüggemann E, Franck U et al (2006) A source study of PM in Saxony by size-segregated characterisation. J Atmos Chem 55:103–130
7. Beuck H, Quass U, Klemm O, Kuhlbusch TAJ (2011) Assessment of sea salt and mineral dust contributions to PM10 in NW Germany using tracer models and positive matrix factorization. Atmos Environ 45:5813–5821
8. Denier van der Gon H, Jozwicka M, Hendriks E et al (2010) Mineral dust as a component of particulate matter. BOP report 500099003 PBL Netherlands Environmental Assessment Agency
9. Gietl JK, Lawrence R, Thorpe AJ et al (2010) Identification of brake wear particles and derivation of a quantitative tracer for brake dust at a major road. Atmos Environ 44:141–146

10. Ebert M, Weinbruch S, Hoffmann P, Ortner HM (2004) The chemical characterization and complex refractive index of rural and urban influenced aerosols determined by individual particle analysis. Atmos Environ 38:6531–6545
11. Worringen A, Ebert M, Weinbruch S (2010) Entwicklung von Methoden zur qualitativen und quantitativen Quellzuordnung von Aerosolpartikeln an einem Verkehrs-Hotspot. Endbericht zum Forschungsvorhaben. TU Darmstadt, Institut für Angewandte Geowissenschaft, Fachgebiet Umweltmineralogie, Darmstadt (Juni 2010)
12. Schauer JJ, Lough GC, Shafer et al (2006) Characterization of metals emitted from motor vehicles. Report no. 133. Health Effects Institute
13. Quass U, Kuhlbusch TAJ, Koch M (2004) Identifizierung von Quellgruppen für die Feinstaubfraktion. IUTA report LP15/2004. Download from http://www.iuta.de/files/feinstaubnrw_2004_abschlussbericht.pdf, Accessed 5 Feb 2012
14. IVU (2003) Ursachenanalyse von Feinstaub (PM10)-Immissionen in Berlin unter Berücksichtigung von Messungen der Staubinhaltsstoffe am Stadtrand in der Innenstadt und in einer Straßenschlucht (Los 4 und 5 Ausbreitungsrechnung und Ursachenanalyse für die urbane und lokale Skala) IVU Umwelt GmbH Senate. Department of Urban Development, Berlin
15. IVU (2012) IMMIS air quality models. Web information http://www.ivu-umwelt.de/front_content.php?idcat=30&changelang=2&idart=23, Accessed 5 Feb 2012
16. TRUMF (2012) http://www.geo.fu-berlin.de/met/ag/trumf/RCG/RCG_Internet_description.pdf, Accessed 5 Feb 2012
17. RIU (2012) http://www.eurad.uni-koeln.de/index_e.html?/modell/eurad_descr_e.html, Accessed 5 Feb 2012
18. Stern R, Builtjes P, Schaap M et al (2008) A model inter-comparison study focussing on episodes with elevated PM10 concentrations. Atmos Environ 42:4567–4588
19. Koo B, Wilson GM, Morris RE, Dunker Yarwood G (2009) Comparison of source apportionment and sensitivity analysis in a particulate matter air quality model. Environ Sci Technol 43:6669–6675
20. Stohl A (1998) Computation accuracy and applications of trajectories – a review and bibliography. Atmos Environ 32:947–966
21. Draxler RR, Hess GD (1998) An overview of the HYSPLIT_4 modeling system of trajectories, dispersion, and deposition. Aust Meteorol Mag 47:295–308
22. Stohl A, Eckhardt S, Forster C et al (2002) A replacement for simple back trajectory calculations in the interpretation of atmospheric trace substance measurements. Atmos Environ 36:4635–4648
23. Stohl A (1996) Trajectory statistics – a new method to establish source-receptor relationships of air pollutants and its application to the transport of particulate sulfate in Europe. Atmos Environ 30:579–587
24. Ashbaugh LL, Malm WC, Zadeh WZ (1985) A residence time probability analysis of sulfur concentrations at grand canyon national park. Atmos Environ 19:1263–1270
25. Zeng Y, Hopke PK (1989) A study of the sources of acid precipitation in Ontario, Canada. Atmos Environ 23:1499–1509
26. Wang YQ, Zhang XY, Draxler RR. (2009) TrajStat: GIS-based software that uses various trajectory statistical analysis methods to identify potential sources from long-term air pollution measurement data. Environ Modell Softw 24:938. Download of product from http://www.meteothinker.com/TrajStatProduct.html, Accessed 5 Feb 2012
27. Quass U, Kuhlbusch TAJ, Hugo A et al (2007) German contribution to the EMEP TFMM assessment report on particulate matter. IUTA-Report Nr. LP 34/2007. Available at http://www.nilu.no/projects/ccc/tfmm/, Accessed 5 Feb 2012
28. Birmili W, Engler C (2011) Studie zur Charakterisierung und Quantifizierung der räumlichen Herkunft der PM10-Belastung an hoch belasteten Orten. Ausarbeitung für das Umweltbundesamt Fachgebiet II 4.2 Förderkennzeichen 31201283. www.env-it.de/luftdaten/download/public/docs/

pollutants/PM10/Studie_zur_Charakterisierung_und_Quantifizierung_der_raeumlichen_Herkunft. pdf. Download 4 Jan 2012
29. Stern R (2006) Der Beitrag des Ferntransports zu den PM10-und den NO_2-Konzentrationen in Deutschland unter besonderer Betrachtung der polnischen Emissionen: Eine Modellstudie. Final report to UBA R&D projects no. 204 42 202/03 and 202 43 270. www.umweltdaten.de/publikationen/fpdf-l/3712.pdf. Accessed 4 Jan 2012
30. IHU (2008) Untersuchung zur Zusammensetzung des Feinstaubs in Schleswig-Holstein und Hamburg. Institut für Hygiene und Umwelt Hamburg/Staatliches Umweltamt Itzehoe
31. Hainsch A (2003) Ursachenanalyse der PM10-Immission in urbanen Gebieten am Beispiel der Stadt Berlin. Dissertation. Technische Universität, Berlin
32. John AC, Kuhlbusch TAJ (2004) Ursachenanalyse von Feinstaub (PM10)-Immissionen in Berlin. Report to the Department on Urban Development of Berlin, IUTA-report LP 09/2004a
33. LAU (2010) Immissionsschutzbericht Sachsen-Anhalt 2010. Landesanstalt für Umweltschutz Sachsen-Anhalt. Download: http://www.mu.sachsen-anhalt.de/lau/luesa/reload.html?berichte-immissionsschutz.html. Accessed 10 Jan 2012
34. Cyrys J, Stolzel M, Heinrich J et al (2003) Elemental composition and sources of fine and ultrafine ambient particles in Erfurt, Germany. Sci Total Environ 305:143–156
35. Yue W, Stolzel M, Cyrys J (2008) Source apportionment of ambient fine particle size distribution using positive matrix factorization in Erfurt Germany. Sci Total Environ 398:133–144
36. TLUG (2008) Web presentation of Thüringer Landesanstalt für Umwelt und Geologie. http://www.tlug-jena.de/umweltdaten/umweltdaten2008/luft/luft_02.html. Accessed 14 Jan 2012
37. Gietl JK, Klemm O (2009) Source identification of size-segregated aerosol in Münster, Germany, by factor analysis. Aerosol Sci Technol 43:828–837
38. Kuhlbusch TAJ, John AC, Romazanowa et al (2003) Identifizierung von PM10-Emissionsquellen im Rahmen der Maßnahmenplanung zur Reduktion der PM10-Immissions belastung in Rheinland-Pfalz. Report to State Environmental Protection Agency of Rhineland-Palatinate IUTA-report LP 06/2003
39. John A, Quass U, Jacobi S et al (2008) Combined Lenschow and PMF source apportionment for the area Frankfurt/Main Germany. Presentation at European Aerosol Conference 2008, Thessaloniki
40. Quass U, John AC, Kuhlbusch TAJ (2009) Quellenzuordnung für Feinstaub in Hessen: Frankfurt/Main und Kleiner Feldberg. IUTA report LP 41/2007
41. Gu J, Pitz M, Schnelle-Kreis J et al (2011) Source apportionment of ambient particles: comparison of positive matrix factorization analysis applied to particle size distribution and chemical composition data. Atmos Environ 45:1849–1857
42. LUBW (2009) Untersuchung von massenrelevanten Inhaltsstoffen in Feinstaub PM10 an drei Messstationen in Baden-Württemberg in den Jahren 2006 und 2007. Dok-No. 72-02/2009. http://www.lubw.baden-wuerttemberg.de/servlet/is/207409/. Accessed 4 Jan 2012
43. Draheim T (2012) Personal communication on chemical composition data from state environment agency of Mecklenburg-Vorpommern
44. Kuhlbusch TAJ, John AC, Quass U (2009) Sources and source contributions to fine particles. Biomarkers 14(S1):23–28
45. Gerwig H (2004) Korngrößendifferenzierte Feinstaubbelastungin Straßennähe in Ballungsgebieten Sachsens. Materialien zur Luftreinhaltung. Sächsisches Landesamt für Umwelt und Geologie, Dresden
46. John AC, Quass U, Kuhlbusch TAJ (2004) Comparison study of the chemical composition of PM10 for days with high mass concentrations in three regions in Germany. Presentation at European Aerosol Conference 2004, Budapest
47. Warneck P (1999) Chemistry of the natural atmosphere. Academic, San Diego
48. Hasager CB, Birmili W, Pappalardo G et al (2010) Atmospheric implications of the volcanic eruptions of Eyjafjallajökull Iceland 2010. Atmos Chem Phys (special issue). http://www.atmos-chem-phys.net/special_issue212.html, Accessed 5 Feb 2012
49. Bruckmann P, Birmili W, Straub W et al (2008) An outbreak of Saharan dust causing high PM10 levels north of the Alps. Gefahrst Reinhalt Luft 68:490–498

50. Jaenicke R (2005) Abundance of cellular material and proteins in the atmosphere. Science 308:73
51. NatAir (2007) Improving and applying methods for the calculation of natural and biogenic emissions and assessment of impactsto the air quality. Final report of the FP6 project. Download from http://natair.ier.uni-stuttgart.de/NatAir_Final_Activity_Report.pdf. Accessed 6 Jan 2012
52. Womiloju TO, Miller JD, Mayer PM et al (2003) Methods to determine the biological composition of particulate matter collected from outdoor air. Atmos Environ 37:4335–4344
53. Lau APS, Lee AKY, Chan CK et al (2006) Ergosterol as a biomarker for the quantification of the fungal biomass in atmospheric aerosols. Atmos Environ 40:249–259
54. Wagener S, Langner M, Hansen U (2012) Spatial and seasonal variations of biogenic tracer compounds in ambient PM10 and PM1 samples in Berlin, Germany. Atmos Environ 47:33–42
55. Manders AMM, Schaap M, Querol X et al (2010) Sea salt concentrations across the European continent. Atmos Environ 44:2434–2442
56. Vester BP, Ebert M, Barnert EB et al (2007) Composition and mixing state of the urban background aerosol in the Rhein-main area (Germany). Atmos Environ 41:6102–6115
57. Putaud JP, Van Dingenen R, Dell'Acqua A et al (2004) Size-segregated aerosol mass closure and chemical composition in Monte cimone (I) during MINATROC. Atmos Chem Phys 4:889–902
58. Holst J, Mayer H, Holst T (2008) Effect of meteorological exchange conditions on PM10 concentration. Meteorol Z 17:273–282
59. Rauterberg-Wulff A (2000) Untersuchungen über die Bedeutung der Staubaufwirbelung für die PM10-Immission an einer Hauptverkehrsstraße. Im Auftrag der Senatsverwaltung für Stadtentwicklung, Umweltschutz und Technologie. TU Berlin (Januar 2000)
60. Schmidt W, Düring I, Lohmeyer A (2011) Einbindung des HBEFA 3.1 in das FIS Umwelt und Verkehr sowie Neufassung der Emissionsfaktoren für Aufwirbelung und Abrieb des Strassenverkehrs. Ingenieurbüro Lohmeyer, Karlsruhe. Download from http://www.umwelt.sachsen.de/umwelt/download/70675_09_10_Endfassung.pdf. Accessed 21 Jan 2012
61. UBA (2010) Handbuch Emissionsfaktoren des Straßenverkehrs, Version 3.1/Januar 2010. Dokumentation zur Version Deutschland erarbeitet durch INFRAS AG Bern/Schweiz in Zusammenarbeit mit IFEU Heidelberg. Hrsg.: Umweltbundesamt, Berlin. http://www.hbefa.net/d/start.html, Accessed 5 Feb 2012
62. Stein G, Wünstel E, Travnicek-Pagaimo W (2009) Reifenabrieb in Feinstaub - Bewertung auf Basis einer neu entwickelten Messmethode In: Reifen - Fahrwerk - Fahrbahn im Spannungsfeld von Sicherheit und Umwelt: VDI-Berichte 2086. VDI-Verlag, Düsseldorf
63. Bukowiecki N, Lienemann P, Hill M et al (2009) Real-world emission factors for antimony and other brake wear related trace elements: size-segregated values for light and heavy duty vehicles. Environ Sci Technol 43:8072–8078
64. Quass U, John A, Beyer M et al (2008) Ermittlung des Beitrages von Reifen-, Kupplungs-, Brems- und Fahrbahnabrieb an den PM10-Emissionen von Strassen. Strassenverkehrstechnik 5:304
65. Bukowiecki N, Gehrig R, Lienemann et al (2009) PM10-Emissionsfaktoren von Abriebspartikeln des Straßenverkehrs (APART). Forschungsauftrag ASTRA 2005/007. Bundesamt für Strassen (August 2009)
66. LUBW (2010) Bestimmung des Beitrags der Holzfeuerung zum PM10-Feinstaub. Dok.-No. 64–01/2010. http://www.lubw.baden-wuerttemberg.de/servlet/is/72944/. Accessed 29 Jan 2012
67. Ehrlich C, Noll G, Kalkoff W-D (2007) Determining PM-emission fractions (PM10, PM2.5, PM1.0) from small-scale combustion units and domestic stoves using different types of fuels including biofuels like wood pellets and energy grain. In: DustConf 2007, Maastricht. http://www.dustconf.com/CLIENT/DUSTCONF/UPLOAD/S4/EHRLICH_.PDF, Accessed 5 Feb 2012
68. UBA (2006) Hintergrundpapier: Die Nebenwirkungen der Behaglichkeit: Feinstaub aus Kamin und Holzofen. Umweltbundesamt, Dessau (9. März 2006)
69. Brandt C, Kunde R, Dobmeier B et al (2011) Ambient PM10 concentrations from wood combustion – emission modeling and dispersion calculation for the city area of Augsburg, Germany. Atmos Environ 45:3466–3474

70. Pacyna EG, Pacyna JM, Fudala J et al (2007) Current and future emissions of selected heavy metals to the atmosphere from anthropogenic sources in Europe. Atmos Environ 41:8557–8566
71. Quass U, Fermann M, Bröker G (2004) The European dioxin Air emission inventory project – final results. Chemosphere 54:1319–1327
72. Ehrlich C, Noll G, Kalkoff W-D et al (2007) PM10, PM2.5 and PM1.0 – emissions from industrial plants – results from measurement programmes in Germany. Atmos Environ 41:6236–6254
73. Kappert W, Bruckmann P, Gladtke D et al (2007) High population density and heavy industry: the challenge for air quality management in North Rhine Westphalia with special focus on the steel production at Duisburg. In: DustConf 2007, Maastricht. http://www.dustconf.org/CLIENT/DUSTCONF/UPLOAD/S2/KAPPERT_.PDF, Accessed 5 Feb 2012
74. Gladtke D, Volkhausen W, Bach B (2009) Estimating the contribution of industrial facilities to annual PM10 concentrations at industrially influenced sites. Atmos Environ 43:4655–4665
75. Bach B, Volkhausen W (2010) The influence of natural and man-made sources on PM10 and PM2.5 concentrations at industrially influenced sites. Gefahrst Reinhalt Luft 70:488–492
76. Monteny G-J (2007) Ammonia emissions in agriculture. Wageningen Academic Pub. ISBN: 908686029X and 9789086860296, 403 pp
77. Dämmgen U (2002) Fine particles and their constituents in Germany - results of denuder filter measurements. In: Hinz T, Rönnpagel B, Linke S (eds) Particulate matter in and from agriculture. Landbauforschung Völkenrode Sonderheft, p 235. http://literatur.vti.bund.de/digbib_extern/zi026867.pdf, Accessed 5 Feb 2012
78. Goossens D, Gross J, Spaan W (2001) Aeolian dust dynamics in agricultural land areas in lower Saxony, Germany. Earth Surf Process Landforms 26:701–720
79. Seedorf J (2004) An emission inventory of livestock-related bioaerosols for lower Saxony, Germany. Atmos Environ 38:6565–6581
80. Lutz M, Rauterberg-Wulff A (2010) Berlin's Low Emission Zone – top or flop? Results of an impact analysis after 2 years in force. In: 14th ETH conference on combustion generated particles, Zurich
81. Donaldson K, Stone V, Clouter A et al (2001) Ultrafine particles. Occup Environ Med 58:211–216
82. Ibald-Mulli A, Wichmann H-E, Kreyling W et al (2002) Epidemiological evidence on health effects of ultrafine particles. J Aerosol Med 15:189–201
83. Janssen NA, Hoek G, Simic-Lawson M, Fischer P, van Bree L, ten Brink H, Keuken M, Atkinson RW, Anderson HR, Brunekreef B, Cassee FR (2011) Black carbon as an additional indicator of the adverse health effects of airborne particles compared with PM10 and PM2.5. Environ Health Perspect 119:1691–1699
84. Birmili W, Weinhold K, Nordmann S et al (2009) Atmospheric aerosol measurements in the German ultrafine aerosol network (GUAN). Part 1. Soot and particle number size distributions. Gefahrst Reinhalt Luft 69:137–145
85. Lohmeyer A et al (2004) Aerosolbudget in einem landwirtschaftlich geprägten Gebiet in Niedersachsen (Atmospheric aerosols in an agriculture-related area of Lower Saxony, Germany). Report for Niedersächsisches Landesamt für Ökologie, 18 p, http://www.lohmeyer.de/literatur/Boesel_Kurzbericht12.02.2004.pdf
86. Ebert M and Weinbruch S (2007) Elektronenmikroskopische Einzelpartikelanalyse atmosphärischer Aerosolpartikel. Report to LANUV NRW
87. Vester BP, Ebert M, Barnert EB, Schneider J, Kandler K, Schutz L et al. (2007) Composition and mixing state of the urban background aerosol in the Rhein-Main area (Germany). Atmos Environ 41:6102–6115
88. Bavarian State Ministry of the Environment and Public Health (2010): Daten+Fakten+Ziele Feinstaub: diffuser Staub -klares Handeln. Download from http://www.bestellen.bayern.de (accession 20.04.2012)
89. Bruckmann P (2010) From Smog to Blue Skies-Remaining Challenges. Conference "Air Quality Management in European Regions–Challenges and success stories" Essen 2010

Air Quality in Urban Environments in the Eastern Mediterranean

A. Karanasiou and N. Mihalopoulos

Abstract The Eastern Mediterranean Basin, EMB, is characterised as an air pollution hotspot, located at a crossroad of air masses coming from Europe, Asia and Africa. The Eastern Mediterranean region is subject to several inputs of natural and anthropogenic pollutants that are generated from numerous regional and local sources. Pollution in the area results from industrial and traffic sources and domestic heating mostly from Europe, Balkans and the Black Sea. In addition dust storms coming mainly from the Sahara desert and to a lesser extent from the Middle East transport high quantities of mineral aerosols which increase significantly the levels of particulate matter. Forest fires and agricultural burning emissions are also affecting the area during the dry season. Moreover, marine aerosols and ship emissions originated from the highly busy shipping routes of the Mediterranean Sea are considered to be an important contributor to the EMB aerosol burden. Finally, other specific features of the Mediterranean Basin such as the high radiation intensity all year long and the high temperature significantly enhance the formation of secondary aerosols.

Keywords Aerosol, Emission sources, Greece, Mediterranean Basin, Particulate matter, Road dust, Sahara dust, Secondary aerosol, Ship emissions

Contents

1 Introduction 220
2 Exceptional Features of the Eastern Mediterranean Basin 221
3 Monitoring Networks in Eastern Mediterranean Basin 221

A. Karanasiou (✉)
Institute of Environmental Assessment and Water Research (IDAEA-CSIC), Barcelona, Spain
e-mail: angeliki.karanasiou@idaea.csic.es

N. Mihalopoulos
Environmental Chemical Processes Laboratory, Department of Chemistry, University of Crete, Voutes Campus, P.O. Box 2208, 71003 Heraklion, Greece

4	Particulate Matter in the Eastern Mediterranean Basin	222
5	Source Contribution in the Eastern Mediterranean Basin	224
	5.1 Sahara Dust	226
	5.2 Road Dust Resuspension	227
	5.3 Forest Fires	228
	5.4 Shipping Emissions	229
	5.5 Secondary Aerosol	230
6	Concluding Remarks	234
References		235

1 Introduction

During the last decades, the Mediterranean basin has experienced a rapid growth in urbanization, vehicle number and use and industrialisation with this being reflected in pollutant emissions to the atmosphere. The air quality limits established by the EU to protect human health are often exceeded. Thus, urban areas have received special attention for their high air pollutant concentrations and the associated degradation of the air quality and public health [1].

The Mediterranean region is divided into three basins. The Western Basin extends from the Iberian Peninsula at the west to the Apennine mountain chain at the east, and from southern France to the northern African coast in the north–south direction. The Central Basin includes the watershed of the Adriatic Sea from the Apennines to the Balkans, and at the south includes the Ionic Sea from the east coast of Tunisia, south of Sicily, across the Libyan Gulf, to an imaginary line extending north–south from the Balkan Peninsula to the Libyan coast. The Eastern Basin extends west from this line to the Israeli and Lebanon coasts and includes the Aegean Sea [2]. The European territory of the Eastern Mediterranean includes two large urban centres: the Greater Athens area (GAA) (>4 million) and the Greater Thessaloniki area (GTA) (>1.5 million) in Greece.

The Eastern Mediterranean Basin, EMB, is characterised as an air pollution hotspot, located at a crossroad of air masses coming from Europe, Asia and Africa. Many different types of aerosols can be encountered within the basin; desert dust transported mainly from the Sahara desert polluted aerosols emitted by urban and industrial activities of continental and Eastern Europe, biomass burning aerosols often produced by seasonal forest fires, marine aerosols and ship emissions originated from the highly busy shipping routes of the Mediterranean Sea. In addition the general climatic conditions in the Eastern Mediterranean such as the low precipitation rate, the high radiation intensity and the high temperature favour the accumulation of air pollutants and the formation of secondary aerosols.

In this chapter the air quality of the EMB is examined in terms of particulate pollution while the characteristic emissions of the basin are explored to further understand the source categories affecting the region.

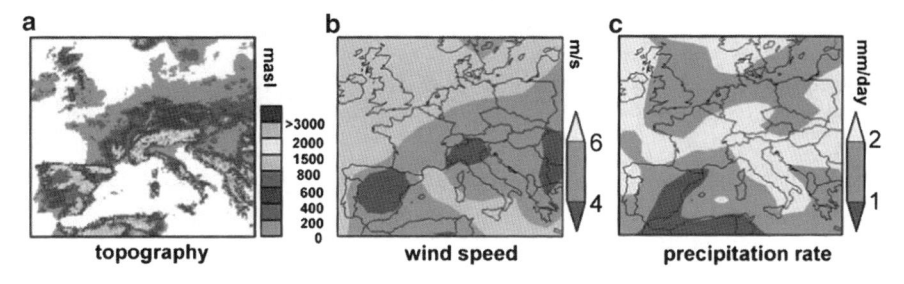

Fig. 1 Topography and mean (1995–2005) values of precipitation rate and 'synoptic component' of wind speed over Europe [3]

2 Exceptional Features of the Eastern Mediterranean Basin

Features such as topography and meteorology play an important role in the transport, dispersion, diffusion and levels of air pollutants. In northern and western regions of Europe (e.g. UK, Germany) the terrain is flat or gently undulating, while the frequent passages of cold fronts and depressions result in high precipitation rates. In contrast, the Mediterranean basin is characterised by a complex topography (Fig. 1), and when coupled with the characteristic synoptic scale patterns, it produces low mean wind speeds (i.e. 'synoptic component' of wind) that hinder the renovation of air masses and favour the accumulation of air pollutants. Because rain is the main mechanism for atmospheric aerosol removal, this has an important consequence: the low precipitation rates of the EMB result in high residence time of airborne particles and consequently high background particulate matter levels. These low rainfall rates coupled with the frequent weak advections of air masses and the frequent air masses recirculation episodes produce regional pollution events [3, 4].

Due to its position the Eastern Mediterranean region is subject to several inputs of natural and anthropogenic pollutants that are generated from several regional and local sources. Dust storms coming mainly from the Sahara desert and to a lesser extent from the Middle East transport high quantities of mineral aerosols which increase significantly the levels of particulate matter [5–7]. In addition local sources (traffic) and long-range transport of air masses from central Europe can cause severe pollution episodes [8]. Moreover, marine aerosols (sea spray) are considered to be an important contributor to the eastern Mediterranean aerosol burden [9]. Finally, other specific features of the Mediterranean Basin such as the high radiation intensity all year long and the high temperature enhance the formation of secondary aerosols [10, 11].

3 Monitoring Networks in Eastern Mediterranean Basin

The national networks have established several monitoring sites conducting measurements of traditional air pollutants including particulate matter. In Greece the Ministry of the Environment has established 15 monitoring stations in the

Greater Athens Area and 8 stations in the Greater Thessaloniki Area covering traffic sites, urban and suburban areas and also industrial areas. In addition other monitoring sites in smaller urban centres (Patras, Volos, Heraklio) have been set. In Cyprus air quality is accessed through a network of nine monitoring stations covering urban, industrial and rural sites.

Except from these national network other remote monitoring stations have been established by EMEP (European Monitoring and Evaluation Programme) at the sites of Finokalia (Grete island, Greece), Aliartos (Central Greece) and Ayia Marina (Cyprus), while a monitoring station of the GAW (Global Atmosphere Watch) programme of WHO (World Health Organization) has been established at Athens suburban area (DEM, Agia Paraskevi). These stations provide continuous measurements of gaseous pollutants (O_3, CO, NO_x) and also measurements of particle properties (optical properties, chemical composition, mass and mass size distribution).

The monitoring network in EMB compared to those established in Central and North Europe or even with the Western Mediterranean Basin (e.g.: Spain) is rather insufficient. Particularly there is a lack of data on continuous or long-term monitoring of the chemical composition of particulate matter [12]. For some substances (e.g. carbonaceous aerosol), the variability is expected to be much larger than can be resolved by integrating the available measurements and the research studies need to be supported by assessment of the local scale variability. In order to understand the temporal evolution (trends) there is also a particular need for aerosol measurements at additional sites with little influence from local and regional emission sources.

4 Particulate Matter in the Eastern Mediterranean Basin

A number of studies (e.g.: [13–22]) have investigated PM concentration levels and their temporal trends in urban sites of Athens. Similarly in the city of Thessaloniki, PM pollution has been the subject of many studies since the early 1990s (e.g., [23–26]). All these studies have reached the conclusion that urban centres in Greece (Athens and Thessaloniki) exhibit high PM concentrations, compared to other European and US cities of the same size. In both cities, all urban stations exceeded the EU annual and 24-h limits several times during the last decade. The long-term monitoring of PM concentrations in the Athens area [14, 16, 27, 28] registered the occurrence of a significant number of PM_{10} exceedances of the limits established by the EU legislation and pointed out the need for abatement strategies.

However, the analysis of the PM_{10} and $PM_{2.5}$ long-term measurements conducted in an urban background site of Athens reveals that despite the great variability of the mass concentration values a descending trend exists. Figure 2 summarises the trend analysis of the 24 h PM_{10} and $PM_{2.5}$ mass concentrations at the urban background station (Agia Paraskevi) within the GAA for the period 2001–2010. The data were obtained from the public air quality database of

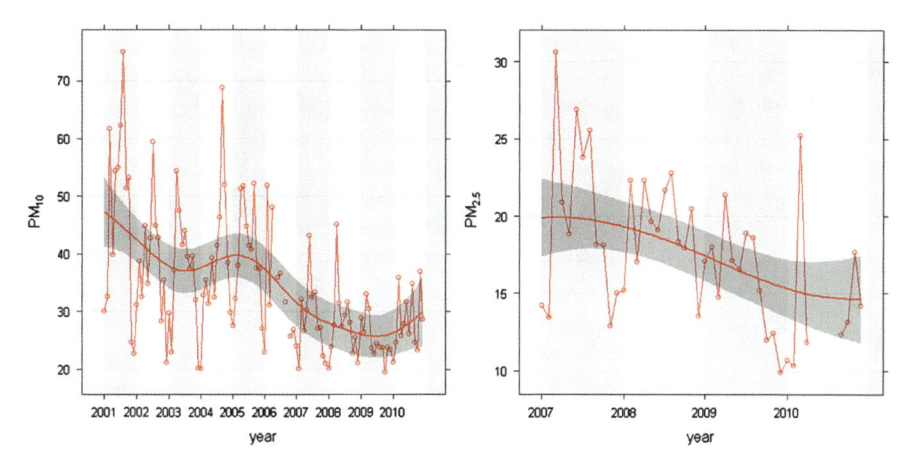

Fig. 2 Trend analysis of PM_{10} during 2001–2010 and $PM_{2.5}$ during 2007–2010 in Athens urban background station Agia Paraskevi

the European Environment Agency EEA (http://www.eea.europa.eu/themes/air/airbase). The decreasing trend observed could be attributed to the effectiveness of the vehicular emission control strategies that have been implemented in the metropolitan area of Athens from the early 1980s. In Athens restrictive measures have been in place since 1983, with odd- and even-registration plate vehicles being banned from the city centre on alternate days, and a total ban enforced when emergency levels of pollution are reached. Although these measures were beneficial at the beginning, the increase in the overall number of private cars offset any positive effects of the strategy. Thus, more permanent measures were introduced, including the elimination of older highly polluted vehicles from the city centre. A retirement plan for old vehicles was introduced in 1991 in an effort to renew the motor vehicles fleet with new cars equipped with catalytic converters. Other major measures were implemented in the early 2000s where the completion of the major infrastructural works in Athens such as the subway, the peripheral roads, the introduction of natural gas and the replacement of old public buses are thought to have improved to a considerable extent the air quality in the area. These later measures were not pollution-responsive, but permanent measures, in order to try and reduce pollution levels on a long-term basis. Another explanation for the reduction in PM levels could be the decline in fuels use as a result of economic crisis, and not only due to the drastic effect of mitigation measures [29].

For the urban area of Thessaloniki the average PM_{10} concentrations were 58 μg m^{-3} during 2001–2010 considerably higher than the annual EU limit value. Interestingly, the PM_{10} concentrations at the urban stations of Thessaloniki are much higher compared to those at the urban stations in Athens. An explanation for this observation is that apart from heavy traffic, the city of Thessaloniki also suffers from intensive industrial activity. It is worth noting that 20% of the industrial activity of the country is located in the area. Nevertheless, despite the high values,

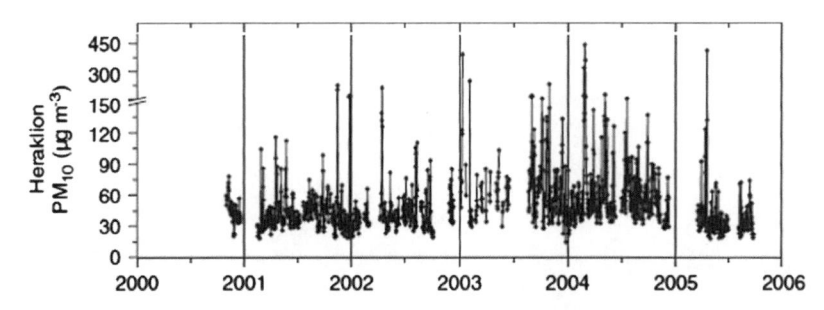

Fig. 3 PM_{10} measurements at Heraklion during the period November 2001–September 2005 [31]

an overall decrease of the PM_{10} annual average concentrations is observed at the urban monitoring stations in Thessaloniki from 2001 to 2010 [30]. Again for Thessaloniki urban area could be argued that the employed abatement strategies combined with the financial crisis led to the reduction of particulate matter.

Results from PM studies conducted in other urban areas in Greece conclude that PM levels have similar values with moderately polluted sites worldwide [31, 32]. Gerasopoulos et al. [31] report on the concentration levels of PM_{10} in the urban area of Heraklion (Crete Island, Greece) during 2001–2005, Fig. 3. In this urban site the average value of PM_{10} is 51 μg/m^3 while after 2003 the exceedances of the established limit values were more frequent with this phenomenon being attributed to the heat wave that influenced central Europe in summer 2003 and to the frequent Saharan dust intrusions.

Partial information is available concerning the aerosol composition over the Eastern Mediterranean since most of the studies conducted in the region report only on specific compounds (e.g: inorganic ions, metals, alkanes, PAHs) and do not provide the complete characterisation of aerosols. In urban sites of Greece water-soluble inorganic ions, mainly sulphate, dominate the PM mass and might contribute up to 60% of the PM_{10} mass followed by mineral dust components and carbonaceous matter [17, 21, 31, 33, 34]. The high sulphate loadings in the EMB are mostly attributed to long-range transport of pollution sources. In fact a 4-year aerosol study carried out at Thessaloniki by means of a Raman lidar indicated higher aerosol optical depth values and backscatter ratio mostly corresponding to air masses originating from the northeast Balkans and Eastern Europe [35].

5 Source Contribution in the Eastern Mediterranean Basin

Table 1 lists a number of studies carried out in urban sites of the Eastern Mediterranean Basin that have used multivariate statistical methods to quantify the mass contribution of sources of particulate matter. In most of these studies, four or five major source categories have been detected. These categories include road/soil dust, traffic emissions, marine aerosol, fuel oil combustion emissions, biomass

Table 1 Mean source contributions (%) to the ambient PM mass in urban areas of Greece (Source: modified table from [66])

Site/PM fraction, study	Road/ soil dust	Traffic	Fuel oil	Biomass burning	Marine aerosol	Secondary aerosol/ unidentified
Thessaloniki/TSP, Samara et al. [36]	7–11	4–5	25–33			54–66
Thessaloniki/fine, Manoli et al. [23]	28	38	14			20
Thessaloniki/coarse, Manoli et al. [23]	57	9	26			8
Thessaloniki/PM$_{10}$, Samara et al. [37]	18–22	45–65	10–35			
Athens/fine, Karanasiou et al. [17]	20	27	12	15	19	7
Athens/coarse, Karanasiou et al. [17]	54	8			16	22
Rhodes Island/PM$_{10}$, Argyropoulos et al. [38]	10–22	32–41		9–12	3–10	17–20

burning and secondary aerosol. As in most urban areas worldwide, traffic is the predominant source especially in the fine fraction contributing up to 65% in PM$_{10}$. Additionally, in all studies there is a significant contribution of road/soil dust originating mainly from deposited dust arising from brake, tyre and road abrasion, construction/demolition, and in minor proportion from windblown soil or regional dust. Nevertheless the contribution of the Sahara dust is not quantified in these studies but is usually included in the mineral dust factor. Not surprisingly, the effect of this mineral source is more important in the PM$_{10}$ and coarse fractions compared to fine particles, although the mineral source was also detected in the finer fractions. Fuel oil combustion during industrial activities has high contribution mainly in Thessaloniki industrial area. Marine aerosol has higher contribution in fine fraction (19%) than in coarse particles. For this source Karanasiou et al. [17] point out the influence of shipping emissions in the urban area of Athens. Biomass burning emissions have rather moderate contribution that does not exceed 15% in fine aerosol fraction. All studies mention that this source probably includes the emissions from forest fires that frequently affect the region mainly during dry periods. Finally as expected, secondary aerosol is also an important source identified in almost all studies due to the distinctive climatological characteristics of the region. However its mass contribution varies significantly probably due to the fact that other sources such as traffic might include a significant proportion of the secondary aerosol.

These characteristic emissions of the Eastern Mediterranean are further explored to understand the source categories affecting the region.

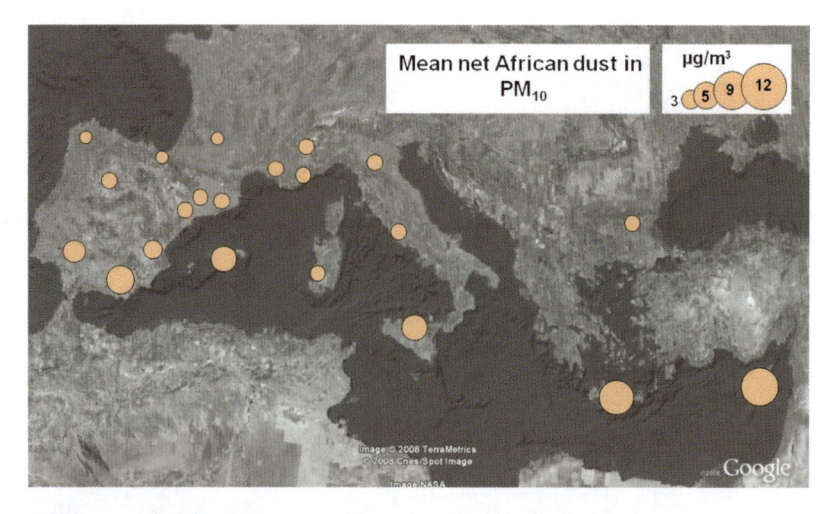

Fig. 4 Mean annual African dust contribution (in $\mu g/m^3$) to PM_{10} levels for the Mediterranean Basin during 2001–2007 at EMEP sites. Source: Querol et al. [5, 6]

5.1 Sahara Dust

The Mediterranean region due to its position is frequently affected by long-range transport of desert dust mainly from the Western and Central Sahara, the most important dust-raising area worldwide. Dust storms moving from Sahara desert to Eastern Mediterranean are more frequent in late spring, summer, and early autumn periods [39]. The concentration of desert dust particles is expected to decrease as moving from the emission source due to particle deposition mechanisms; therefore, the contribution of dust to PM_{10} concentration might vary significantly depending on the location. Thus, the PM_{10} African dust levels have a clear increasing trend from the north to south and western to eastern of the Mediterranean Basin (Fig. 4). Overall the Eastern Mediterranean Basin suffers from more intensive African dust intrusions compared to the Western Mediterranean Basin [5, 6, 69].

These dust outbreaks may greatly increase the ambient air levels of PM recorded in air quality monitoring networks. Indeed in Athens during the period 2001–2005 for approximately 50% of the days with daily PM_{10} exceedance were recorded intrusions of desert dust having either low or high mass contribution [40, 41]. Similarly, several major Saharan dust events are identified in the region of Crete and Cyprus where the PM_{10} limit value exceedances are regularly coinciding with the transport of air masses from the Sahara desert [5, 6, 31, 67].

The air masses that reach Greece leading to the most severe dust events usually originate from Western Sahara region. The spatial and temporal evolution of a Saharan dust outbreak during 30–31 August 2003 has been studied in detail using coordinated measurements by three lidar systems over Greece located at Thessaloniki, Athens and Finokalia, Crete. Although the air masses over the studied areas had a different route and thus different spatial evolution, the dust particles originated from the same source located in the Western Saharan region. However, the Sahara dust particles are usually strongly diluted by mixing with marine and urban aerosols as they are moving from the Sahara desert towards continental Greece [42].

The transported desert dust particles normally lie in the coarse size fraction and thus, the $PM_{2.5}/PM_{10}$ ratio is relatively low (<0.4) differing significantly from the typical values in polluted urban areas of the Eastern Mediterranean Basin [16]. During these dust outbreaks the contribution of mineral dust to PM_{10} mass concentrations might reach 60–80%, while the average mineral dust contribution to PM_{10} ranges from 13% to 19% [31, 43].

The mineralogical composition of Sahara dust particles shows the predominance of aluminosilicates (clays). Illite is also present in many cases while quartz particles are rare. Scanning Electron Microscopy (SEM) results on dust composition transported over different regions in the Mediterranean Basin have shown that Al-rich clay minerals such as illite and kaolinite are very common in PM_{10} for Cyprus and dominant for Crete. Dust particles are also very rich in calcium which is distributed between calcite, dolomite and sulphates and Ca–Si particles (e.g. smectites) whereas iron oxides are often detected [43].

5.2 Road Dust Resuspension

As previously mentioned, in the big urban centres of Eastern Mediterranean Basin the concentrations of PM_{10} are frequently above the threshold limit values. However, the quantification of vehicles exhaust emissions suggests that solely the latter cannot possibly explain the observed high concentrations.

The PM generated by traffic can be divided into two categories according to its mode of formation. Fuel combustion is the primary mechanism by which particles are formed; however, there are a number of other processes, involving mechanical abrasion and corrosion, which can also result in PM being released directly to the atmosphere. These processes include: tyre wear, brake wear, clutch wear and road surface abrasion. The abrasion and corrosion processes can also lead to the deposition of particles on the road surface. The material which collects on the road surface, referred to as 'road dust', may also contain exhaust particles, and material from a range of sources that are not related to road transport (e.g. mineral and biogenic material, and material from industrial/domestic activity). Road dust may subsequently be suspended or resuspended in the atmosphere as a result of tyre shear, vehicle-generated turbulence, and the action of the wind. In central and northern Western Europe, frequent precipitation (causing dust washout and immobilisation) and lower dust deposition reduce the amount of road dust available for resuspension but this does not occur that frequent in the Mediterranean Basin. Consequently the relevance of the resuspended road dust is estimated to be high for the Mediterranean region.

Research studies on the quantification of the road dust contribution on ambient PM levels are scarce for Eastern Mediterranean countries. In addition, emission inventories for the region hardly include these non-exhaust PM_{10} emissions (tirewear, break-lining and road abrasion).

Only a few source apportionment studies report on the contribution of road dust by means of receptor models. In Athens urban area road dust might contribute up to 50% to coarse fraction and up to 20% to fine particle mass [17]. The road dust contribution of 28% to PM_3 and 57% to PM_{10} was found in the central area of Thessaloniki while for the greater area of Thessaloniki a road dust average contribution to PM_{10} up to 22% was calculated [23, 37].

While exhaust emissions are expected to go down due to the continuous introduction of better technologies in the circulating fleet, non-exhaust sources are expected to rise as they depend on the number of circulating vehicles, which in the case of Greece in the last decade raises at a constant rate [44].

5.3 Forest Fires

Many studies have shown that in North Europe major biomass burning emissions are mostly linked with wood smoke from fireplaces and stoves, whereas in Southern Europe wildfires can be the most important biomass combustion source. The Mediterranean region is frequently under the influence of this phenomenon, especially during dry periods. Although wildfires can be a major contributor of particulate matter into the atmosphere, forest fire emissions are poorly quantified in the literature, due to the difficulties induced in estimating their temporal and spatial distribution.

With respect to forest fires, Greece faces one of the Europe's most severe problems during summer. Fires burning near urban environments may produce particularly dense pollution plumes that carry a mixture of anthropogenic and biomass burning pollutants. From late August to early September 2007 the forest fires in Evia and Peloponnese (Greece mainland) affected the air quality in areas far away from the fire sources. In addition to local emission sources, these forest fires were major contributors to airborne particle mass levels in Athens urban area. Their average contribution to PM_{10} concentrations in Athens was calculated 28 $\mu g\ m^{-3}$ while the average contribution from traffic emissions and other local sources was 33 $\mu g\ m^{-3}$ [45].

A mesoscale air quality modelling system has been used to quantify the contribution of forest fire emissions during July 2000 when widespread forest fires occurred in the Greek mainland. The forest fire emissions were the largest contributors to the air pollution problem in regions tens of kilometres away from the fire source during the studied period. These emissions were calculated to cause an increase in the average PM_{10} concentration, organic aerosol mass, and gaseous concentration of several pollutants, among them CO, NO_x, and NH_3. An average contribution of 50% to the PM_{10} concentration over the region around the burnt area and downwind of the fire source (approximately 500 km) was calculated with a maximum of 80% [32].

Sciare et al. [46] by performing long-term (5-year) measurements of elemental carbon (EC) and organic carbon (OC) in bulk aerosols in the Mediterranean Basin

(Crete Island) showed that long-range transport of agricultural waste burning from European countries surrounding the Black Sea occurred twice a year (March–April and July–September). The contribution of biomass burning to the concentrations of EC and OC was shown to be rather small (20% and 14%, respectively, on a yearly basis), although this contribution could be much higher on a monthly basis and showed important seasonal and interannual variability.

The study of Amiridis et al. [47] demonstrated that the smoke plume from the fires in Athens during the summer of 2009 was distributed homogenously within the Planetary Boundary Layer of Athens, reaching in some cases free troposphere heights (2–4 km). Both columnar aerosol retrievals and surface measurements revealed the dominance of fine over coarse mode. Regarding PM_{10} and $PM_{2.5}$ urban background levels showed an increase of about 100% and 150% during the fires, demonstrating the dominance of fine particles near the surface due to the fire plume. The signature of the smoke plume was characterised by the high concentration levels of carbonaceous aerosol both EC and OC. During the forest fires, more OC than EC is emitted, resulting in a relatively higher OC/EC ratio. At Athens urban background site the average OC/EC ratio ranged between 10 and 14 during the forest fires while in normal conditions was about 4 [47].

5.4 Shipping Emissions

About 70% of the emissions from oceangoing shipping occur within 400 km of the coastlines along the main trade routes. Thereby, ship emissions can have an impact on the air quality in coastal regions and may partly offset the decline of land-based sources and coastal pollution due to national control measures [48]. In coastal urban areas, emissions from commercial shipping (passenger and cargo) may also constitute a relevant source of PM and SO_2 emissions [49]. According to global estimates, shipping emits between 0.9 and 1.7 million tonnes of total suspended particles (TSP) annually that mainly contain organic carbon, sulphate and inorganic constituents (V, Ni, Ca, Fe) [50].

The Mediterranean Basin is highly influenced by shipping emissions due to its extremely busy shipping routes. These emissions may account for 2–4% of the mean annual ambient air PM_{10} levels (25% primary particles and 75% secondary particles) and for 14% of the mean annual $PM_{2.5}$ in Mediterranean urban areas [51]. It has been estimated that 54% of the total sulphate aerosol column burden over the Mediterranean in summer originates from ship emissions [52].

Greece is found to be a major and growing contributor of exhaust emissions from domestic and international shipping. Through 1984–2008, the ship emission inventory for Greece had an almost fourfold increase. In particular, the PM emissions from domestic and international shipping have an increasing trend that reached 24 million tonnes in 2008 (Fig. 5). Greece contributes to the European and Mediterranean emission inventory from shipping with 7.3% and 14.1%, respectively [53].

Studies on ship emissions have been carried out mainly in harbour areas and coastal cities in the Western Mediterranean Basin by aerosol chemical analysis and

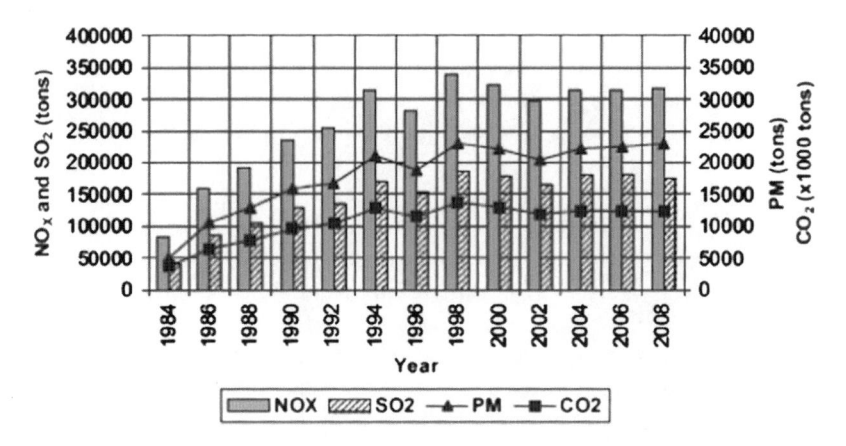

Fig. 5 Fuel-based emission inventory from domestic and international shipping during the period 1984–2008 [53]

source apportionment techniques [54, 55]. As fuel oil is enriched in V and Ni content with respect to the crust, these metals were generally used as markers to identify the fuel oil combustion source. Measurements of aerosol chemical composition made on the island of Lampedusa, south of Sicily, revealed the influence of heavy fuel oil combustion emissions due to the elevated Ni and V soluble fraction and high V and Ni to Si ratios. Back trajectories analysis suggested that air masses prevalently came from the Sicily channel region, where intense ship traffic occurs [56].

For the Eastern Mediterranean Basin the impact of shipping emissions on the urban air quality has been hardly investigated. Only a few studies report on the contribution to ambient PM levels by shipping emissions. Recently, Tzannetos [57] calculated that the $PM_{2.5}$ emissions due to ship activity within the passenger port of Piraeus constitute 15% of the total emissions within the greater area of Athens. Karanasiou et al. [17] attributed the relatively high sulphate concentrations in the urban aerosol of Athens to the emissions from ships from the port of Piraeus.

5.5 Secondary Aerosol

5.5.1 Secondary Inorganic Aerosol

Sulphates (SO_4^{2-}), ammonium (NH_4^+) and nitrates (NO_3^-) are the main secondary inorganic aerosol ions as they account for about two thirds of the total ionic mass in PM_1 and for about 50% in PM_{10} in Athens [21]. These ions represent two different major source categories: fuel combustion and vehicular circulation. Theodosi et al. [21] studied the spatial variability of these ions in 2 sites within GAA: Lykovrisi (LYK) and Goudi (GOU). LYK is a moderately populated municipality, in the northern part of the GAA, 10 km from the city centre; GOU is located downtown Athens (3 km from the city centre) and is influenced by traffic emissions. High spatial homogeneity has been observed for all three ions at these two locations as indicated by the significant correlations with slopes close to 1, denoting relatively small contribution from local sources. Indeed for the PM_1 fraction, the correlations (r) between LYK and GOU were 0.93 for SO_4^{2-} and NH_4^+ with slope of 1.1, while 0.81 for NO_3^- with slope of 1.0. For PM_{10} the correlations (r) were 0.94 for SO_4^{2-} with slope of 0.9 and 0.81 with slope of 0.9 for NO_3^-.

The temporal variability for each of the main ions is also studied and compared to a regional background site (Finokalia; FKL) to assess the importance of regional and local sources.

Non-sea-salt sulphate (nss-SO_4^{2-}): nss-SO_4^{2-} presents a prominent peak in winter (only in GAA) and summer (in all sites; Fig. 6a, b). The summer peak could be related to enhanced photochemistry, lack of precipitation, low air mass renovation at regional scale or the increment of the summer mixing layer depth favouring the regional mixing of polluted air masses [58]. Note also that during summer air masses are almost exclusively originating from Central/Eastern Europe which bring higher levels of SO_2 than the rest of the wind sectors [7, 59]. The secondary maxima of SO_4^{2-} concentration commonly recorded during the winter could concur with the anticyclonic pollution episodes as also indicated by the high nitrate levels [5, 6].

In the case of nss-SO_4^{2-} there is a clear decreasing gradient from urban to suburban and natural sites (5.3, 5.0 and 4.2 µg m^{-3} in PM_1 for GOU, LYK and FKL, respectively). In addition a clear seasonal variation is observed in the GAA/FKL ratio regarding nss-SO_4^{2-}. Indeed during the warm season the GAA/FKL nss-SO_4^{2-} ratio ranges between 1.1 and 1.25, indicating that sulphur levels above Greece are largely controlled by long-range transport and processes evolving at a large spatial scale. On the other hand, during winter the GAA/FKL nss-SO_4^{2-} ratio significantly increases ranging from 1.4 to 1.5. This behaviour indicates significant contribution from local anthropogenic sources (combustion of sulphur-rich diesel for domestic heating) within the GAA during the cold-season.

Nitrate (NO_3^-): As expected higher contributions of nitrate are found at the urban and suburban sites compared to the background site due to the presence of local sources of NO_x in conjunction with thermodynamic conditions producing stable ammonium nitrate (PM_{10}: 2.7, 2.5 and 1.7 µg m^{-3}; PM_1: 0.8, 0.8 and 0.1 µg m^{-3}, for LYK, GOU and FKL, respectively; Fig. 6c, d). In the GAA, NO_3^- presents strong seasonal variability in both PM_1 and PM_{10}, with higher values during colder months, which, as in the case of SO_4^{2-}, are likely to originate from local pollution sources and especially vehicular traffic. On the contrary, no clear seasonal trend is observed at FKL. The summer minimum of NO_3^- in the GAA, which is more prominent in the fine mode, is due to instable ammonium nitrate formation during that period [68, 69]. No ammonium nitrate formation occurs during the warm season due to high temperature in agreement with Eleftheriadis et al. [15].

Nitrate (NO_3^-) behaves differently at FKL and the GAA. At FKL on average, about 94% of particulate nitrate (NO_3^-) is associated with coarse particles, strongly

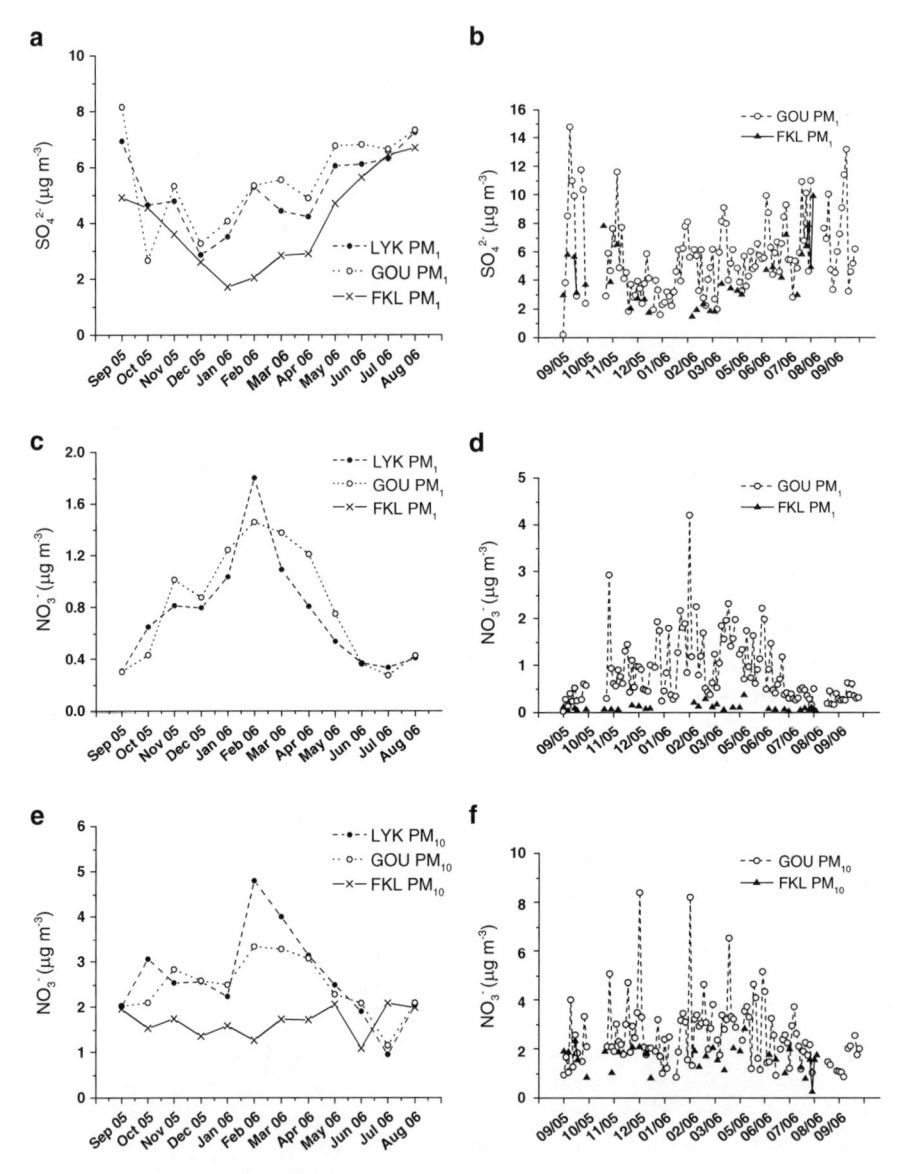

Fig. 6 Seasonal (**a, c** and **e**) and daily (**b, d** and **f**) variations (μg m^{-3}) for SO_4^{2-} in PM_1 and NO_3^- at both PM_1 and PM_{10} samples [21]

indicating that it is mainly chemically bounded with alkaline ions [60]. In GAA as in FKL, the most likely formation pathway for particulate nitrate (NO_3^-) in the coarse mode (60% of the total nitrate) is the reaction of gaseous nitric acid or some other nitrogen compounds with sea salt and mineral dust particles [61]. In addition in the GAA, a significant portion of NO_3^- (about 30%) was found in fine mode, indicating ammonium nitrate formation mainly for the winter period.

By comparing the NO_3^- levels in PM_{10}, the difference between GAA and FKL minimised during the warm season (2.1 and 1.8 μg m^{-3}, respectively). On the other hand, local anthropogenic sources within the GAA dominate during the cold season as for PM_{10} the GAA/FKL ratio is reaching values up to 2. Similar trend also existed for NO_3^- in the PM_1 fraction. However, the GAA/FKL ratio during winter was much pronounced and reached values as high as 8 that is almost double the factor of 4–5 during the warm season. This observation is in agreement with ammonium nitrate formation in the GAA during winter.

5.5.2 Secondary Organic Aerosol

Grivas et al. [62] were the first to report continuous measurements of EC and OC, at an urban location in central Athens, Greece, for an 8-month period (January–August). Average concentrations of 2.2 μgC m^{-3} and 6.8 μgC m^{-3} were observed, for EC and OC, respectively. The contribution of carbonaceous compounds (EC plus organic matter) to PM_{10} was calculated at 26% in agreement with previous estimates [21, 63]. The seasonal variability of EC was found to be limited, while OC mean concentrations were significantly higher (by 23%), during the warm months (May–August). EC and OC produced a bimodal diurnal cycle, with the morning rush hour traffic mode prevailing. However, midday-to-afternoon presence of secondary organic aerosol (SOA) was strongly indicated. The temporal variation of EC, OC and their correlation patterns with primary and secondary gaseous pollutants, suggested that, although primary emissions affected both fractions, SOA formation is an important factor to be accounted for, especially during the photochemical season. Secondary organic carbon was estimated using the EC tracer method and orthogonal regression on OC, EC hourly concentration data. The average contributions of secondary organic carbon (SOC) to OC were calculated at 20.9% for the cold period and 30.3% for the warm period. The SOC diurnal variations suggested photochemical formation throughout the year, intensified during summer months, with the correlation coefficient between SOC and the sum of oxidants ($NO_2 + O_3$) reaching up to 0.84.

Theodosi et al. [64] studied the spatial variability of carbonaceous aerosols along several locations around the Eastern Mediterranean, including big cities (Athens and Istanbul), urban background (Sinop, Erdemli), rural background (Aegina and Pentcli) and regional background sites (Gokceada and Finokalia, Fig. 7). In this work the concentration of SOC was estimated from the following equation:

$$SOC = OC_{total} - (OC/EC)_{minimum} \times EC \qquad (1)$$

By using an OC/EC minimum ratio of 0.3–0.4 suggested by Pio et al. [65] and based on the tunnel studies, SOC percentage contribution can be estimated for all

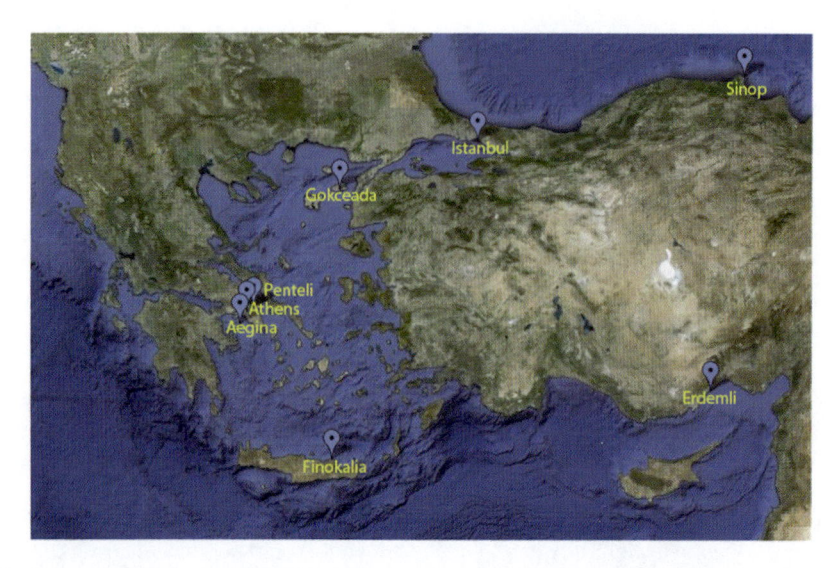

Fig. 7 Location of the sampling sites with annual measurements of carbonaceous aerosols

studied locations. As expected higher SOC values were found at regional or rural background locations reaching from up to 90% at Finokalia down to 45% in Athens and Istanbul. A clear seasonal variability in SOC contribution is observed only in urban or urban background sites with summer time SOC values being 1.5 times higher compared to winter. At rural and regional background sites receiving most of the time processed air masses no such variability can be seen, indicating very fast organic aerosol processing (within few hours after the emission). By using the OC/EC values reported at Grivas et al. (2011) paper and the above-mentioned procedure, SOC% contribution for Athens can reach much higher values (up to 70% during summer) than those reported by Grivas et al. 2011.

6 Concluding Remarks

The Mediterranean Basin has unique characteristics concerning the regional atmospheric circulation, regional climate and also air quality. Air quality in the area is strongly affected by various sources such as local sources and long-range transport of African dust or distance anthropogenic pollution sources over the region. Previous studies have shown that a significant number of PM exceedances registered in urban centres are associated with regional pollution or natural dust transport. Furthermore, the Mediterranean Basin is a region highly affected by climatic changes. Even though the overview of the particulate matter levels during the last decade shows some improvements that have decreased the exceedances and the human exposure, the region is characterised by poor air quality.

Although some effort has recently been paid on characterising particulate matter and quantifying its sources, still further information needs to be gained for the East

Mediterranean Basin. Additional studies of the finer particle fractions and chemical characterisation will elucidate the sources and impact of particle pollution in the area. Particularly there is a lack of data on carbonaceous aerosol that organic and elemental carbon observations could help to distinguish the contribution of secondary versus primary sources.

References

1. Kanakidou M, Mihalopoulos N, Kindap T, Im U, Vrekoussis M, Gerasopoulos E, Dermitzaki E, Unal A, Kocak M, Markakis K, Melas D, Kouvarakis G, Youssef AF, Richter A, Hatzianastassiou N, Hilboll A, Ebojie F, von Savigny C, Ladstaetter-Weissenmayer A, Burrows J, Moubasher H (2011) Megacities as hot spots of air pollution in the East Mediterranean. Atmos Environ 45:1223–1235
2. Millán MM, Sanz MJ, Salvador R, Mantilla E (2002) Atmospheric dynamics and ozone cycles related to nitrogen deposition in the western Mediterranean. Environ Pollut 118:167–186
3. Rodriguez S, Querol X, Alastuey A, de la Rosa J (2007) Atmospheric particulate matter and air quality in the Mediterranean: a review. Environ Chem Lett 5:1–7
4. Kallos G, Kotroni V, Lagouvardos K, Papadopoulos A (1998) On the long-range transport of air pollutants from Europe to Africa. Geophys Res Lett 25:619–622
5. Querol X, Pey J, Pandolfi M, Alastuey A, Cusack M, Pérez N, Moreno T, Viana M, Mihalopoulos N, Kallos G, Kleanthous S (2009) African dust contributions to mean ambient PM10 mass-levels across the Mediterranean Basin. Atmos Environ 43:4266–4277
6. Querol X, Alastuey A, Pey J, Cusack M, Pérez N, Mihalopoulos N, Theodosi C, Gerasopoulos E, Kubilay N, Koçak M (2009) Variability in regional background aerosols within the Mediterranean. Atmos Chem Phys 9:4575–4591. doi:10.5194/acp-9-4575-2009
7. Sciare J, Bardouki H, Moulin C, Mihalopoulos N (2003) Aerosol sources and their contribution to the chemical composition of aerosols in the Eastern Mediterranean Sea. Atmos Chem Phys 3:291–302
8. Lelieveld J, Berresheim H, Borrmann S, Crutzen PJ et al (2002) Global air pollution crossroads over the Mediterranean. Science 298:794–799
9. Eleftheriadis K, Colbeck I, Housiadas C, Lazaridis M, Mihalopoulos N, Mitsakou C, Smolik J, Zdimal V (2006) Size distribution, composition and origin of the submicron aerosol in the marine boundary layer during the eastern Mediterranean "SUB-AERO" experiment. Atmos Environ 40:6245–6260
10. Kouvarakis G, Tsigaridis K, Kanakidou M, Mihalopoulos N (2000) Temporal variations of surface regional background ozone over Crete Island in southeast Mediterranean. J Geophys Res 105:4399–4407
11. Millan MM, Salvador R, Mantilla E, Kallos G (1997) Photo-oxidant dynamics in the Mediterranean basin in summer: results from European research projects. J Geophys Res 102:8811–8823
12. Tørseth K, Aas W, Breivik K, Fjæraa AM, Fiebig M, Hjellbrekke AG, Lund Myhre C, Solberg S, Yttri KE (2012) Introduction to the European Monitoring and Evaluation Programme (EMEP) and observed atmospheric composition change during 1972–2009. Atmos Chem Phys 12:5447–5481
13. Aleksandropoulou V, Eleftheriadis K, Diapouli E, Torseth K, Lazaridis M (2012) Assessing PM_{10} source reduction in urban agglomerations for air quality compliance. J Environ Monit 14:266–278
14. Chaloulakou A et al (2003) Measurements of PM10 and PM2.5 particle concentrations in Athens, Greece. Atmos Environ 37:649–660
15. Eleftheriadis K, Balis D, Ziomas I, Colbeck I, Manalis N (1998) Atmospheric aerosol and gaseous species in Athens, Greece. Atmos Environ 32:2183–2191

16. Grivas G, Chaloulakou A, Kassomenos P (2008) An overview of the PM10 pollution problem, in the Metropolitan Area of Athens, Greece, Assessment of controlling factors and potential impact of long-range transport. Sci Total Environ 389:165–177
17. Karanasiou A, Siskos PA, Eleftheriadis K (2009) Assessment of source apportionment by Positive Matrix Factorization analysis on fine and coarse urban aerosol size fractions. Atmos Environ 43:3385–3395
18. Karanasiou A, Sitaras IE, Siskos PA, Eleftheriadis K (2007) Size distribution and sources of trace metals and n-alkanes in the Athens urban aerosol during summer. Atmos Environ 41:2368–2381
19. Pateraki S, Assimakopoulos VD, Bougiatioti A, Kouvarakis G, Mihalopoulos N, Vasilakos C (2012) Carbonaceous and ionic compositional patterns of fine particles over an urban Mediterranean area. Sci Total Environ 424:251–263
20. Scheff P, Valiozis C (1990) Characterization and source identification of respirable particulate matter in Athens, Greece. Atmos Environ 24:203–211
21. Theodosi C, Grivas G, Zarmpas P, Chaloulakou A, Mihalopoulos N (2011) Mass and chemical composition of size-segregated aerosols (PM1, PM2.5, PM10) over Athens, Greece: local versus regional sources. Atmos Chem Phys. doi:10.5194/acp-11-11895-2011
22. Vardoulakis S, Kassomenos P (2008) Sources and factors affecting PM10 levels in two European cities: Implications for local air quality management. Atmos Environ 42:3949–3963
23. Manoli E, Voutsa D, Samara C (2002) Chemical characterization and source identification/apportionment of fine and coarse air particles in Thessaloniki, Greece. Atmos Environ 36:949–961
24. Samara C et al (1990) Characterization of airborne particulate matter in Thessaloniki, Greece. 1. Source related heavy-metal concentrations within TSP. Toxicol Environ Chem 29:107–119
25. Tsitouridou R, Samara C (1993) First results of acidic and alkaline constituents determination in air particulates of Thessaloniki, Greece. Atmos Environ Part B 27:313–319
26. Chrysikou LP, Samara C (2009) Seasonal variation of the size distribution of urban particulate matter and associated organic pollutants in the ambient air. Atmos Environ 43:4557–4569
27. Chaloulakou A, Kassomenos P, Grivas G, Spyrellis N (2005) Particulate matter and black smoke concentration levels in central Athens, Greece. Environ Int 31:651–659
28. Grivas G, Chaloulakou A, Samara C, Spyrellis N (2004) Spatial and temporal variation of PM_{10} mass concentrations within the greater area of Athens, Greece, Water Air Soil Pollut 158:357–371
29. Cusack M, Alastuey A, Pérez N, Pey J, Querol X (2012) Trends of particulate matter ($PM_{2.5}$) and chemical composition at a regional background site in the Western Mediterranean over the last nine years (2002–2010). Atmos Chem Phys Discuss 12:10995–11033
30. Triantafyllou E, Biskos G (2012) Overview of the temporal variation of PM10 mass concentrations in the two major cities in Greece: Athens and Thessaloniki, Global Nest article in press
31. Gerasopoulos E, Kouvarakis G, Babasakalis P, Vrekoussis M, Putaud J-P, Mihalopoulos N (2006) Origin and variability of particulate matter (PM10) mass concentrations over the Eastern Mediterranean. Atmos Environ 40:4679–4690
32. Lazaridis M, Latos M, Aleksandropoulou V, Hov Ø, Papayannis A, Tørseth K (2008) Contribution of forest fire emissions to atmospheric pollution in Greece. Air Qual Atmos Health 1:143–158
33. Siskos PA, Bakeas EB, Lioli I, Smirnioudi VN, Koutrakis P (2001) Chemical characterization of PM2.5 aerosols in Athens-Greece. Environ Technol 22:687–695
34. Terzi E, Argyropoulos G, Bougatioti A, Mihalopoulos N, Nikolaou K, Samara C (2010) Chemical composition and mass closure of ambient PM10 at urban sites. Atmos Environ 44:2231–2239
35. Amiridis V, Balis DS, Kazadzis S, Bais A, Giannakaki E, Papayannis A, Zerefos C (2005) Four-year aerosol observations with a Raman lidar at Thessaloniki, Greece, in the framework of European Aerosol Research Lidar Network (EARLINET). J Geophys Res 110:D21203. doi:10.1029/2005JD006190
36. Samara C, Kouimtzis TH, Katsoulos GA (1994) Characterisation of airborne particulate matter in Thessaloniki, Greece Ð Part III: comparison of two multivariate modelling approaches for

the sources of apportionment of heavy metal concentrations within total suspended particles. Toxicol Environ Chem 44:147–160
37. Samara C et al (2003) Chemical mass balance source apportionment of PM10 in an industrialized urban area of Northern Greece. Atmos Environ 37:41–54
38. Argyropoulos G, Manoli E, Kouras A, Samara C (2012) Concentrations and source apportionment of PM_{10} and associated major and trace elements in the Rhodes Island, Greece. Sci Total Environ 432:12–22
39. Middleton NJ, Goudie AS (2001) Saharan dust: sources and trajectories. Trans Inst Br Geogr NS 26:165–181
40. Kallos G, Astitha M, Katsafados P, Spyrou C (2007) Long-range transport of anthropogenically and naturally produced particulate matter in the Mediterranean and North Atlantic: current state of knowledge. J Appl Meteorol Climatol 46:1230–1251
41. Mitsakou C, Kallos G, Papantoniou N, Spyrou C, Solomos S, Astitha M, Housiadas C (2008) Saharan dust levels in Greece and received inhalation doses. Atmos Chem Phys 8:11967–11996
42. Papayannis A, Mamouri RE, Amiridis V, Kazadzis S, Perez C, Tsaknakis G, Kokkalis P, Baldasano JM (2009) Systematic lidar observations of Saharan dust layers over Athens, Greece in the frame of EARLINET project (2004e2006). Ann Geophys 27:3611–3620
43. Remoundaki E, Bourliva A, Kokkalis P, Mamouri RE, Papayannis A, Grigoratos T, Samara C, Tsezos M (2011) Composition of PM10 during a Saharan dust transport event over Athens, Greece. Sci Total Environ 409:4361–4372
44. Markakis K, Poupkou A, Melas D, Zerefos C (2010) A GIS based methodology for the compilation of an anthropogenic PM10 emission inventory in Greece. Environ Pollut Res 1(2):71–81
45. Liu Y, Kahn RA, Chaloulakou A, Koutrakis P (2009) Analysis of the impact of the forest fires in August 2007 on air quality of Athens using multi-sensor aerosol remote sensing data, meteorology and surface observations. Atmos Environ 43(21):3310–3318
46. Sciare J, Oikonomou K, Favez O, Markaki Z, Liakakou E, Cachier H, Mihalopoulos N (2008) Long-term measurements of carbonaceous aerosols in the Eastern Mediterranean: evidence of long-range transport of biomass burning. Atmos Chem Phys 8:5551–5563
47. Amiridis V, Zerefos C, Kazadzis S, Gerasopoulos E, Eleftheratos K, Vrekoussis M, Stohl A, Mamouri RE, Kokkalis P, Papayannis A, Eleftheriadis K, Diapouli E, Keramitsoglou I, Kontoes C, Kotroni V, Lagouvardos K, Marinou E, Giannakaki E, Kostopoulou E, Giannakopoulos C, Richter A, Burrows JP, Mihalopoulos N (2012) Impact of the 2009 Attica wild fires on the air quality in urban Athens. Atmos Environ 46:536–544
48. Eyring V, Isaksen ISA, Berntsen T, Collins WJ, Corbett JJ, Endresen O, Grainger RG, Moldanova J, Schlager H, Stevenson DS (2010) Transport impacts on atmosphere and climate: shipping. Atmos Environ 44:4735–4771
49. Miranda AI, Marchi E, Ferretti M, Millán MM (2008) Chapter 9 Forest fires and air quality issues in Southern Europe. Dev Environ Sci 8:209–231
50. Moldanová J, Fridell E, Popovicheva O, Demirdjian B, Tishkova V, Faccinetto A, Focsa C (2009) Characterisation of particulate matter and gaseous emissions from a large ship diesel engine. Atmos Environ 43:2632–2641
51. Viana M, Amato F, Alastuey A, Querol X, Saúl G, Herce-Garraleta D, Fernandez-Patier R (2009) Chemical tracers of particulate emissions from commercial shipping. Environ Sci Technol 43:7472–7477
52. Marmer E, Langmann B (2005) Impact of ship emissions on the Mediterranean summertime pollution and climate: a regional model study. Atmos Environ 39:4659–4669
53. Tzannatos E (2010) Ship emissions and their externalities for Greece. Atmos Environ 44:2194–2202
54. Mazzei F, D'Alessandro A, Lucarelli F, Nava S, Prati P, Valli G, Vecchi R (2008) Characterization of particulate matter sources in an urban environment. Sci Total Environ 401:81–89
55. Pandolfi M, Gonzalez-Castanedo Y, Alastuey A, da la Rosa JD, Mantilla E, de la Campa AS, Querol X, Pey J, Amato F, Moreno T (2011) Source apportionment of PM10 and PM2.5 at

multiple sites in the strait of Gibraltar by PMF: impact of shipping emissions. Environ Sci Pollut Res 18:260–269
56. Becagli S, Sferlazzo DM, Pace G, di Sarra A, Bommarito C, Calzolai G, Ghedini C, Lucarelli F, Meloni D, Monteleone F, Severi M, Traversi R, Udisti R (2012) Evidence for heavy fuel oil combustion aerosols from chemical analyses at the island of Lampedusa: a possible large role of ships emissions in the Mediterranean. Atmos Chem Phys 12:3479–3492
57. Tzannatos E (2010) Ship emissions and their externalities for the port of Piraeus – Greece. Atmos Environ 44:400–407
58. Mihalopoulos N, Kerminen VM, Kanakidou M, Berresheim H, Sciare J (2007) Formation of particulate sulfur species (sulfate and methanesulfonate) during summer over the Eastern Mediterranean: a modelling approach. Atmos Environ 41(32):6860–6871
59. Kouvarakis G, Bardouki H, Mihalopoulos N (2002) Sulfur budget above the Eastern Mediterranean: relative contribution of anthropogenic and biogenic sources. Tellus 54B:201–212
60. Mamane Y, Gottlieb J (1992) Nitrate formation on sea salt and mineral particles—a single particle approach. Atmos Environ 26A:1763–1769
61. Metzger S, Mihalopoulos N, Lelieveld J (2006) Importance of mineral cations and organics in gas–aerosol partitioning of reactive nitrogen compounds: case study based on MINOS results. Atmos Chem Phys 6:2549–2567
62. Grivas G, Cheristanidis S, Chaloulakou A (2012) Elemental and organic carbon in the urban environment of Athens. Seasonal and diurnal variations and estimates of secondary organic carbón. Sci Total Environ 414:535–545
63. Sillanpaa M, Hillamo R, Saarikoski S, Frey A, Pennanen A, Makkonen U et al (2006) Chemical composition and mass closure of particulate matter at six urban sites in Europe. Atmos Environ 40(Suppl 2):212–223
64. Theodosi C, Bougiatioti A, Smouliotis D, Zarmpas P, Kocak M, Mihalopoulos N (2012) Carbonaceous aerosols over the Mediterranean and Black Sea, Proceedings of the 11th International Conference on Meteorology Climatology and Atmospheric Physics, Athens 29/5-1/6 2012, Advance in meteorology, climatology and atmospheric physics, Springer Atmospheric Sciences, Helmis CG, Natsos PT (eds) Volume 1, doi 10.1007/978-3-642-291172-2, 2012, pp 1233–1238
65. Pio C, Cerqueira M, Harrison RM, Nunes T, Mirante F, Alves C, Oliveira C, Sanchez de la Campa A, Artíñano B, Matos M (2011) OC/EC ratio observations in Europe: re-thinking the approach for apportionment between primary and secondary organic carbon. Atmos Environ 45:6121–6132
66. Eleftheriadis K, Diapouli E, Gini MI, Vasilatou V, Samara C, Argyropoulos G (2011) Source apportionment of airborne particulate matter for three urban centers in Greece. In: European aerosol conference, 4–9 September 2011
67. Kopanakis I, Eleftheriadis K, Mihalopoulos N, Lydakis-Simantiris N, Katsivela E, Pentari D, Zarmpas P, Lazaridis M (2012) Atmos Res 106:93–107
68. Harrison RM, Pio CA (1983) Size-differentiated composition of inorganic atmospheric aerosols of both marine and polluted continental origin. Atmos Environ 17:1733–1738
69. Querol X, Alastuey A, Ruiz CR, Artiñano B, Hanssonc HC, Harrison RM, Buringh E, ten Brink HM, Lutz M, Bruckmannh P, Straehl P, Schneider J (2004) Speciation and origin of PM10 and PM2.5 in selected European cities. Atmos Environ 38:6547–6555

Anthropogenic and Natural Constituents in PM10 at Urban and Rural Sites in North-Western Europe: Concentrations, Chemical Composition and Sources

Ernie Weijers and Martijn Schaap

Abstract This study focuses on north-western region of Europe discussing questions like the following: Which anthropogenic and natural constituents build up the particulate matter? To what extent do they contribute to the total mass? And where do these constituents originate? To answer, we elaborated data sets containing chemical information of PM recently becoming available in the Netherlands, Germany and Belgium.

The chemical composition of PM10 shows a considerable conformity in these countries. Always, secondary inorganic aerosols (SIA) are the major constituent ($\pm 40\%$) followed by the carbonaceous compounds ($\pm 25\%$). Contributions of sea salt and mineral dust vary between 10% and 15% depending on presence and distance of respective sources. The unidentified mass is some 15% indicating that the composition of PM10 in this region is fairly well known.

PM10 concentrations and constituents appear systematically higher at urban sites. Urban increments have been measured for most chemical constituents. Nearby (anthropogenic) sources and reduced dispersion in the urbanised areas are the main determining factors here. The observed increment for SIA is caused by more nitrate and sulphate. It is explained by depletion of chloride stabilising part of the nitrate and sulphate in the coarse mode. The question then arises how to assign the coarse mode nitrate (and sulphate) in the mass closure exercise as they replace the chloride.

Important for the national and European air pollution policy is how much of the measured particulate matter is of anthropogenic origin. A simple assessment indicates that 20–25% of PM10 is of natural origin; hence, the majority of PM10

E. Weijers (✉)
Energy research Centre of the Netherlands (ECN), P.O. Box 1, 1755 ZG Petten, The Netherlands
e-mail: weijers@ecn.nl

M. Schaap
TNO, P.O. Box 80015, 3508 TA Utrecht, The Netherlands
e-mail: martijn.schaap@tno.nl

in the north-western-European region is of anthropogenic origin. The uncertainty in this analysis is considerable, and the result is indicative.

A chemical transport model (LOTOS-EUROS) was used to obtain a detailed source apportionment. In total, 75% of the modelled PM10 mass could be explained. The important contributions to PM10 come from agriculture, on- and off-road transport and natural sources (sea salt). Secondary contributions are derived from power generation, industrial processes and combustion as well as households. Of the modelled part, 70–80% of PM10 over the Netherlands is anthropogenic. The increase in source contribution going from low to high PM levels is proportional for most sectors, except for agriculture and transport, which become more important mainly due to the more than proportional rise in ammonium nitrate concentrations. Sea-salt concentrations decline with rising PM10. The same was found for Spain, but here, the impact of Saharan dust on PM episodes is clearly recognisable and much larger than in north-western Europe. Natural sources in Spain contribute about half of the modelled PM10 concentrations. Significant anthropogenic sources are similar to those in north-western Europe.

Keywords Anthropogenic contribution, Chemical composition, Chemical transport modelling, Natural contribution, North-western Europe, PM10, Source apportionment, Spain

Contents

1	Introduction and Scope	240
2	Concentrations and Chemical Composition at Urban and Rural Sites	241
	2.1 Data Sets	241
	2.2 Handling	241
	2.3 Chemical Composition	243
	2.4 The Urban Increment	245
3	Anthropogenic and Natural Contributions to PM in North-Western Europe	246
4	Modelled Origin of Particulate Matter in the Netherlands and Spain	249
5	Concluding Remarks	254
References		256

1 Introduction and Scope

Although air quality in Europe has improved substantially over the last decades, it still poses a significant threat to human health [1]. Short- and long-term health effects have been described, but so far there is no conclusive evidence which component(s) or property can be held responsible. As a consequence, the European air quality guidelines keep a focus on particulate mass, and, next to the PM10 guideline, the Directive 2008/50/EC introduced additional objectives with respect to PM2.5. Although various abatement measures have been implemented to reduce the levels of particulate matter, many European countries still have problems adhering to the guidelines; in particular, this concerns the maximum number of exceedances of the daily limit value for PM10. The heavily populated regions in the north-west of Europe is an example where problematic "hot spots" are found in urbanised and industrialised agglomerations.

In addition to the strength and location of local, regional and continental emission sources, levels and chemical composition of ambient PM depend on climatology, trajectories, rain scavenging potential, recirculation of air masses, dispersive atmospheric conditions and geography (proximity to coast or arid zones, topography, soil cover). These factors largely differ over Europe. Reduction measures in one region may therefore not be optimal for other regions. In the design of a meaningful abatement strategy, one needs therefore to take into account the changing characteristics of PM, or more specifically, one should assess which anthropogenic and natural constituents build up the particulate matter, to what extent they contribute to the total mass and where they come from.

To answer these questions, we elaborated three PM data sets bearing chemical information that recently became available in the Netherlands, Germany and Belgium. In addition, a first-order quantitative estimate is given of the ratio between natural and anthropogenic PM10 mass; this ratio defines the "playing field" of policy-decision makers. To end, a modelling exercise is described comparing the major emission sources for north-western Europe (here defined as the three abovementioned countries) with those of the Iberian Peninsula.

2 Concentrations and Chemical Composition at Urban and Rural Sites

2.1 Data Sets

During the last decade, a considerable number of studies have focussed on the speciation of PM in different regions of Europe (e.g. [2–7, 29]). Recently, dedicated measurement campaigns were carried out in north-western Europe: "CHEMKAR" (Belgium) by the Flemish Environmental Agency [8], "BOP" in the Netherlands within the framework of the Netherlands Research Programme on Particulate Matter [9] and in North Rhine-Westphalia IUTA (Germany).

The respective data sets are used here to illustrate the general chemistry of PM10 in north-western Europe. Always, urban and rural sites have been compared. Some data features are given in Table 1. The components of interest are sulphate, nitrate, ammonium, elemental carbon, organic carbon, sodium, chloride and elements.

2.2 Handling

Chemical analyses of PM samples usually provide a major part of the total particulate mass collected on a filter. Certain tracers or combination of tracers are

Table 1 Data sets

Country	Name and type of site	Mass fractions	Period
Belgium (CHEMKAR)	Zelzate (ub), Borgerhout (ub) Houtem(rb), Aarschot (rb)	PM10	September 2006– September 2007
The Netherlands (BOP)	Schiedam (ub), Hellendoorn (rb)	PM10, PM2.5, PM10-2.5	January–August 2008
Germany (IUTA/LANUV)	Styrum (ub), Eiffel (rb)	PM10	April–September 2008

ub urban background, *rb* rural background

used to estimate the contributions from specific sources like sea salt or mineral dust. In order to make a sound comparison between the three sets, contributions were (re)calculated using the same algorithms. The presence of *secondary inorganic aerosol* (SIA) is calculated as the sum of NO3, SO4 and NH4 and results from direct measurements. Sulphate concentrations are corrected for a small amount of sea salt.

Carbonaceous compounds in PM contain other elements (e.g. oxygen) augmenting the organic mass. Organic matter (OM) concentrations are usually calculated from organic carbon, but the conversion factor remains rather uncertain. Factors vary between 1.2 and 2 and probably vary between lower values near sources and higher values after processing in aged air masses. Here, a factor of 1.4 is used which seems most common in literature. Elemental carbon (EC) can be measured straightforward. In the case of sea salt (SS), there are two tracers: sodium and chloride. Several algorithms are in practice; one of them calculates the SS part from both Na and Cl. However, in the case of chloride, reactions with HNO3 may occur in the atmosphere as well as on the filter thereby releasing HCl (under the formation of NaNO3). In addition, the presence of chloride may suffer from the evaporation of NH4Cl (from quartz filters). For these reasons, sodium is selected as the sole tracer: SS = 3.26*Na (the factor follows from the composition of sea water).

Various algorithms are also in use to estimate the contribution of mineral dust (MD). With MD is meant all fugitive windblown and mechanically resuspended dust with a composition comparable to the earth's crust. Since chemical analyses of PM samples measure elements directly, the approach here is to sum over those elements known to be present in the earth's crust: Al, Si, CO3, Ca, Fe, K, Mn, Ti and P [10]. Weights were first recalculated to correct for their oxidised form (e.g. Si is usually present as SiO_2). MD is a parameter difficult to estimate. The use of other algorithms in the estimation of MD results in different values, e.g. the one formulated by Denier van der Gon et al. [11]. Also, local anthropogenic sources may contribute (e.g. metallurgical industry).

Finally, the concentrations of all constituents were summated and compared with the gravimetrically measured mass to establish the unaccounted mass denoted unknown. The assumption here is that the various data sets accurately describe the PM characteristics. The filtration devices employed at the sites were all equivalent to the reference method. However, corresponding studies use different procedures with respect to filter handling, data treatment and selection, analytical techniques, etc. In addition, temporal variation, local site characteristics, artefacts like volatilisation and particle-bound water as well as varying distances from major sources will affect results in different ways. This will affect the absolute concentrations that have been measured. Putaud et al. [2] extensively describe the accuracy of the analytical techniques common in Europe. While ion chromatography (SIA) performs well (<10%), an accurate determination of EC, OC and MD remains a challenge. EC concentrations by various techniques differ by a factor of 4, and OC by 30%. The uncertainty in the determination of mineral dust amounts can reach 100% for reasons mentioned above.

Only PM10 is discussed as data on PM2.5 are much less available. In our region, the average mass contribution of PM2.5 to PM10 is about 60–70%, and the relative distribution of the different chemical parts in PM2.5 usually resembles that of PM10. All components are present in both the fine and coarse fraction. Whereas SIA, EC and OM dominate more in the fine fraction, SS and MD contribute more to the coarse mode.

2.3 Chemical Composition

Figure 1 shows the chemical distributions for the urban and regional background sites in the three PM10 data sets selected here. Common characteristics can be observed: always, the major constituent is the SIA, followed by the carbon-containing components (EC+OM). Smaller, but non-negligible, contributions arise from MD and SS. At least 81% of the particulate mass was explained in the mass balances. The highest closure was seen at the urban background site Borgerhout (87%) which seems due to relatively a high amount of carbonaceous material. Relatively, there is more SIA present at the rural sites: 42–44% versus 34–41% at urban sites. However, differences between urban and rural sites are not very large stressing the role of SIA in building up background levels [31, 32]. In north-western Europe, agricultural activities (like livestock and soil fertilising) frequently take place yielding high emissions of ammonia in rural areas. Combined with the NOx emissions from intensified traffic and SO2 emissions from industry, the formation of ammonium nitrate and ammonium sulphate aerosols is favoured [12].

The sum of EC and OM contributes another 20–27% (at urban sites) and 18–24% (rural) to the PM mass, with OM dominating EC at most sites. Like in the case of SIA, differences between rural and urban sites appear modest which is probably caused by a considerable natural (biogenic) input. A study by ten Brink et al. [13] on the presence of ^{14}C in PM filter samples revealed that at least 64% of the organic carbon measured at an urban Dutch site was contemporary (i.e. emissions from biogenic material and biomass wood combustion).

A gradient for sea salt is observed as expected. Near the North Sea (Houtem (rural background), Schiedam (urban background)), the marine contribution can be as high as 16–18%. Further inland, some 6–10% is measured. In Germany, it declines to some 4%. On average, the sea-salt contribution in this region is a substantial 10% and is due to the dominance of transport of clean marine air from the West diluting anthropogenic emissions onshore and transporting the pollution further eastward over the European continent.

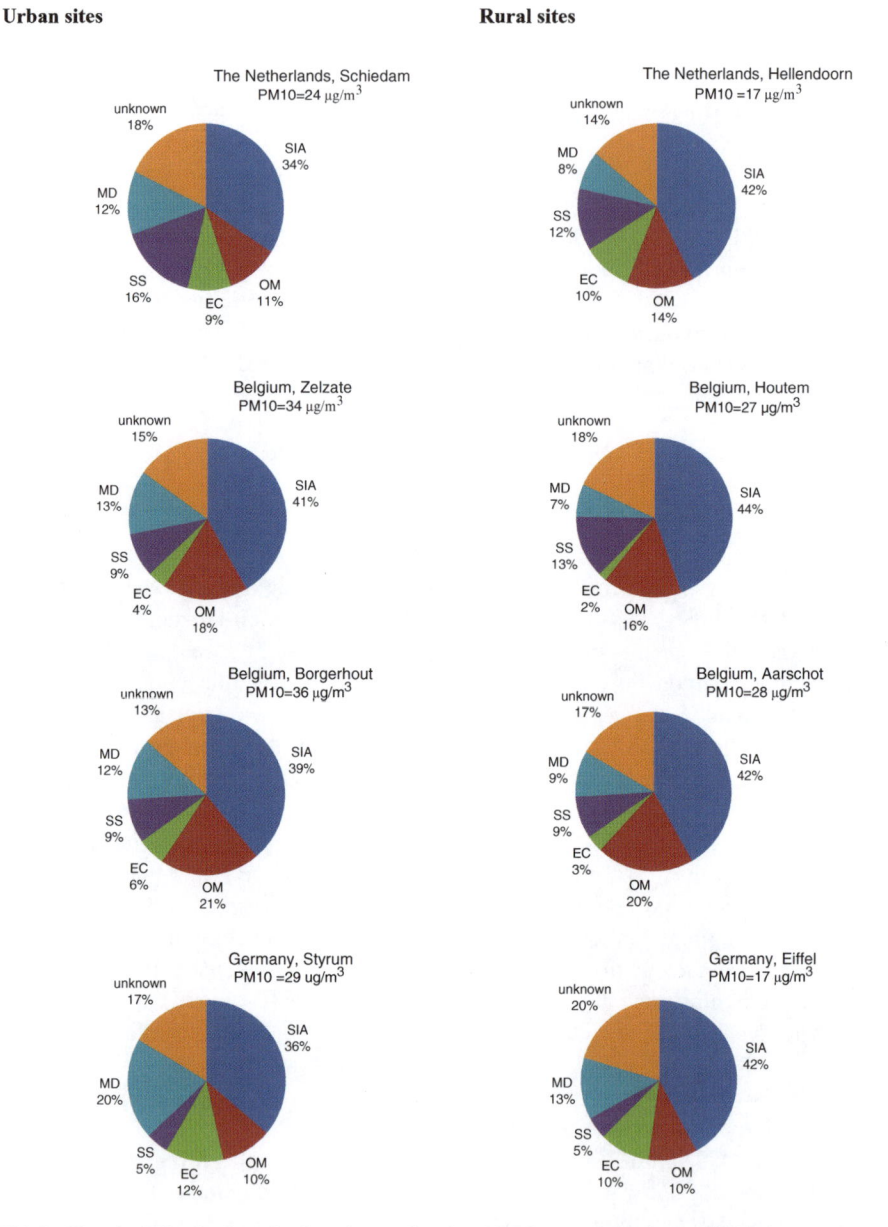

Fig. 1 Chemical distributions for the urban and regional background sites in the PM10 data sets as derived from BOP (the Netherlands), CHEMKAR (Belgium) and IUTA/LANUV (Germany)

The fourth constituent, mineral dust, appears comparable in Belgium and the Netherlands: 12–13% at urban sites and 7–9% at rural sites. A deviant level of MD is observed at the both German sites (20% and 13%, respectively) and is attributed to industrial emissions (steel industry) in this region.

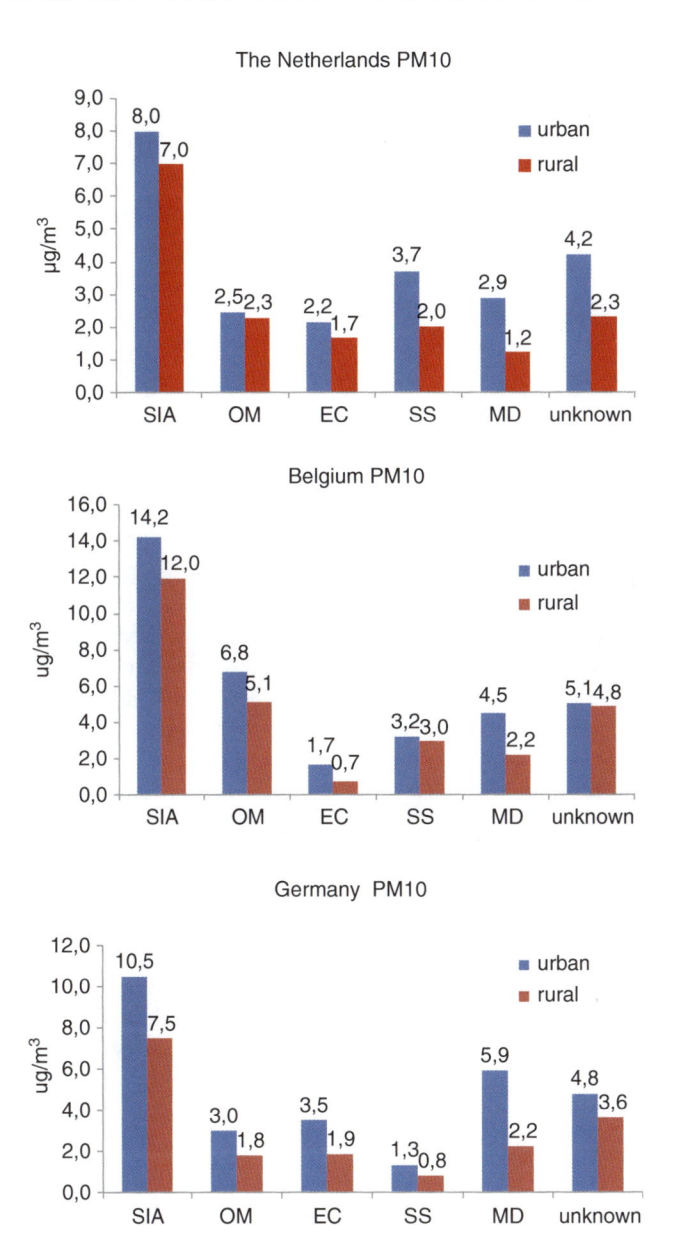

Fig. 2 Chemical distributions for the urban and regional background sites in the PM10 data sets as derived from BOP (the Netherlands), CHEMKAR (Belgium) and IUTA/LANUV (Germany)

2.4 The Urban Increment

In Fig. 2, the absolute concentrations at the urban and rural sites are compared. For the Belgian data set, we averaged over the two urban and two rural sites.

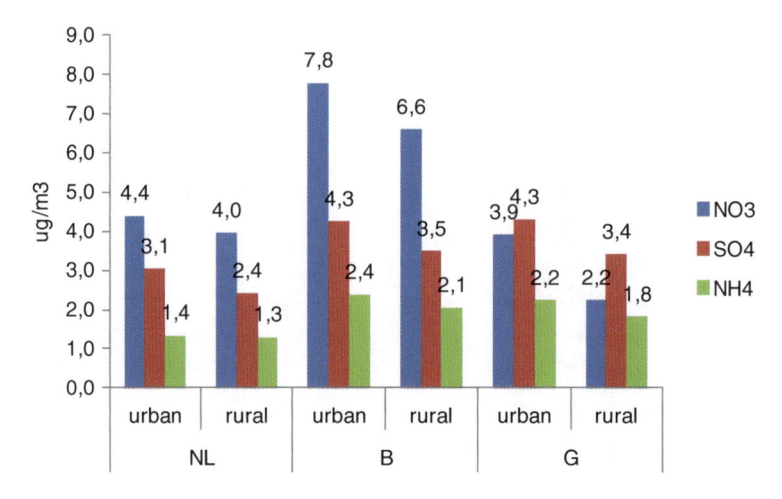

Fig. 3 The urban increment of the SIA components

Like in most cities in Europe, a rise in PM10 concentrations is measured when moving from rural to urban sites. Clearly, a comparison like this depends on the selection of sites, in particular distance between sites and surroundings affect results like these. Here, the increase in PM10 is between 7 and 11 μg/m^3. No single constituent (or emission source) can be held solely "responsible": all (including the unknown part) appear higher at the urban sites, but there is no systematic pattern observed here. In Belgium and Germany, the increment is largely due to more SIA, OM+EC and MD (between 2 and 4 μg/m^3). In the Dutch data set, however, the unknown part contributes most. The increase of sea salt at the urban site Schiedam is due to its location close to the North Sea. The rise of MD is substantial at all sites.

The urban increment observed for SIA systematically returns for each of its component (Fig. 3).

It is seen that the increment of SIA is predominantly caused by nitrate (the Netherlands and Belgium) and sulphate (Germany), while the change for ammonium is modest. Levels of nitrate and sulphate are usually higher in urban or industrial areas in Europe [14, 15]. In marine and coastal atmospheres, nitric acid is converted into particulate nitrate (NaNO$_3$) through the reaction with sea-salt particles resulting in the release of HCl: NaCl + HNO3 → NaNO3 + HCl. In contrast to NH$_4$NO$_3$, NaNO$_3$ is a non-volatile compound under atmospheric conditions; therefore, partitioning of nitric acid into the sea salt is irreversible. A similar reaction applies for sulphuric acid (and sulphate). Most of the nitrate here is found in the fine mode as ammonium nitrate (and ammonium sulphate).

3 Anthropogenic and Natural Contributions to PM in North-Western Europe

Important for air pollution policy in Europe is the contribution to PM10 that is of anthropogenic origin. It is this fraction that can be targeted by national and European abatement strategies. Below, a first-order rather pragmatic assessment is given of the natural versus anthropogenic contributions to PM in north-western Europe. It is based on the prevalence of major constituents in the chemical mass closure while incorporating specifics of the region considered.

Sea spray emissions are the most important natural source of primary sulphate. Applying the sodium-to-sulphate ratio in seawater learns that roughly 5% of the particulate sulphate is of marine origin. Though sulphate may be emitted directly into the atmosphere [17], its major source is the oxidation of sulphur dioxide. In north-western Europe, anthropogenic SO_2 emissions are derived mostly from the combustion of sulphur-containing fuels for power generation and international shipping. In Europe, natural emissions for SO_2 include volcanoes, but given location, height and distance to the Netherlands, their contribution is probably low (i.e. not more than 1% to ground level sulphate). Contributions of other biogenic sources of SO_2 like oxidation of DMS, COS and H_2S are also of minor importance [16]. Wildfire emissions of SO_2 are commonly neglected in studies directed to wild land fire emissions. In the approximation here 5% is used as a conservative estimate. To estimate how much of the nitrate is natural, emissions from soil, biomass burning and lightning should be considered. Studies on NO emissions from soil report emissions between 59 and 190 kton in the EU15 ([17, 30]). Relative to the total inventoried emissions, the central value is 4%. From these emissions, about two-third is attributed to agricultural soils and one-third to forest soils. Soil NO_x emissions largely depend on the nutrient input (through fertilisation or atmospheric deposition) implying that the abovementioned contribution is only partly natural. Lightning depends linearly on the amount of convective precipitation. In Europe, corresponding emissions are estimated to be about 1% (65 kton) of the total inventoried emissions. It mainly takes place between 1 and 5 km altitude and mostly in southern Europe (Friedrich et al. 2008). Wild fires estimates range between 20 and 50 kton (Friedrich et al. 2008; [18]) and occur mostly in countries around the Mediterranean. Given the short life time of NOx and the rather short transport distance of nitrate, we assume that lightning and wild land fires can be neglected as sources in north-western Europe. In the approximation here, the natural fraction of NO_x emissions and therewith nitrate is taken between 0% and 5%.

Ammonium in aerosols originates from the neutralisation of sulphuric and nitric acid by ammonia. Ammonia is emitted by different sources, most notably animal manure, traffic and application of fertiliser. In general, emissions are for the largest part (80–95%) associated with agricultural activities [19]. Erisman et al. [20] estimated the natural emissions at about 10% of the total emissions in Europe. This percentage includes contributions from wild animals and wetlands. We assume a similar percentage for ammonium in north-western Europe.

The main source of elemental carbon is the (incomplete) combustion of fossil fuels. Wild land fires are occasional sources of vast amounts of carbonaceous particles. Though wild land fires may seem natural, the vast majority of present-day fires are due to human behaviour. Hence, elemental carbon is almost exclusively anthropogenic.

Organic matter contains numerous chemical constituents of which only a small fraction has been identified. Organic carbon is released by the incomplete combustion of fuels but in addition originates from farming activities (stables, harvesting) as well as from nature. Anthropogenic combustion sources as well as (small) stable emissions have been inventoried. Half of the observed concentrations of OC [21, 22] can be explained. Hence, a significant (unknown) contribution is from anthropogenic sources like secondary formation, wildfires, harvesting, abrasion processes and biological material (fungal spores and plant debris). As mentioned previously, ^{14}C analysis suggests that some 70% of the OC mass in the Netherlands is of living material where biogenic SOA, agriculture and other landscaping activities may have contributed. The complicating factor in the estimation of the natural part of OM is the uncertainty related to its secondary component (oxidation of VOCs yields products with low vapour pressures that may condense on existing aerosol) as the major formation routes are not well known. As a consequence, the ratio of natural to anthropogenic SOA is under debate. It is postulated that biogenic sources are a major contributor to atmospheric SOA (e.g. [22]). Assuming that the unexplained OC using present-day emission inventories is for the largest part SOA, an upper limit of around 50% for natural OM is obtained. As a substantial part of the SOA may very well be anthropogenic, it is postulated that half of the SOA as natural, leading to a lower limit of 25% for natural OM.

Sources of mineral dust in total PM10 are wind erosion of bare soils, agricultural land management, resuspension of road dust from paved an unpaved roads, road wear, handling of materials and building and construction activities. Only wind erosion may contribute to the natural fraction. Saharan dust is regularly transported to countries around the Mediterranean Sea. In north-western Europe, dust transport from the Sahara occurs once or twice a year and is therefore not very significant. Korcz et al. [23] and Schaap et al. [24] show that windblown dust at the European continent is a rather small source compared to traffic resuspension and agricultural land management. More importantly, the windblown dust source strength from soils other than arable land is low. Windblown dust emissions are strongly related to anthropogenic changes in surface vegetation cover and are regarded as mostly anthropogenic. Consequently, the total mineral dust concentration in air is expected to be anthropogenic for a large part. In the approximation, 10% is assumed as a conservative estimate for the natural contribution to MD.

The emission of sea salt is mainly dependent on wind speed. It is considered the second largest contributor in the global aerosol budget, as a vast area of the earth consists of sea. The aerosols consist mainly of sodium chloride. Other constituents of atmospheric sea salt reflect the composition of sea water (magnesium, sulphate, calcium and potassium). Sea salt is the only pure natural aerosol component.

The anthropogenic contribution of the unknown part is by definition unknown. Water accounts to some extent for the unknown fraction. As most of that water may be associated with SIA, it is likely that a significant part of the unknown fraction is anthropogenic. It is assumed that the anthropogenic and natural parts of the unknown fraction resemble those of the defined mass.

The natural fraction in the total mass can now be obtained by adding the relative natural mass contributions per constituent. For instance, 10% of PM10 is associated with sea salt adding to 10% of natural PM10 (see Table 2). Note that this fraction represents the source attribution and therewith fresh sea salt as discussed in the

Table 2 Analysis of the natural contribution (%), for all components, and for PM10 as a whole, at the rural background site (Hellendoorn)

Composition	PM10 (%)	Natural contribution (%)	Rural background PM10 (%)
NO$_3$	20	0–5	0–1
SO$_4$	14	5	1
NH$_4$	8	10	1
MD	10	20	2
SS	10	100	10
EC	6	0	0
OM	15	25–50	4–8
Sum known			18–23
Unknown	15	–[a]	3–3
Total			21–26

For each constituent, the relative contributions to PM10 are listed as well as the estimated natural contributions. The last column provides the resulting natural contribution, expressed in a percentage of PM10 mass (with a low-high estimate for the natural contribution in NO$_3$ and OC)
[a]The natural contribution is assumed to be the same as the total natural percentage of PM mass without the unknown fraction

previous section and would be lower when one reflects the chloride loss in this calculation. Nitrate contributes 20%, of which 0–5% is assumed to be natural resulting in a contribution of 0–1% natural PM10 for the low and high case, respectively. Hence, nitrate contributes very little to the natural fraction. Adding all contributions, the estimated natural fraction of PM10 is between 21% and 26%. Hence, between one-fifth and one-quarter of PM10 is estimated to be of natural origin. The uncertainty in such a simple analysis is quite large. Hence, we rounded all data to the closest interval of a half per cent. As such, the uncertainty in the summation is around 2–3%. It is concluded that the most important natural contributions originate from sea salt (100%) and organic material (with an upper limit of 50%).

4 Modelled Origin of Particulate Matter in the Netherlands and Spain

So far, we have used composition measurements and mass closure studies to interpret PM distributions and origin. Furthermore, a simple analysis on the natural fraction of PM was described. Detailed speciation data sets such as used in this study enable to use more elaborate statistical approaches to identify source origins, such as positive matrix factorisation [25]. Examples of these techniques are given for Germany by Quass in this volume [32]. However, these methods are only able to distinguish between a limited number of source categories. Furthermore, they are typically not able to provide a source apportionment for secondary components.

Complementary to experimental data, a chemical transport model (CTM) can be used to obtain a more detailed source apportionment. CTMs provide calculations of the evolution of the air pollution situation across a region based on emission inventories and atmospheric process descriptions. Here we use the LOTOS-EUROS CTM to investigate the origin of particulate matter in north-western Europe. The LOTOS-EUROS model has been equipped with a module that tracks the contribution of different source sectors throughout a model simulation [26]. Hence, for each modelled process, e.g. advection and chemistry, the concentration change is calculated as well as change in source contributions. The module has been used to assess the origin of PM in the Netherlands by Hendriks et al. [27]. Here, we summarise the results for the Netherlands assuming it to be representative for north-western Europe (i.e. Belgium and western Germany) and also provide results for Spain to contrast the two regions.

To assess the source contributions to modelled PM10, two simulations were performed. A simulation across Europe at 0.5° longitude × 0.25° latitude (about 28 × 28 km) resolution provided the results which were used as boundary conditions for a simulation at a resolution of 0.125° longitude × 0.0625° latitude (7 × 7 km) centred over the Netherlands. The simulations were performed for 2007, 2008 and 2009 to average out meteorological variability. Labels were applied to distinguish Dutch and foreign emissions sources specified to SNAP (Selected Nomenclature for sources of Air Pollution) level 1, which uses ten main sectors:

1. Power generation: combustion in energy and transformation industries
2. Other combustion: nonindustrial combustion mainly households
3. Industrial combustion in manufacturing industry
4. Industrial process emissions
5. Extraction and distribution of fossil fuels and geothermal energy
6. Solvent use
7. Road transport
8. Other transport: other mobile sources and machinery
9. Waste treatment and disposal
10. Agriculture

Natural emissions and PM originating from the initial conditions, aloft conditions and PM coming from regions outside the model domain were tracked as well.

In Fig. 4, the modelled PM10 distribution is shown for Europe. For the Netherlands, modelled PM10 concentrations range from 13 µg/m^3 in the North and 18 µg/m^3 in the South to 22 µg/m^3 in the densely populated and industrialised western part of the country. Comparison to observations shows that present-day chemistry transport models are not able to fully explain the observed particulate matter mass. LOTOS-EUROS misses about 40% of the measured PM10 mass, of which a large part can be explained. The model underestimates nitrate concentrations by about 25%, whereas total carbon concentrations are underestimated by about 60%. The latter is largely attributed to organic matter and corresponds to up to 4–5 µg/m^3 of OM. The reason for the underestimation of organic carbon is that

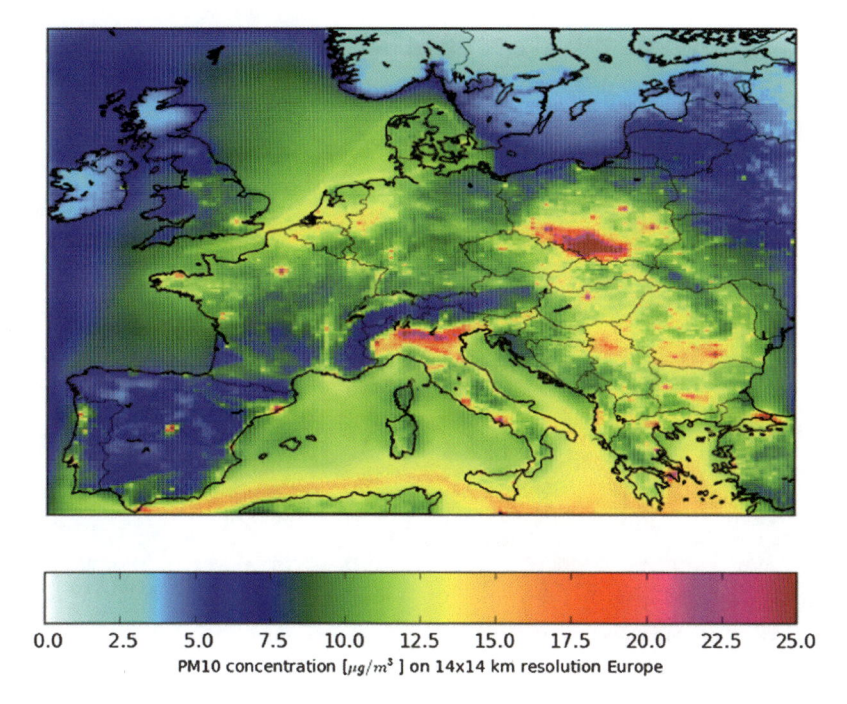

Fig. 4 Modelled distribution of PM10 across Europe for 2007–2009 excluding mineral dust

the model does not include a relevant process description. The uncertainties in the process description including secondary organic aerosol formation and the semi-volatile nature of OM are so large that they are not considered robust enough to incorporate in a source apportionment study. Model to observation comparison shows no significant biases for sulphate, ammonium and sea salt. Hence, the source apportionment presented here covers about 60% of the observed PM mass.

The source apportionment is performed for each component separately. Figure 5 shows the source attribution per sector for the most important PM components averaged across the Netherlands. Some components are dominated by a few sectors. Agriculture is by far the most important source of ammonium, causing over 90% of the emissions and concentration of this substance in the Netherlands. The mineral dust concentration in the Netherlands originates for about 25% from outside the model domain which is mainly associated with a few desert dust episodes. The remaining mineral dust is equally divided between agriculture and road transport. The unspecified primary particulate mass is dominated by industrial process emissions. Nitrate, sulphate and EC concentrations originate mainly from sectors in which fuel combustion is important (e.g. transport, industrial combustion and power generation). For the other components, the sector origin is more diffuse.

Summing all separate contributions allows to assess the origin of the modelled total mass. The most important contributions to total PM10 mass in this region are associated with agriculture, on-road and off-road transport and natural sources. Together these explain about 75% of the modelled mass. Secondary contributions

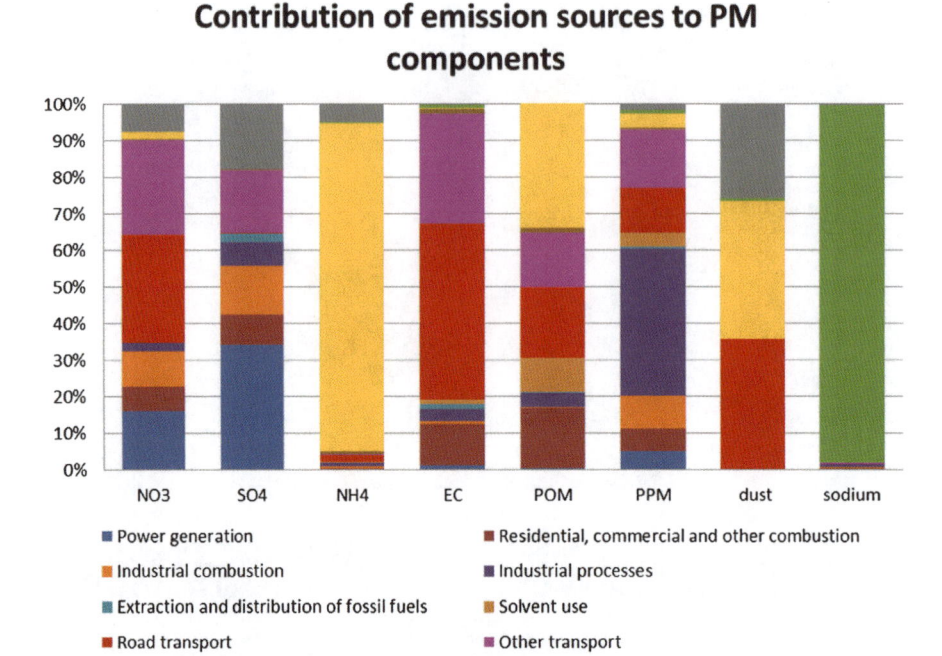

Contribution of emission sources to PM components

Fig. 5 Annual average origin of modelled PM10 components in the Netherlands [27]

are derived from power generation, industrial processes and combustion as well as households. Waste treatment, solvent use and extraction of fossil fuels do not contribute significantly. Of the modelled part, 70–80% of PM10 over the Netherlands is anthropogenic, which is in agreement with the analysis described in the previous section.

To investigate the differences in source contributions between peak episodes and periods with lower modelled PM concentrations, all days were categorised based on the average modelled concentration of PM10. The source attribution as function of PM10 concentration in the Netherlands is shown in Fig. 6. The natural contribution is highest when the total modelled PM concentration is low. Low concentrations are associated with westerly winds, resulting in transport of sea salt from the North Sea and Atlantic Ocean to the Netherlands. High PM concentrations occur mainly with easterly winds and/or stable stagnant conditions, during which the influx of sea salt is much smaller. The increase in concentration going from low to high PM levels is proportional for most sectors, except for agriculture and transport, which become more important mainly due to the more than proportional rise in ammonium nitrate concentrations. In the high concentration range also the impact of a few desert dust episodes is visible.

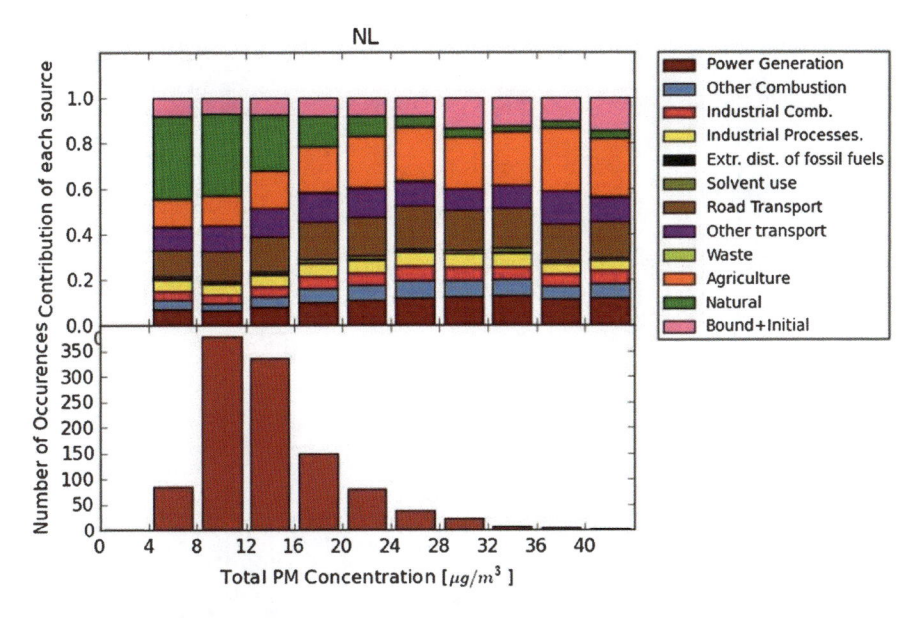

Fig. 6 Modelled source attribution on a sectorial basis for the Netherlands as function of total modelled PM10 mass. For each concentration bin, the number of occurrences is given

To contrast the situation in north-western Europe, we also present results for Spain. Before addressing the source attribution, we need to highlight an important difference between the two regions. In north-western Europe, the population density is generally high, and cities are located close to each other. In Spain, this is not the case. Major cities with high PM10 concentrations are located far from each other, and away from the coast population density is very low. Hence, large differences in concentrations are present between the rural background and urban concentration levels.

Figure 7 presents the source attribution as function of modelled PM10 concentration averaged across the whole Spanish mainland. Hence, the apportionment is dominated by the rural areas. This feature also explains the relatively low modelled concentrations, also in comparison to north-western Europe. Note that the emissions from the ocean are depicted as natural, whereas those of dust are classified as boundary. This is due to the fact that a global model simulation incorporating desert dust is used to provide the influx at the model domain boundaries. Whereas sea-salt concentration decreases with PM in Spain, the impact of desert dust on PM episodes is clearly recognisable and, for obvious reasons (Saharan desert sand), much larger than in north-western Europe. Summarised, the natural sources (sea salt and desert dust) contribute about half of the modelled PM10 concentrations across the whole of Spain. Anthropogenic sources that contribute significantly are similar to those in north-western Europe and comprise agriculture, transport and industrial combustion. Zooming into large urban agglomerations such as Madrid and Barcelona, the importance of anthropogenic sources is significantly larger than for the country as a whole and cause regional maxima modelled PM10 (Fig. 4). It is especially the transport sector that significantly gains importance.

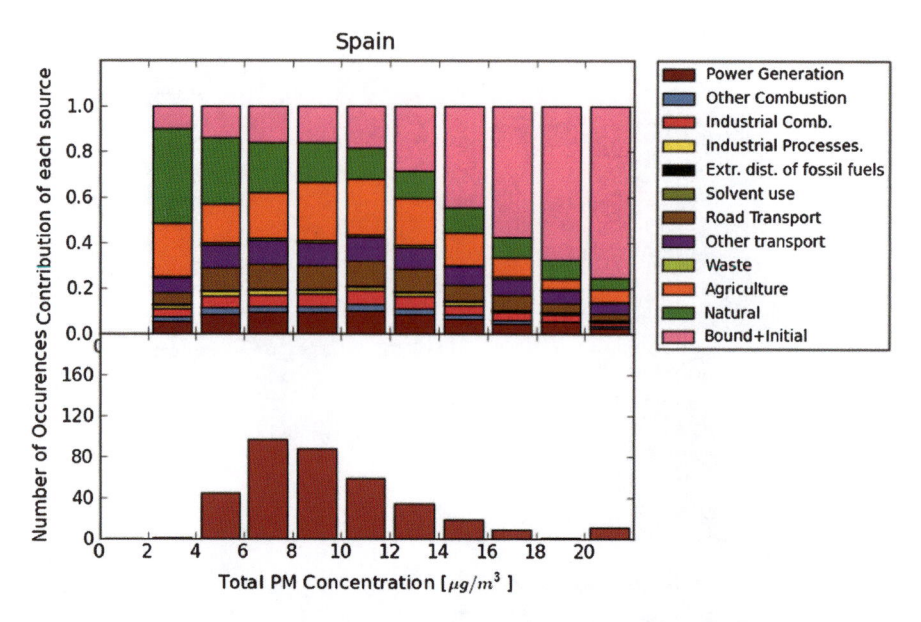

Fig. 7 Modelled source attribution on a sectorial basis for the Spain as function of total modelled PM10 mass. For each concentration bin, the number of occurrences is given. Note that desert dust is incorporated as boundary conditions in the simulations, explaining the large contribution of boundary conditions for Spain

5 Concluding Remarks

The chemical composition of PM10 on various measurement sites in the Netherlands, Germany and Belgium shows a considerable conformity. Always, SIA is the major constituent (±40%) followed by the carbonaceous compounds (±25%). Contributions of sea salt and mineral dust vary between 10% and 15% depending on location (distance to the North Sea) or presence of local sources. The unidentified mass is only in the order of 15% showing that the composition of PM10 in the region is rather well known.

PM10 concentrations are systematically higher at urban sites. Roughly, every constituent in PM10 appears higher in the urban area. The reduced dispersion in urban areas and the presence of dominating (anthropogenic) sources are the main reasons. However, the locations and distance between the rural and urban site may also have influenced the analysis. For instance, the higher urban concentrations of sea salt are induced by a closer proximity to the coast, though a small gradient in salt may be explained by the resuspension of road salt in winter. Elementary carbon in cities primarily originates from diesel emissions. The organic mass is a very complex group with anthropogenic and natural sources. Although Robinson et al. [28] showed that photo-oxidation of diesel emissions rapidly generates organic aerosol ten Brink et al. [13], indicated that only one-third of the OC in a Dutch city was due to fossil fuel combustion. Hence, the origin of OM remains unclear.

Mineral dust sources in urban areas include road dust resuspension and demolition and construction activities. There is no apparent reason why the unknown part should also be higher or lower at urban or rural sites. A possible cause may be the amount of water attached to SIA which appears increased. Hygroscopic salts on particles, like ammonium nitrate and ammonium sulphate in the fine fraction, and sodium nitrate and sodium sulphate in the coarse fraction attract water erroneously, increasing the PM mass.

The reason why SIA is higher in urban areas is less obvious as these are secondary aerosols. The observed increment is predominantly caused by more nitrate and sulphate. The reaction of nitric acid and sulphuric acid with the sea-salt aerosol in a marine urbanised environment follows an irreversible reaction scheme. In essence, the chloride depletion stabilises part of the nitrate and sulphate in the coarse mode and may partly explain part of the observed increment. However, it also raises the question how to assign the coarse mode nitrate in the mass closure. The sea salt and nitrate contributions cannot simply be added any more as nitrate replaces chloride. Reduction of NO_x emissions may cause a reduction of coarse mode nitrate, which is partly compensated by the fact that chloride is not lost anymore. A reduction would yield a net result of $((NO_3-Cl)/NO_3 = (62-35)/62=)$ 27/62 times the nitrate reduction (where the numbers are molar weights of the respective components), and this factor could be used to scale back the coarse nitrate fraction in the chemical mass balance. A similar reasoning may be valid for the anthropogenic sulphate in the coarse fraction. Corrections like these are uncommon in current mass closure studies, and consequences will have to be explored in more detail.

Important for the national and European air pollution policy is the question how much of the measured particulate matter is of anthropogenic origin. It is this fraction that can be targeted by national and European abatement strategies. A pragmatic assessment of the natural versus anthropogenic contributions to PM was given here. The estimation is that between 20% and 25% of PM10 is of natural origin for countries like Germany, Belgium and the Netherlands. Hence, the majority of PM in the north-western European region is of anthropogenic origin. The uncertainty in such an analysis is considerable, and the result should be taken as indicative.

A CTM was used to obtain a more detailed source apportionment for the Netherlands. The model explains about 60% of the observed PM10 mass concentration. Application of a dedicated source apportionment module showed that the origin of the individual species may differ considerably. The most important contributions to total PM10 mass in north-western Europe are associated with agriculture, on-road and off-road transport as well as natural sources (sea salt). Together these explain about 75% of the modelled mass. Secondary contributions are derived from power generation, industrial processes and combustion as well as households. Of the modelled part, 70–80% of PM10 over the Netherlands is anthropogenic. The increase in source contribution going from low to high PM levels is proportional for most sectors, except for agriculture and transport, which become more important mainly due to the more than proportional rise in ammonium nitrate concentrations. Sea-salt concentrations decline with rising PM10.

Whereas sea-salt concentrations also decrease with rising PM in Spain, the impact of Saharan dust on PM episodes is clearly recognisable and much larger than in north-western Europe. The natural sources (sea salt and desert dust) contribute about half of the modelled PM10 concentrations across the whole of Spain, according to our modelling exercise. Anthropogenic sources that contribute significantly are similar to those in north-western Europe.

Acknowledgements The use of the "CHEMKAR" data in this study was kindly permitted by the Flemish Environmental Agency [8]. The "BOP" campaign was organised by a consortium of the Dutch research institutes (ECN, TNO and RIVM) within the framework of the second Netherlands Research Programme on Particulate Matter (sponsored by the Ministry of Infrastructure and Environment I&M). The German air quality campaign (executed by IUTA) was sponsored by the North Rhine-Westphalia State Agency for Nature, Environment and Consumer Protection (LANUV) and the North Rhine-Westphalia Ministry of Environment (MKULNV). The LOTOS-EUROS modelling exercise presented here was partly funded by the 7th Framework Programme of the European Commission EnerGEO and by the second Netherlands Research Programme on Particulate Matter.

References

1. EEA (2007) Air pollution in Europe 1990–2004, EEA report 2/2007, Copenhagen
2. Putaud J-P, Raesa F, Van Dingenen R, Bruggemann E, Facchini M, Decesari S, Fuzzi S, Gehrig R, Hueglin C, Laj P, Lorbeer G, Maenhaut W, Mihalopoulos N, Mueller K, Querol X, Rodriguez S, Schneider J, Spindler G, ten Brink H, Torseth K, Wiedensohler A (2004) A European aerosol phenomenology - 2: chemical characteristics of particulate matter at kerbside, urban, rural and background locations in Europe. Atmos Environ 38:2579–2595
3. Putaud J-P, Van Dingenen R, Alastuey A, Bauer H, Birmili W, Cyrys J, Flentje H, Fuzzi S, Gehrig R, Hansson HC, Harrison RM, Herrmann H, Hitzenberger R, Hüglin C, Jones AM, Kasper-Giebl A, Kiss G, Kousa A, Kuhlbusch TAJ, Löschau G, Maenhaut W, Molnar A, Moreno T, Pekkanen J, Perrino C, Pitz M, Puxbaum H, Querol X, Rodriguez S, Salma I, Schwarz J, Smolik J, Schneider J, Spindler G, ten Brink H, Tursic JJ, Viana M, Wiedensohler A, Raes F (2009) A European aerosol phenomenology - 3: physical and chemical characteristics of particulate matter from 60 rural, urban, and kerbside sites across Europe. Atmos Environ 44:1308–1320
4. Viana M, Kuhlbusch TAJ, Querol X, Alastuey A, Harrison RM, Hopke PK, Winiwarter W, Vallius M, Szidat S, Prévôt ASH, Hueglin C, Bloemen H, Wåhlin P, Vecchi R, Miranda AI, Kasper-Giebl A, Maenhaut W, Hitzenberger R (2008) Source apportionment of particulate matter in Europe: a review of methods and results. J Aerosol Sci 39:827–849
5. Sillanpää M, Hillama R, Saarikoski S, Frey A, Pennanen A, Makkonen U, Spolnik Z, Van Grieken R, Braniš M, Brunekreef B, Chalbot M-C, Kuhlbusch T, Sunyer J, Kerminen V-M, Kulmala M, Salonen RO (2006) Chemical composition and mass closure of particulate matter at six urban sites in Europe. Atmos Environ 40:212–223
6. Salvador P, Artı B, Querol X, Alastuey A, Costoya M (2007) Characterisation of local and external contributions of atmospheric particulate matter at a background coastal site. Atmos Environ 41:1–17
7. Mazzei F, D'Alessandro A, Lucarelli F, Nava S, Prati P, Valli G, Vecchi R (2008) Characterization of particulate matter sources in an urban environment. Sci Total Environ 41:81–89

8. VMM (2009): Chemkar PM10: Chemische karakterisatie van fijn stof in Vlaanderen, 2006–2007. http://www.vmm.be/publicaties/2009/CK_PM10_TW.pdf/view (in Dutch with English summary)
9. PBL, Resultaten op hoofdlijnen en beleidsconsequenties, Beleidsgericht Onderzoeksprogramma Fijn Stof, PBL-rapport 500099013, 2010
10. Querol X, Alastuey A, Rodrıguez S, Plana F, Ruiz CR, Cots N, Massague G, Puig O (2001) PM10 and PM2.5 source apportionment in the Barcelona Metropolitan Area, Catalonia, Spain. Atmos Environ 35(36):6407–6419
11. Denier van der Gon H, Jozwicka M, Hendriks E, Gondwe M, Schaap M (2010) Mineral dust as a constituent of particulate matter. PBL report 500099003, Bilthoven, the Netherlands
12. Weijers EP, Sahan E, Ten Brink HM, Schaap M, Matthijsen J, Otjes RP, Van Arkel F (2010) Contribution of secondary inorganic aerosols to PM10 and PM2.5 in the Netherlands; measurements and modelling results. PBL Report 500099006, Bilthoven, the Netherlands
13. ten Brink HM, Weijers E.P, Röckmann T, Dusek U (2010) ^{14}C analysis of filter samples for source apportionment of PM in the Netherlands, ECN Report E–10-005, 2010
14. Perez N, Castillo S, Pey J, Alastuey A, Viana M, Querol X (2008) Interpretation of the variability of regional background aerosols in the Western Mediterranean. Sci Total Environ 407:527–540
15. Drechsler S, Uhrner U, Lumpp R (2006) Sensitivity of urban and rural ammonium nitrate particulate matter to precursor emissions in Southern Germany. Workshop on Contribution of Natural Sources to PM Levels in Europe, JRC Ispra, 12–13 October 2006
16. Bates TS, Lamb BK, Guenther A, Dignon J, Stoiber RE (1992) Sulfur emissions to the atmosphere from natural sources. J Atmos Chem 14:315–337
17. Simpson D, Winiwarter W, Börjesson G, Cinderby S, Ferreiro A, Guenther A, Hewitt N, Janson R, Khalil MAK, Owen S, Pierce T, Puxbaum H, Shearer M, Skiba U, Steinbrecher R, Tarrason L, Öquist MG (1999) Inventorying emissions from nature in Europe. J Geophys Res 104:8113–8152
18. Hoelzemann J, Schultz MG, Brasseur GP, Granier C, Simon M (2004) Global wildland fire emission model (GWEM): evaluating the use of global area burnt satellite data. J Geophys Res 109:D14S04
19. van der Hoek KW (1998) Estimating ammonia emission factors in Europe: summary of the work of the UNECE ammonia expert panel. Atmos Environ 32:315–316
20. Erisman JW, Sutton MA, Galloway J, Klimont Z, Winiwarter W (2009) How a century of ammonia synthesis changed the world. Nat Geosci 1:636–639
21. Schaap M, Spindler G, Schulz M, Acker K, Maenhaut W, Berner A, Wieprecht W, Streit N, Mueller K, Brüggemann E, Putaud J-P, Puxbaum H, Baltensperger U, ten Brink HM (2004) Artefacts in the sampling of nitrate studied in the "INTERCOMP" campaigns of EUROTRAC-AEROSOL. Atmos Environ 38:6487–6496
22. Simpson D, Yttri KE, Klimont Z, Kupiainen K, Caseiro A, Gelencsér A, Pio CA, Puxbaum H, Legrand M (2007) Modeling carbonaceous aerosol over Europe: analysis of the CARBOSOL and EMEP EC/OC campaigns. J Geophys Res Atmos 112:D23S14
23. Korcz M, Fudała J, Kliś C (2009) Estimation of windblown dust emissions in Europe and its vicinity. Atmos Environ 43:1410–1420
24. Schaap M, Manders AMM, Hendriks ECJ, Cnossen JM, Segers AJS, Denier van der Gon HAC, Jozwicka M, Sauter FJ, Velders GJM, Matthijsen J, Builtjes PJH (2009) Regional modelling of particulate matter for the Netherlands, PBL Report 500099006, Bilthoven, the Netherlands
25. Kuhlbusch TAJ, John AC, Quass U (2010) Sources and source contributions to fine particles. Biomarkers 14:23–28
26. Schaap M, Kranenburg R, Huibregtse JN, Segers AJA, Hendriks C (2012) Development of a source apportionment module in LOTOS-EUROS. TNO report TNO-060-UT-2012-00161, 2012
27. Hendriks C, Kranenburg R, Kuenen J, van Gijlswijk R, Denier van der Gon H, Schaap M (2012) The origin of ambient Particulate Matter concentrations in the Netherlands, TNO report 060-UT-2012-00474, Utrecht, the Netherlands

28. Robinson AL, Donahue NM, Shrivastava MK, Weitkamp EA, Sage AM, Grieshop AP, Lane TE, Pierce JR, Pandis SN (2007) Rethinking organic aerosols: semi volatile emissions and photochemical aging. Science 315:1259–1262
29. Almeida SM, Pio CA, Freitas MC, Reis MA, Trancoso MA (2005) Source apportionment of fine and coarse particulate matter in a sub-urban area at the Western European Coast. Atmos Environ 39:3127–3138
30. Friedrich R (2007) Improving and applying methods for the calculation of natural and biogenic emissions and assessment of impacts to the air quality. Final project activity report 2007. http://natair.ier.uni-stuttgart.de/. Accessed March 2011
31. Weijers EP, Schaap M, Nguyen L, Matthijsen J, Denier van der Gon HAC, ten Brink HM, Hoogerbrugge R (2011) Anthropogenic and natural constituents in particulate matter in the Netherlands. Atmos Chem Phys 11:1–14
32. Quass U, John AC, Kuhlbusch TAJ (2012) Source apportionment of airborne dust in Germany: methods and results. Hdb Env Chem, DOI 10.1007/698_2012_182

Particulate Matter and Exposure Modelling in Europe

A.I. Miranda, J. Valente, J. Ferreira, and C. Borrego

Abstract Recent estimates delivered by the European Environment Agency indicate that exposure to atmospheric particulate matter (PM) causes approximately three million deaths per year in the world. Exceedances of PM thresholds have been reported by the majority of the European Union member countries, mainly in urban agglomerations where human exposure is also higher.

Health effects of air pollution, namely of PM levels, are the result of a chain of events, going from the release of pollutants that lead to an atmospheric concentration, over the personal exposure, uptake, and resulting internal dose to the subsequent health effect.

The focus of this chapter will be on PM ambient concentrations as input for population exposure estimation, which are key variables for exposure models, and are generally measured at air quality monitoring stations or estimated by an adequate air quality modelling system. This chapter critically overviews PM modelling activities going on in Europe in the scope of exposure assessment, identifying advantages and gaps and recommending future uses and developments.

Keywords Exposure, Health effects, Modelling, Particulate matter

Contents

1 Introduction 260
2 Air Quality Modelling 261
3 Exposure Modelling 263
4 PM Exposure Modelling Applications in Europe 266
5 New Developments 268
6 Final Comments 270
References 271

A.I. Miranda (✉), J. Valente, J. Ferreira, and C. Borrego
CESAM & Department of Environment and Planning, University of Aveiro, 3810 - 193 Aveiro, Portugal
e-mail: miranda@ua.pt

1 Introduction

Particles in the atmosphere arise from natural sources, such as windborne dust, sea spray and volcanoes, and from anthropogenic activities, such as combustion of fuels. Emitted directly as particles (primary aerosol) or formed in the atmosphere by gas-to-particle conversion processes (secondary aerosol), atmospheric aerosols are generally considered to be the particles that range in from a few nanometres to tens of micrometres in diameter [1].

Increased understanding of the health issues associated with high concentrations of particulate matter (PM) have contributed to the identification of PM as the reason for one of the most critical air pollution problems nowadays. Estimates from the European Environment Agency [2] indicate that exposure to PM causes approximately three million deaths per year in the world. PM exceedances to the European Directive 2008/50/EC [3] thresholds have been reported by the majority of the European Union (EU) member countries, mainly in urban agglomerations, where human exposure is also higher [4]. Fine particulates ($PM_{2.5}$) are considered to be responsible for increased mortality over Europe. Anthropogenic $PM_{2.5}$ levels are expected to be responsible for a loss of 10 months of life expectancy in some regions of Europe by 2020, in spite of application of the current legislation devoted to air pollution control [5]. It is also recognised that adverse effects from PM long-term exposure occur whatever the concentration levels are [6, 7].

Health effects of air pollution, namely of PM levels in the air, are the result of a chain of events, going from the release of pollutants leading to an ambient atmospheric concentration, over the personal exposure, uptake, and resulting internal dose to the subsequent health effect (Fig. 1).

The conditions for these events vary considerably and have to be accounted for, in order to ensure a proper assessment [8]. Ambient concentrations and exposure are the stages of the chain that are mainly covered within this chapter.

Human exposure refers to the individual contact (not uptake) with a pollutant concentration. It is, then, important to distinguish between concentration and exposure. According to Sexton and Ryan [9] concentration is a physical characteristic of the environment at a certain place and time (amount of material per unit volume of air), while the term exposure stands for the interaction between the environment and a living subject. For exposure to take place two events need to occur simultaneously: pollution concentration at a particular time and place, and the presence of a person in that same place and time. Thus exposure has the dimension of "(mass) × (time)/(volume)".

Exposure studies can be carried out aiming to estimate exposure of one individual (personal exposure) or of a larger population group (population exposure), through direct or indirect methods. Direct methods are measurements made by personal portable exposure monitors. The personal exposure monitoring devices that people carry with them must be lightweight, silent, highly autonomous and

Emissions → Ambient Concentrations → Exposure → Dose → Health Effects

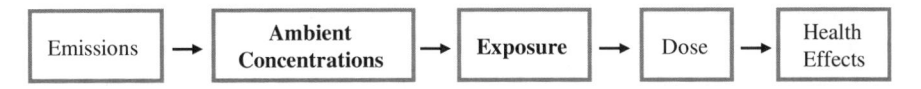

Fig. 1 Chain of events: from emission of pollutants to health effects

1 week is about the maximum time that any population representative sample of individuals will comply with personal exposure measurements [8, 9].

When using indirect methods, the exposure is determined by combining information about the time spent in specific locations, called microenvironments, and the pollutants concentrations at these same places. A microenvironment is defined as a three-dimensional space where the pollution concentration at some specified time is spatially uniform or has constant statistical properties [8]. The microenvironment can be interior of a car, inside a house, inside an office or school, outdoors, etc. Thus total exposure is the sum of the exposure in all microenvironments during the time of interest.

A constraint for using the indirect method is that the residence time of the person (termed time–activity pattern) needs to be known together with the pollution concentrations in each of the microenvironment at the time the person is present.

This chapter will focus on PM ambient concentrations, which are key variables for exposure models, and are generally obtained by direct measurements in air quality monitoring stations. However, depending on the location and dimension of the region to be studied, monitoring data could not be sufficient to characterise PM levels or to perform population exposure estimations. Numerical models complement and improve the information provided by measured concentration data. These models simulate the changes of pollutant concentrations in the air using a set of mathematical equations that translate the chemical and physical processes in the atmosphere.

In this scope the purpose of the present work is to critically overview PM modelling activities, going on mainly in Europe, aiming to identify advantages and gaps and to recommend future uses and developments.

2 Air Quality Modelling

Air quality models use mathematical and numerical techniques to simulate the physical and chemical processes that affect air pollutants, namely PM, as they disperse and react in the atmosphere. Based on meteorological and emission data inputs, these models are designed to characterise primary PM emitted directly into the atmosphere and, in some cases, secondary PM formed as a result of complex chemical reactions within the atmosphere.

Seigneur [10] provides an overview of the current status of air quality models simulating PM levels. Holmes and Morawska [11], more recently, have performed a detailed review of modelling tools regarding the dispersion of particles in the atmosphere.

Dispersion PM air quality models are usually simple; they estimate the concentration of PM at specified ground-level receptors, considering only the dispersion and not the chemical transformation processes. They simulate the atmospheric transport, the turbulent atmospheric diffusion, and some are also able to simulate the ground deposition. The most simple dispersion models use the Gaussian approach (ISC3, AERMOD, AUSPLUME are examples of Gaussian models currently in use to estimate PM concentration values). Notwithstanding the simplicity of this approach, Gaussian models have already been used to estimate PM concentrations for exposure calculation. For example, some authors [12] evaluated population exposure to PM using information from a multiple-source emission and a Gaussian dispersion modelling system.

Chemical aerosol models simulate the changes of PM in the atmosphere using a set of mathematical equations characterising the chemical and physical processes in the atmosphere. They became widely recognised and routinely utilised tools for regulatory analysis and attainment demonstrations by assessing the effectiveness of control strategies. The simulation of the dynamics of multicomponent atmospheric aerosols is an impressive problem that includes new particle formation by homogeneous heteromolecular nucleation, gas-to-particle conversion, coagulation and dry deposition [13]. Most of the current chemical aerosol models have adopted the three-dimensional Eulerian grid modelling mainly because of its ability to better and more fully characterise physical processes in the atmosphere and predict the species concentrations throughout the entire model domain [14].

The choice of an appropriate model is heavily dependent on the intended application. In particular, the science of the model must match the pollutant(s) of concern. If the pollutant of concern is fine PM, the model chemistry must be able to handle reactions of nitrogen oxides (NO_x), sulphur dioxide (SO_2), volatile organic compounds (VOC), ammonia, etc. Reactions in both the gas and aqueous phases must be included, and preferably also heterogeneous reactions taking place on the surfaces of particles. Apart from correct treatment of transport and diffusion, the formation and growth of particles must be included, and the model must be able to track the evolution of particle mass as a function of size. The ability to treat deposition of pollutants to the surface of the earth by both wet and dry processes is also required.

Air quality PM models require a considerable volume of data. The specific needs reflect the science incorporated in the model, but typically include the following: emissions for all sources and for each of the chemical species treated by the model; geophysical data as topography, land use category, vegetation type and additional data for some local scale modelling, like building geometry; meteorology to drive the transport and dispersion in the model; and initial and boundary conditions taken from typical or averaged values measured, or previously modelled, for the region of interest. Figure 2 presents the typical structure of an off-line air quality modelling system, including the inputs and outputs usually considered.

Output data are usually the temporal and spatial distribution of PM concentration values and sophisticated modelling approaches are available, which allow assessing PM at high spatial and temporal resolution. These PM concentration results will be the basis, with time–activity profiles, to exposure estimation.

Particulate Matter and Exposure Modelling in Europe

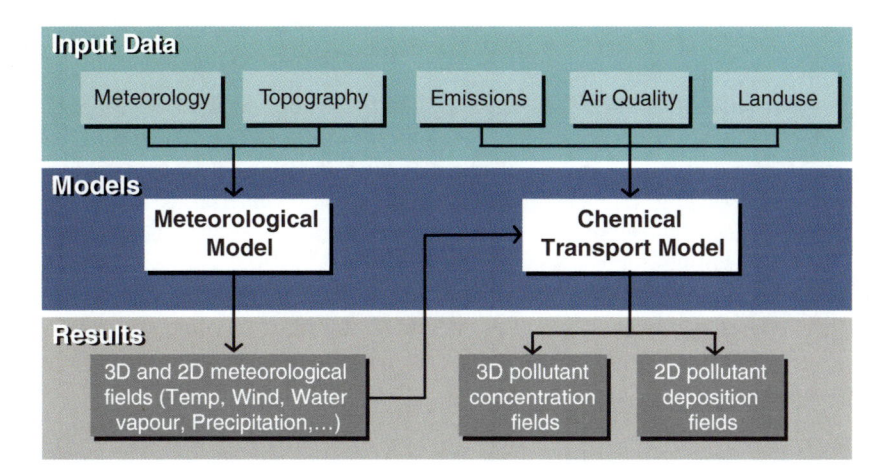

Fig. 2 Structure of an off-line air quality modelling system, showing the inputs and outputs of a meteorological and a chemical-transport dispersion model

Because of the effects of uncertainty and its inherent randomness, it is not possible for an air quality model to ever be "perfect", and there is always a base amount of scatter that cannot be removed [15]. Uncertainties in PM modelling have been estimated in some works. For instance, Borrego et al. [16] evaluated PM_{10} hourly concentration results over Berlin for the year 2002, which were obtained using two different scale model approaches. At the regional/urban scale, the applied Eulerian chemistry-transport model was able to simulate PM_{10} with a satisfactory uncertainty level, according to the quality objectives defined by the EU Directive and to the statistical quality parameters suggested by Borrego et al. [17]. At local scale, a Canyon Plume Box model has been used to simulate PM_{10} concentrations. The uncertainty of PM_{10} model results exhibited also a reasonable performance regarding the model quality objectives and statistical parameters application.

In summary, notwithstanding the need to always assess PM modelling uncertainties PM air quality models are widely applied in Europe for various purposes, such as to help decision makers on the development of policies and air quality management systems for protection of ecosystems and human health or for air quality forecast, and consequently human exposure and health effects prevention.

3 Exposure Modelling

PM concentration fields coming from air quality models are estimations of outdoor microenvironments that combined with gridded population and microenvironments information can be used for exposure modelling and estimation of doses and health effects, integrating the source to dose assessment chain (recall Fig. 1).

The general approach for exposure estimation can be expressed by Hertel et al. [8] equation:

$$\exp_i = C_i \times t_i, \qquad (1)$$

where $\exp_i$ is the total exposure for the person i over the specified period of time, C_i is the pollutant concentration in a given location, and t_i is the time spent by the person i in that specific location. As a result, the exposure value is expressed in concentration × time (e.g. µg m^{-3} h), and thus can be interpreted as the mean pollutant concentration value to which the individual has been exposed during a given period of time (e.g. 1 h).

For exposure assessment, a combination of the spatial distribution of air quality, namely PM levels, and an individual location on time (for personal exposure) and population activity data and density (for population exposure) is required. In what concerns time–activity data strategies are different when defining personal or population exposure. For personal exposure, what is needed is an individual time–activity pattern. This can be obtained, for example, by personal interview or using GPS tracking, since the exact place of the individual at each moment is needed. For population exposure the used information is usually aggregated by population class, and is focused on mobility of the population, work/school–home displacements, presented, for example, as origin–destination matrixes (number of people and time spent in displacement by means of transport). Those matrices allow the calculation of the number of people that enters and leaves the modelling domain. Numerical models are needed for the PM mapping along the time, once the monitoring networks are able to assess the air quality in the single stations of the monitoring network, and not the whole area of interest.

According to IPCS [18] an exposure model is a conceptual or mathematical representation of the exposure process, designed to reflect real-world human exposure scenarios and processes. There are many different ways to classify exposure models. A consensus appears to be developing around the following classification scheme proposed by the World Health Organization [19], which has been adopted in this chapter: (a) mechanistic or empirical and (b) deterministic or stochastic (probabilistic). Table 1 lists these model categories. However, alternative classifications may be considered as well.

Kousa et al. [20] classified exposure models as statistical, mathematical and mathematical-stochastic models. Statistical models are based on the historical data and capture the past statistical trend of pollutants [21]. The mathematical modelling, also called deterministic modelling, involves application of emission inventories, combined with air quality and population activity modelling. The stochastic approach attempts to include a treatment of the inherent uncertainties of the model [22].

Mathematical exposure models applied to urban areas have been presented by Jensen [23], Kousa et al. [20] and Wu et al. [24]. The model presented by Jensen [23] is based on the use of traffic flow computations and the operational street pollution model (OSPM) for evaluating outdoor air pollutants concentrations in urban areas. The activity patterns of the population have been evaluated using

Table 1 Model categories [19]

	Mechanistic	Empirical
Deterministic	Mathematical constructs of physical/chemical processes that predict fixed outputs for a fixed set of inputs	Statistical models based on measured input and output values (e.g. regression models that relate air concentrations and blood levels of a chemical or ambient pollutant concentration with personal exposures)
Stochastic	Mathematical construction of physical/chemical processes that predict the range and probability density distribution of an exposure model outcome (e.g. predicted distribution of personal exposures within a study population)	Regression-based models, where model variables and coefficients are represented by probability distributions, representing variability and/or uncertainty in the model inputs and parameters

various administrative databases and standardised time–activity profiles. The modelling system uses a GIS in combining and processing the concentration and population data activity. The model was applied to evaluate population exposure in one specific municipality in Denmark.

An individual exposure model (IEM) was developed by Wu et al. [24] to retrospectively estimate the long-term average exposure of individual children from Southern California to several pollutants, including PM_{10} and $PM_{2.5}$. In the IEM model, pollutant concentration due to both local mobile source emissions and meteorologically transported pollutants were taken into account by combining a Gaussian line source model (CALINE4) with a regional air quality model (SMOG). Information from the Southern California Children's Health Study (CHS) survey was used to group each child into a specific time–activity category, for which corresponding time–activity profiles were sampled.

Once the outdoor concentration has been calculated by air quality models, the indoor pollutant concentration can also be modelled based on an understanding of the ways in which indoor air becomes exchanged with outdoor air, together with the deposition or decay dynamics of the pollutants, and with indoor emission source rates characteristics. Several methodologies exist to estimate indoor air pollution concentrations from outdoor modelled concentrations. These include a variety of empirical approaches based on: statistical evaluation of test data and a least-square regression analysis; deterministic models based on a pollutant mass balance around a particular indoor air volume; or a combination of both approaches. Most of the current available studies (e.g. [24–29]) are based on experimental data, resulting from measurements of outdoor and indoor concentrations for different microenvironments in order to establish a relation between indoor and outdoor (I/O) PM concentrations. Morawska and Congrong [30] presented a review of studies conducted in different countries concluding that in the absence of known indoor sources the I/O ratios range from 0.50 to 0.98. These values show that firstly the contribution of outdoor air as a source of indoor particles is very significant, and secondly that is rather consistent across all the studies.

4 PM Exposure Modelling Applications in Europe

A number of PM exposure modelling studies based on air quality modelling results have already been performed in Europe. Hänninen et al. [31] present the results of the estimation of children (<15 years) and elderly (> = 65 years) exposure in Turin city area for a specific time period. An Eulerian mesoscale model was used to obtain PM_{10} ambient concentrations in a 1 km × 1 km horizontal resolution grid, which were then used to estimate the indoor concentrations necessary for exposure calculations [31]. For the studied episode, average exposure levels experienced by both population groups (43.5 and 45.8 µg m^{-3}) were lower than the levels recorded at urban monitoring sites (111.3 µg m^{-3}), as could be expected due to the substantial fraction of time spent indoors. Within the whole modelling area, average exposures ranged from 20 to 95 µg m^{-3}.

Exposure to PM_{10} air concentrations was estimated for the municipality of Porto (the second largest Portuguese city) based on the application of a mesoscale air quality and exposure modelling system [32]. Data on population, time–activity patterns, microenvironments characterisation and input/output empirical relations were gathered. Figure 3 presents the annual averages of simulated PM_{10} concentration and individual exposure fields for the study domain. The spatial distribution of exposure levels follows the concentration field; however, exposure annual averages (varying between 40 and 70 µg m^{-3} h) are much lower than outdoor average concentrations, which reach 90 µg m^{-3}. The results have also shown important differences between outdoor and indoor concentrations, stressing the need to include indoor concentrations quantification in the exposure assessment.

Some other studies were performed relating human exposure in urban areas based on ambient PM air concentrations determined with computational fluid dynamics (CFD) modelling applications. The three-dimensional CFD model MISKAM has been successfully implemented to provide better assessment of exposure to traffic-related air pollutants in urban areas [33].

Borrego et al. [26] describe a methodology (schematically shown in Fig. 4) to estimate population exposure to traffic-related PM in urban areas based on the estimation of ambient pollutant concentrations with the CFD model VADIS. This methodology combines information on concentrations at different microenvironments and population time–activity pattern data. A downscaling from a mesoscale meteorological and dispersion model to a local scale air quality model was done to define the boundary conditions for the local scale application. Simple I/O relations were used by Borrego et al. [26] to determine PM_{10} indoor concentrations from outdoor concentrations. The prime objective was the quantification of an integrated exposure expressed as an accumulated population exposure index (APEI). The APEI index was defined as the daily accumulated exposure over

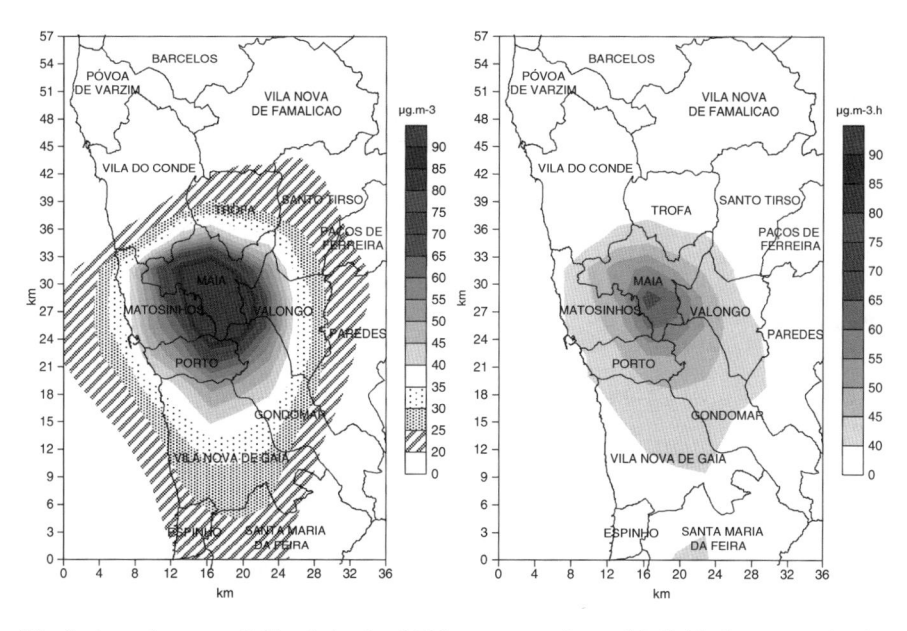

Particulate Matter and Exposure Modelling in Europe 267

Fig. 3 Annual average fields of simulated PM_{10} concentration and individual exposure for the Metropolitan Area of Porto study domain

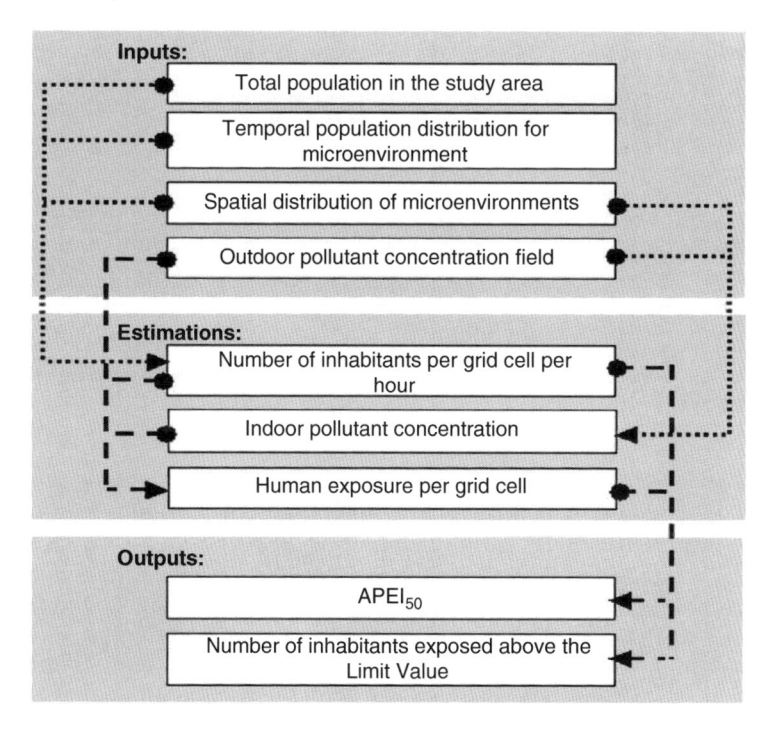

Fig. 4 Schematic representation of the exposure model [26]

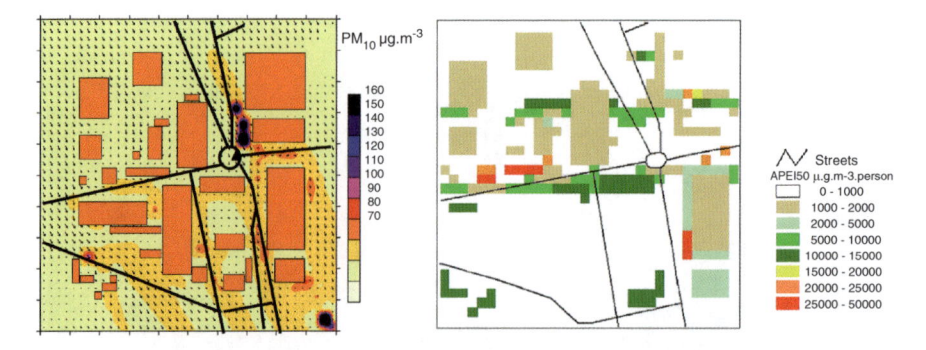

Fig. 5 PM$_{10}$ simulated field and exposure results for an urban area in Lisbon [26]

the pollutant concentration threshold and weighted by the number of inhabitants exposed. Therefore, it is calculated as a sum of the positive differences between the hourly mean PM$_{10}$ concentration and a threshold, multiplied by the number of inhabitants exposed, and then integrated over 24 h. Results of a PM$_{10}$ hourly field predicted by the VADIS local dispersion model and the population exposure expressed in terms of the accumulated index (APEI50) are presented in Fig. 5.

This methodology was a first approach to estimate population exposure, calculated as the total daily values above the thresholds recommended by the European Commission Directive for long- and short-term health effects. Obtained results revealed that in Lisbon city centre a large number of persons are exposed to PM levels exceeding the legislated daily limit value of 50 μg m^{-3}.

McNabola et al. [34] also used a CFD to investigate whether pedestrians using the boardwalk would have a lower air pollution exposure than those using the adjoining footpath along the road. The results show considerable reductions in pedestrian exposure to traffic derived PM along the boardwalk as opposed to footpath.

5 New Developments

Aerosol modelling tools registered a fast development along the last years. Nowadays, and besides the complexity of the formation mechanism of aerosols and the great number of individual chemical species involved, it is possible to simulate them and to estimate population exposure based on chemical transport model results. Modelling organic aerosol is among the most demanding aspects of air quality simulations because the formation processes and evolution are poorly understood and, in spite of the recent improvements in air quality models, organic aerosols can be underestimated by a large amount [35, 36]. Improvements still have to be done, regarding the chemical and physical processes, especially in the case of secondary organic compounds (SOA), which processes continue to be involved in a high level of uncertainty.

Also special care should be taken to reduce uncertainties on emission data and measurements. The validation of an aerosol model requires the analysis of the aerosol chemical composition for the main particulate species (ammonium, sulphate, nitrate and secondary organic aerosol). To find data to perform this kind of more complete evaluation is not always easy. The same applies to emissions data. The lack of detailed information regarding the chemical composition of aerosols obliges modellers to use previously defined aerosols components distributions, which are found in the literature. Present knowledge in emission processes is yet lacunal, especially concerning suspension and resuspension of deposited particles [37].

Moreover, the large diversity of environments and sources of aerosols makes difficult to obtain a general assessment of models performance for a given model all over a large region like Europe, simultaneously at urban and regional scales. Northern Europe undergoes a very different climate than southern Europe, and sources mostly result from combustion in large cities and industrial areas, while southern Europe may be significantly influenced by wind-blown dust.

The issue of modelling scales and objective is very important. City-Delta exercise in which mesoscale models were used to calculate PM concentrations in urban areas allowed concluding that finer-scale models show better performance for PM_{10} in the cities than large-scale models. However, these improvements are limited because these models do not generally use small-scale meteorology, and still have limited vertical resolution [37]. A further increase in horizontal resolution, aiming to reach urban scale, may be necessary to increase further the skill. Dynamical downscaling, from regional till local scale, could be a way to increase the simulation resolution, taking into account the influence of large-scale phenomena in micro-scale.

Improvements of aerosol modelling performance have been focused on data assimilation of PM and aerosols species. However, they have so far mainly been indirectly assimilated via aerosol optical depth (AOD), either using satellite data alone, or in combination with ground-based data. This is confounded by a number of difficulties due to the complex character of the different constituents of PM. It is thus not possible to use such assimilation systems yet to assess the PM concentrations in cities or urban areas. Before using more sophisticated techniques of data assimilation it is of vital importance that the model is thoroughly evaluated and validated, using proper model input data and measurements. This part of the work should not be underestimated, as it often forms the key to success in combining monitoring and modelling. Ideally, the model should show little or no bias as compared to the measurements, i.e., the model should not underestimate or overestimate concentrations systematically, or as an average, as compared to the measurements. Most data assimilation methods work best if there are little or no bias between the model and the measurements. There should also be a reasonable good time correlation (perhaps 0.5 or higher) between the two before attempting to use such methods.

The ensemble approach should also be further investigated as a way to improve PM estimates. The Air Quality Model Evaluation International Initiative (AQMEII—http://aqmeii.jrc.ec.europa.eu/aqmeii2.html) is developing and testing innovative model evaluation methodologies, focusing on ensemble, to improve knowledge about relevant processes and to increase confidence in model performance for better support of policy development.

Exposure modelling is a promising approach to exposure assessment, using ambient air concentration modelling and the respective spatial and temporal distribution, combined with population statistics, to get a statistical assessment. Aiming at personal exposure information may become more accurate, using traffic models' information on optimised daily trips (work, school, shopping) or even tracking individuals mobile phone positions. Technology for exposure assessment is available; it is a matter of costs, data handling and data privacy regulations to access such information. Anyway, a statistical assessment may prove sufficient for many purposes.

Indoor environments are (in the absence of indoor sources) assumed to be much less exposed to PM than outdoor air, due to the filtering effect of the building. Nevertheless, some studies indicate that high indoor values are also found [28]. Still, data show that ambient concentration levels reflect also the temporal variation that is seen indoors, at least for PM. In this sense, the use of ambient data as a proxy for the overall exposure has to be further analysed, since it can introduce high inaccuracy in exposure calculations. How can the exposure to indoor sources (cooking, cleaning, candle lights, smoking) be assessed and what are the health implications of those sources are still questions to investigate, taking into consideration PM outdoor–indoor penetration and the dependency between lifetime indoor and the particles size, composition, physical properties (e.g. volatility, hygroscopicity) and removal processes (absorption/adsorption onto indoor surfaces). Moreover, it is clear that the fraction of time people spend indoors changes between seasons and between places (especially northern vs. southern Europe) and also depends on work habits that may change with time (mechanisation of outdoor labour, but also leisure activities). Finally, influence of the outdoor environment to indoor air quality or the indoor particle formation will strongly depend on age and cultural habits, partly again triggered by environmental conditions. Together with the need to look into other metrics than PM alone, this indicates the importance of better understanding the extent people are exposed to particulate matter – with exposure as the potential to inhale and retain a certain dose of material.

6 Final Comments

Air quality numerical models are useful tools for the mapping of PM, once the monitoring networks are able to assess the air quality in single stations, and not a whole area of interest. Mathematical models which simulate the evolution of both gaseous and aerosol species are recent due to the complexity and variability of the processes in which particulate matter is involved [38].

Nowadays, highly sophisticated modelling approaches are available, which allow assessing PM at high spatial and temporal resolution, as needed for human exposure estimation. Thus no new models need to be developed (models predicting transport and transformation of aerosols in the atmosphere are available). Instead, methods need to be devised which are able to reduce uncertainty of modelled outputs. The respective results made available for a certain use allows understanding if answers to specific user questions can or cannot be supplied reliably.

Therefore, before being used for policies and health evaluation, PM air quality models must be evaluated, a process which can now be carried out over long time periods [37, 39] due to the increase in computer power and memory. All models are useful and the choice of an appropriate model is heavily dependent on the intended application: the type and dimension of the area, and the final goal of the study (air quality management, exposure and health estimations, etc.).

As shown along this chapter, a reliable air quality model is a valuable tool for human exposure studies, once modelled concentrations at different spatial scales and time resolutions allow to better characterising the air quality at the microenvironments visited by a target population, rather than monitoring values that are site and time specific. Moreover, air quality and exposure modelling approach considers the contribution of indoor environments, where people spend most of their time, to the exposure estimation.

During the last two decades or so several chemistry-transport models have been developed in Europe and elsewhere. They are already widely applied for PM exposure and health-related issues, at local, urban and regional scales, but an effort is still needed to take more advantage of this third generation models (Chemical Transport Models including aerosol chemistry) on epidemiological studies.

Acknowledgements The authors would like to acknowledge the financial support of the 3rd European Framework Program and the Portuguese Ministry of Science, Technology and Higher Education, through the Foundation for Science and Technology (FCT), for the Post-Doc grants of J. Ferreira (SFRH/BPD/40620/2007) and J. Valente (SFRH/BPD/78933/2011) and for the funding of research project INSPIRAR (PTDC/AAC-AMB/103895/2008), supported in the scope of the Competitiveness Factors Thematic Operational Programme (COMPETE) of the Community Support Framework III and by the European Community Fund FEDER.

References

1. Seinfeld JH, Pandis SN (2006) Atmospheric chemistry and physics: from air pollution to climate change. Wiley, Hoboken
2. EEA (2003) Europe's environment: the third assessment. European Environment Agency, Copenhagen
3. EC (2008) Directive 2008/50/EC of the European Parliament and of the Council of 21 May 2008 on ambient air quality and cleaner air for Europe. European Commission, Brussels
4. EEA (2005) Environment and health. European Environment Agency, Copenhagen

5. Amann M, Bertok I, Cabala R, Gyarfas F, Heyes C, Klimont Z, Schopp W, Wagner F (2005) A further emission control scenario for the Clean Air for Europe (CAFE) Program. IIASA, Laxenburg
6. Pope CA, Burnett RT, Thun MJ, Calle EE, Krewski D, Ito K, Thurston GD (2002) Lung cancer, cardiopulmonary mortality, and long-term exposure to fine particulate air pollution. JAMA 287:1132–1141
7. WHO (2006) Health risks of particulate matter from long-range transboundary air pollution. World Health Organization, Copenhagen
8. Hertel O, De Leeuw F, Raaschou-Nielsen O, Jensen S, Gee D, Herbarth O, Pryor S, Palmgren F, Olsen E (2001) Human exposure to outdoor air pollution – IUPAC Technical Report. Pure Appl Chem 73:933–958
9. Sexton K, Ryan PB (1988) Assessment of human exposure to air pollution: methods, measurements and models. In: Watson AY, Bates RR, Kennedy D (eds) Air pollution, the automobile and public health. National Academic Press, Washington
10. Seigneur C (2001) Current status of air quality models for particulate matter. J Air Waste Manage Assoc 51:1508–1521
11. Holmes NS, Morawska L (2006) A review of dispersion modelling and its application to the dispersion of particles: an overview of different dispersion models available. Atmos Environ 40:5902–5928
12. Johansson C, Hadenius A, Johansson PA, Jonson T (1999) Shape. The Stockholm study on health effects of air pollution and their economic consequences. Part I. NO_2 and particulate matter in Stockholm-concentrations and population exposure. Swedish National Road Administration, Stockholm
13. Zannetti P (1990) Air pollution modelling. Van Nostrand Reinhold, New York
14. Reid N, Misra PK, Amman M, Hales J (2007) Air quality modeling for policy development. J Toxicol Environ Health A 70:295–310
15. Chang JC, Hanna SR (2004) Air quality model performance evaluation. Meteorol Atmos Phys 87:167–196
16. Borrego C, Monteiro A, Costa AM, Martins H, Miranda AI, Builtjes P, Lutz M (2007) Estimation of modelling uncertainty for air quality assessment: Berlin case. In: Sixth international conference of urban air quality, Limassol
17. Borrego C, Monteiro A, Ferreira J, Miranda AI, Costa AM, Carvalho AC, Lopes M (2008) Procedures for estimation of modelling uncertainty in air quality assessment. Environ Int 34:613–620
18. IPCS (2004) IPCS Risk Assessment Terminology. WHO, Geneva
19. WHO (2005) Principles of characterizing and applying human exposure models. WHO, Copenhagen
20. Kousa A, Kukkonen J, Karppinen A, Aarnio P, Koskentalo T (2002) A model for evaluating the population exposure to ambient air pollution in an urban area. Atmos Environ 36:2109–2119
21. Goyal P, Chan AT, Jaiswal N (2006) Statistical models for the prediction of respirable suspended particulate matter in urban cities. Atmos Environ 40:2068–2077
22. Burke JM, Zufall MJ, Ozkaynak H (2001) A population exposure model for particulate matter: case study results for PM2.5 in Philadelphia, PA. J Expo Anal Environ Epidemiol 11:470–489
23. Jensen SS (1999) A geographic approach to modelling human exposure to traffic air pollution using GIS. National Environmental Research Institute, Denmark
24. Wu J, Lurmann F, Winer A, Lu R, Turco R, Funk T (2005) Development of an individual exposure model for application to the Southern California children's health study. Atmos Environ 39:259–273
25. Baek S-O, Kim Y-S, Perry R (1997) Indoor air quality in homes, offices and restaurants in Korean urban areas – indoor/outdoor relationships. Atmos Environ 31:529–544
26. Borrego C, Tchepel O, Costa AM, Martins H, Ferreira J, Miranda AI (2006) Traffic-related particulate air pollution exposure in urban areas. Atmos Environ 40:7205–7214

27. Chau CK, Tu EY, Chan DWT, Burnett J (2002) Estimating the total exposure to air pollutants for different population age groups in Hong Kong. Environ Int 27:617–630
28. Martins PC, Valente J, Papoila AL, Caires I, Araújo-Martins J, Mata P, Lopes M, Torres S, Rosado-Pinto J, Borrego C, Annesi-Maesano I, Neuparth N (2011) Airways changes related to air pollution exposure in wheezing children. Eur Respir J 39:246–253
29. Monn C (2001) Exposure assessment of air pollutants: a review on spatial heterogeneity and indoor/outdoor/personal exposure to suspended particulate matter, nitrogen dioxide and ozone. Atmos Environ 35:1–32
30. Morawska L, Congrong H (2003) Particle concentration levels and size distribution characteristics in residential and non-industrial workplace environment. In: Morawska L, Salthammer T (eds) Indoor environment: airborne particles and settled dust. Wiley-VCH, Weinheim
31. Hänninen O, Karppinen A, Valkama I, Kauhaniemi M, Kukkonen J, Kousa A, Aarnio P, Sokhi R, Skouloudis A, Jantunen M (2005) Recommendations and best practices for population exposure assessment in the context of air quality modelling. FUMAPEX Project Deliverable 2005. Danish Meteorology Institute, Copenhagen
32. Borrego C, Lopes M, Valente J, Tchepel O, Miranda AI, Ferreira J (2008) The role of PM10 in air quality and exposure in urban areas. In: Brebbia CA, Longhurst JWS (eds) Air pollution 2008 conference. WIT Trans Ecol Environ 116:511–520
33. Lohmeyer A, Eichhorn J, Flassak T, Kunz W (2002) WinMISKAM 4.2 microscale flow and dispersion model for built up areas, recent developments. In: Eleventh international symposium transport and air pollution, Graz
34. McNabola A, Broderick BM, Gill LW (2008) Reduced exposure to air pollution on the boardwalk in Dublin, Ireland. Measurement and prediction. Environ Int 34:86–93
35. Aksoyoglu S, Keller J, Barmpadimos I, Oderbolz D, Lanz VA, Prévôt ASH, Baltensperger U (2011) Aerosol modelling in Europe with a focus on Switzerland during summer and winter episodes. Atmos Chem Phys 11:7355–7373
36. Hodzic A, Jimenez JL, Madronich S, Aiken AC, Bessagnet B, Curci G, Fast J, Lamarque JF, Onasch TB, Roux G, Ulbrich IM (2009) Modeling organic aerosols during MILAGRO: application of the CHIMERE model and importance of biogenic secondary organic aerosols. Atmos Chem Phys Discuss 9:12207–12281
37. Vautard R, Builtjes PHJ, Thunis P, Cuvelier C, Bedogni M, Bessagnet B, Honoré C, Moussiopoulos N, Pirovano G, Schaap M, Stern R, Tarrason L, Wind P (2007) Evaluation and intercomparison of Ozone and PM10 simulations by several chemistry transport models over four European cities within the CityDelta project. Atmos Environ 41:173–188
38. Nenes A, Pandis SN, Pilinis C (1998) ISORROPIA: a new thermodynamic equilibrium model for multiphase multicomponent inorganic aerosols. Aquat Geochem 4:123–152
39. Monteiro A, Miranda AI, Borrego C, Vautard R, Ferreira J, Perez AT (2007) Long-term assessment of particulate matter using CHIMERE model. Atmos Environ 41:7726–7738

Part II
Future Air Quality Monitoring Strategies and Research Directions

Air Pollution Monitoring Strategies and Technologies for Urban Areas

Thomas A.J. Kuhlbusch, Ulrich Quass, Gary Fuller, Mar Viana, Xavier Querol, Klea Katsouyanni, and Paul Quincey

Abstract Current ambient air quality monitoring is solely based on fixed monitoring sites, not always reflecting the exposure and effects on humans. This article reviews the current situation in Europe, the USA and Asia and discusses the main differences and similarities. Based on the analysis of the relation between monitoring techniques and strategies as well as the analysis of current trends and developments in monitoring strategies, new concepts and directions of future urban monitoring networks and strategies are presented and discussed.

T.A.J. Kuhlbusch (✉)
IUTA e.V., Air Quality & Sustainable Nanotechnology, Bliersheimer Strasse 60, 47229 Duisburg, Germany

Center for Nanointegration Duisburg-Essen (CENIDE), 47057 Duisburg, Germany

King's College London, Environmental Research Group, Franklin Wilkins Building, 150 Stamford Street, London SE1 9NH, UK
e-mail: tky@iuta.de

U. Quass
IUTA e.V., Air Quality & Sustainable Nanotechnology, Bliersheimer Strasse 60, 47229 Duisburg, Germany

G. Fuller
King's College London, Environmental Research Group, Franklin Wilkins Building, 150 Stamford Street, London SE1 9NH, UK

M. Viana and X. Querol
Institute of Environmental Assessment and Water Research (IDÆA-CSIC), Lluis Solé i Sabarís S/N, 08028 Barcelona, Spain

K. Katsouyanni
Department of Hygiene and Epidemiology, University of Athens, 75 Mikras Asias Avenue, 115 27 Athens, Greece

P. Quincey
National Physical Laboratory, Hampton Road, Teddington TW11 0LW, UK

Keywords Air quality regulation, Future developments, Technology trends, Urban air quality monitoring

Contents

1   Introduction 278
2   Current Legislative Monitoring Requirements in Different Parts of the World 280
    2.1   USA 280
    2.2   Asia 282
    2.3   Europe 283
    2.4   Comparison of the Air Quality Monitoring Strategies 284
3   How Do Air Quality Monitoring Technologies Influence Strategies? 286
4   Recent Technological Developments 289
5   Possible Future Monitoring Strategies 291
References 294

1 Introduction

Current strategies for monitoring urban air pollution arise from the abundant evidence of the harm that air pollution causes to human health (see the extensive review in [1]) and the broader environment. Types of harm include the acidification of soils and water [2, 3], eutrophication [4], and the erosion and soiling of buildings and structures (e.g. [5]). Globally air pollution may have both positive and negative effects including significant climatic implications [6] and their role as an important source of essential nutrients and minerals, e.g. for the Amazon region in South America [7]. The strategy chosen for air quality monitoring depends on the impact to be monitored, as well as on the available measurement techniques and financial considerations.

For example, monitoring long range transport of pollutants for the purpose of assessing eutrophication and acidification led to the set-up of the European Monitoring and Evaluation Programme [8] in 1979. The focus on investigating long range transport of acidic substances led to EMEP establishing remote background sites in Europe and standardised measurements for, e.g. sulphate and nitrate in particles and rainwater. This monitoring combined with modelling allows trans-boundary transport of these pollutants to be assessed and reduction measures to be established.

The main rationale for ambient air quality monitoring in urban areas concerns the negative health effects of pollutants on humans. The linkage between air pollutants and health effects has been established for many decades, though evidence from urban air pollution effects became most infamously known by the assessment of the London smog episode in 1952 [9]. This event and other investigations published before and afterwards lead to the establishment of legislations to improve ambient air quality. Table 1 gives an example of the development of air quality-related regulations in the UK and the European Union.

All air quality-related legislation is based on the presumption of a direct line connecting emissions, ambient air quality, exposure, uptake, dose and effect. Pursuing this idea, regulation starts at the point of emission. Whilst the emissions

Table 1 History of air pollution in the UK (based on [10])

Year and legislation	Focus/purpose
1273	Use of coal prohibited in London as being "prejudicial to health"
1306 – Royal Proclamation	Prohibiting craftsmen from using sea-coal in their furnaces
1845 – Railway Clauses Consolidated Act	Required railway engines to consume their own smoke
1847 – The Improvement Clauses Act	Contained a section dealing with factory smoke
1863 – Alkali, etc. Works Regulation Act	Required that 95% of the offensive emissions should be arrested
1866 – The Sanitary Act	Empowered sanitary authorities to take action in cases of smoke nuisances
1875 – The Public Health Act	Contained a section on smoke abatement from which legislation to the present day has been based
1946	First smokeless zone and prior approval legislation
1956 – Clean Air Act	Introduced smoke control areas, controlled chimney heights
1970 – EC Directive 70/220/EEC	Air pollution measures against gases from positive ignition engines of motor vehicles (emissions of CO and hydrocarbons)
1972 – EC Directive 72/306/EEC	Measures against diesel engine emissions limiting black smoke emissions from heavy duty vehicles
1974 – Control of Pollution Act	Regulation of motor fuels. Limits amount of sulphur in fuel oil
1975 – EC Directive 75/716/EEC	Sulphur content of certain liquid fuels. Amended in 1987 EC Directive 87/219/EEC: (1) The motor and oil fuel
1978 – EC Directive 78/611/EEC	Lead content of petrol limited to 0.4 gl^{-1}
1979 – International Convention on Long Range Transboundary Pollution	Introduced to control the transboundary effects of acid rain and to limit emission of acidifying pollutants
1980 – EC Directive 80/779/EEC	Air quality limit values and guide values for sulphur dioxide and suspended particles
1982 – EC Directive 82/884/EEC	Limit value for lead in the air
1984 – Directive 84/360/EEC	Common framework directive on combating pollution from industrial plants
1989 – EC Directive 89/427/EEC	Limit values and guide values of air quality for sulphur dioxide and suspended particulates. Harmonised measurement methods
1990 – Environmental Protection Act	Smaller emission sources under air pollution control by local authorities for the first time
1991 – The Road Vehicles Regulations	Set standards for in service emissions of CO and hydrocarbons to be included in the MOT test
1992 – EC Directive 92/72/EEC	Harmonised procedure for monitoring, exchange of information and public warnings for ozone
1995 – The Environment Act	New statutory framework for local air quality management with obligation to publish a National Strategy
1996 – EC Directive 96/62/EC	Framework for controlling levels of SO_2, NO_2, particulate matter, Pb, O_3, benzene, CO and other

(continued)

Table 1 (continued)

Year and legislation	Focus/purpose
	hydrocarbons leading to Daughter Directives 1–4: 1999/30/EC, 2000/69/EC, 2002/3/EC and 2004/107/EC
1997 – The National Air Quality Strategy	Published with commitments to achieve new air quality objectives throughout the UK by 2005
2008 – EC Directive 2008/50/EC	Merges existing legislation into one directive (excl. 4th daughter directive), sets new objectives for PM2.5 (fine particles) including the limit value and exposure-related objectives, and gives possibility to discount natural sources of pollution

from large industrial sources can be measured directly, the sheer multitude of individual air pollution sources (e.g. cars on the road or home heating systems) means that they cannot be assessed individually and instead regulatory assessment is extended to the measurement of ambient air quality. Emission monitoring, even though not the focus of this article, is hence intrinsically linked to ambient air quality monitoring to ensure that ambient limit values are achieved and allow efficient abatement strategies for air quality improvements when needed.

Extending the ideas of Chen et al. [11], the purpose of ambient air quality monitoring in urban areas can be described as:

1. Determining whether legal limit or target values are met at the monitoring sites
2. Deriving population-exposure-related values for health impact assessments
3. Providing information to check emissions inventories and pollutant evolution, allowing effective abatement strategies to be planned and executed
4. Providing real-time information allowing rapid response to air quality deterioration
5. Providing reliable data with which to validate air quality models and new monitoring methods
6. Providing timely information on air quality to all interested parties including the general public

Monitoring strategies can be summarised in terms of spatial representativity (i.e. siting criteria, including fixed or mobile and numbers of sites), time resolution and measurement accuracy. Network design will of course be constrained by the available technologies and cost. Currently the main focus for monitoring networks is to satisfy the requirements of air quality legislation, as described in the next section.

2 Current Legislative Monitoring Requirements in Different Parts of the World

2.1 USA

The US Environmental Protection Agency (EPA) is required to set up National Ambient Air Quality Standards for substances possibly harmful for humans or the environment. This requirement is formulated in the Clean Air Act, last amended in

Table 2 Air pollutants regulated by the National Ambient Air Quality Standards as of October 2011 [12]

Pollutant (final rule cite)	Primary/secondary	Avg. time	Level	Form
Carbon monoxide	Primary	8 h	9 ppm	Not more than one exceedance per year
		1 h	35 ppm	
Lead	Primary + secondary	3 month average	0.15 $\mu g/m^3$	Not to be exceeded
Nitrogen dioxide	Primary	1 h	100 ppb	8th percentile, averaged over 3 years
	Primary + secondary	Annual	53 ppb	Annual mean
Ozone	Primary + secondary	8 h	0.075 ppm	Annual 4th highest daily max. 8-h concentration, avg. over 3 years
Particles $PM_{2.5}$	Primary + secondary	Annual	15 $\mu g/m^3$	Annual mean, avg. over 3 years
		24 h	35 $\mu g/m^3$	98th percentile, avg. over 3 years
PM_{10}	Primary + secondary	24 h	150 $\mu g/m^3$	Max. one exceedance per year, over 3 years
Sulphur dioxide	Primary	1 h	75 ppb	99th percentile of 1-h daily max. concentrations, averaged over 3 years
	Secondary	3 h	0.5 ppm	Not to be exceeded more than once per year

1990, which also differentiates between primary standards providing public health protection, including protecting the health of "sensitive" populations such as asthmatics, children and the elderly, and secondary standards providing public welfare protection, including protection against decreased visibility and damage to animals, crops, vegetation and buildings. The criteria air pollutants, applicable throughout the USA, are listed with their limit values, called "Ambient Air Quality Standards" in the USA, in Table 2.

Two main networks are differentiated in the USA: the state and local air monitoring stations (SLAMS) and the national air monitoring stations (NAMS). The two networks mainly differ in their requirements with respect to area of coverage with more, smaller-scale representative monitoring sites within the SLAMS. The objectives, primarily for SLAMS, as laid out in CFR 40 [13] are to:

(a) Provide air pollution data to the general public in a timely manner
(b) Support compliance with ambient air quality standards and emissions strategy development
(c) Support air pollution research studies, specifically providing data for researchers working on health effects assessments and atmospheric processes, or for monitoring methods development

Both networks, SLAMS and NAMS, use the same siting criteria as laid out by the US EPA, differentiating micro, middle, neighbourhood, urban, regional and national to global scale [13]. This differentiation is solely based on their spatial scale of representativeness. Representativeness can be seen as an area surrounding the measurement site where air pollutant concentrations are reasonably similar to those measured at the monitoring site.

Based on the objectives and scales of spatial representativeness needed, CFR 40 [13] differentiates six site types:

(a) Sites located to determine the highest concentrations expected to occur in the area covered by the network,
(b) Sites located to measure typical concentrations in areas of high population density,
(c) Sites located to determine the impact of significant sources or source categories on air quality,
(d) Sites located to determine general background concentration levels,
(e) Sites located to determine the extent of regional pollutant transport among populated areas; and in support of secondary standards and
(f) Sites located to measure air pollution impacts on visibility, vegetation damage or other welfare-based impacts.

Each site type matches a certain need of spatial representativeness. For example, sites located to determine local hot-spots are at places where the highest concentrations are expected and will mostly be microscale monitoring sites, while those measuring air pollution impacts on visibility and vegetation damage have to be representative over regional to national scale.

CFR 40 [13] reflects this not only in general form but also with respect to specific pollutants. Ozone, a typical large-scale secondary pollutant (formed in the gas phase in polluted air) accordingly has only to be measured at urban and larger scales during the "ozone season", as derived from seasonal measurements for each federal state and listed in CFR 40 [13]. In contrast, the main areas required to monitor for CO and NO_2 are at the micro- and middle-scale where the highest concentrations and exposure may occur. It is interesting to note that microscale (near road) NO_2 measurements according to CFR 40 [13] are required to be conducted "within 50 m of target road segments in order to measure expected peak concentrations". A microscale site typically represents an area impacted by the plume with dimensions extending up to approximately 100 m. The distance from a source is a very critical parameter when compliance to limit values is required.

The current strategy and approach for air quality monitoring networks in the USA is therefore best described as pursuing the aims of ensuring the achievement of air quality limit values, air quality control, and allowing rapid intervention to prevent air quality deterioration.

2.2 Asia

Air quality standards and monitoring strategies across Asia are diverse and exhibit many differences between the countries. For brevity only the standards and monitoring concept for Japan are presented in any detail. For Taiwan, India, China and

Table 3 Japanese air quality standards (based on [15])

Pollutant		Avg. time	Level	Form
Carbon monoxide		24 h	10 ppm	Not to be exceeded
		1 h	20 ppm	For any consecutive 8 h period
Lead				No limit value
Nitrogen dioxide		24 h	0.04–0.06 ppm	Values to be within or below that zone
Ozone[a]		1 h	0.06 ppm	Not to be exceeded
Particle pollution	$PM_{2.5}$	Year	15 µg/m^3	Not to be exceeded
		24 h	35 µg/m^3	Annual 98th percentile value
	PM_{10}			The daily average for hourly values shall not exceed 0.10 mg/m^3, and hourly values shall not exceed 0.20 mg/m^3
Sulphur dioxide		24 h	0.04 ppm	Not to be exceeded
		1 h	0.1 ppm	Not to be exceeded

[a]Defined as oxidising substances (e.g. O_3 and PAN, peroxyacetyl nitrate) capable of isolating iodine from neutral potassium iodide, excluding nitrogen dioxide

other Asian countries, please refer to the corresponding webpages [14]. The Japanese air quality standards are summarised in Table 3.

The Japanese "Air Pollution Control Law" dates back to 1968 and was designed to promote comprehensive air pollution control measures. Subsequent revisions have included extensions of regulatory objects, nationwide regulation and enforced standards, e.g. those for specific dust (asbestos) in 1989, vehicle fuel in 1995, harmful air pollutants in 1996 and volatile organic compounds (VOC) in 2004.

Article 1 of the recent Japanese Air Pollution Control Law [16] states:

> The purposes of this Law are as follows.
> One is to protect the *public health* and *preserve the living environment* with respect to air pollution, by controlling emissions of soot, smoke and particulate from the business activities of factories and business establishments; by controlling emissions of particulate while buildings are being demolished; by promoting various measures concerning hazardous air pollutants; and, by setting maximum permissible limits for automobile exhaust gases, etc.
> The other is to help victims of air pollution-related health damage by providing a *liability regime* health damage caused by air pollution from business activities.

It is interesting to note that this law clearly indicates a liability regime for those who were harmed which is not the case in the corresponding laws of the USA and the European Union.

2.3 Europe

Air Quality is regulated at the European Union level by Directive 2008/50/EC [17] with currently one additional directive (2004/107/EC) [18] covering arsenic, Cd, Hg, Ni and polycyclic hydrocarbons (PAH) in air. The European directives include more pollutants than those listed by the US EPA as "criteria air pollutants common throughout the United States"; specifically As, Ni, PAH, specifically benzo(a)pyrene), Hg and Cd.

The Directive requires air quality monitoring for the protection of human health and the environment. This is specifically addressed in the Directive preamble which states that air pollution has to be reduced to such a level that any harm to humans is kept at a minimum and takes special account of the needs of sensitive subpopulations, such as children and elderly people.

Monitoring of additional, non-regulated substances is also requested at regional background locations (EMEP sites) to (a) compare rural concentrations to those in other areas, e.g. urban concentrations, (b) identify and quantify regional sources and their contributions and (c) allow the assessment of the trans-national movement of air pollutants. Substances requested for this type of regional background station are the anions and cations SO_4^{2-}, NO_3^-, Mg^{2+}, Ca^{2+}, NH^{4+}, Cl^-, Na^+, K^+, plus elemental carbon and organic carbon, in the $PM_{2.5}$ particle size fraction.

Table 4 summarises the air quality standards stipulated in the 2008 EC Directive. Additionally, benzene is regulated not to exceed 5 µg/m^3 as an annual average.

The European Directive [17], like the US regulations, differentiates separate site types for the protection of human health and the environment. Three main site types are defined for health purposes:

- Hot-spot monitoring: Publicly accessible areas in urban environments where highest concentrations of a pollutant are to be expected, including areas influenced by traffic or industrial emissions,
- Urban area monitoring: Measurements conducted at a location representative of a larger area to represent population exposure,
- Rural area monitoring: A sampling location which is not influenced by industry or city areas within the surrounding 5 km.

A single site type is defined for the protection of the environment; these are regional background sites designed to represent an area of at least 1,000 km^2. Regional background monitoring sites should be at least 5 km from significant sources and more than 20 km away from any conurbation.

In addition to the above site criteria, microscale requirements are set out within the European directive [17]. Traffic-related hot-spot sampling locations shall not be more than 10 m away from the edge of the traffic lane and should be representative of a road length of 100 m. The Directive [17] additionally gives further recommendations including distances from walls and sampling height.

2.4 Comparison of the Air Quality Monitoring Strategies

A comparison of the air quality strategies of the USA, Japan and Europe reveals many similarities.

Table 4 Air pollutants regulated in Europe (based on [17])

Pollutant		Avg. time	Level	Form
Carbon monoxide		8 h	10 mg/m^3	Not to be exceeded
Lead		Annual	0.5 μg/m^3	Not to be exceeded
Nitrogen dioxide		Annual	40 μg/m^3	Not to be exceeded
		1 h	200 μg/m^3	Not to be exceeded more than 18 times a year
		Annual	30 μg/m^3	As NO$_x$, for environmental protection: Not to be exceeded at regional sites
Ozone		8 h	120 μg/m^3	For health protection: Not to be exceeded more than 25 times per year, averaged over 3 years
		AOT40	18,000 μg/m^3*h	For environmental protection: Sum of differences between 8 am and 8 pm of hourly value exceeding 80 μg/m^3 between May and July over 3 years
Particles	PM$_{2.5}$	Annual	25 μg/m^3	To be reached in steps with no exceedance in 2015
		AEI[a]	20 μg/m^3	Not to be exceeded
	PM$_{10}$	Annual	40 μg/m^3	Not to be exceeded
		24 h	50 μg/m^3	Not to be exceeded more than 35 times a year
Sulphur dioxide		24 h	125 μg/m^3	Not to be exceeded more than 3 times a year.
		1 h	350 μg/m^3	Not to be exceeded more than 24 times a year
		Annual	20 μg/m^3	For environmental protection: Not to be exceeded at regional sites

[a]Average Exposure Indicator: the representative mean of urban background monitoring stations in a country, calculated as a rolling 3-year average

All strategies

- Contain elements of emission monitoring (not further discussed in detail)
- Assume a link between emissions and ambient air quality
- Define site types depending on distance to major sources of air pollution
- Rely on a combination of emission control and ambient air quality monitoring
- Include limit values for the safety of the population and the environment
- Stress the importance of providing timely information to the public

Strong similarities can also be seen between the lists of regulated air pollutants in Tables 2, 3 and 4. There are some air pollutants beyond those listed in the tables that are also required to be monitored, including volatile organic compounds (VOCs) and polycyclic aromatic hydrocarbons (PAHs).

Still significant differences exist in

- The area of liability

European law does not state anything concerning liability in the case of harm or damage by air pollution, whereas specific sections are dedicated to this question in Japanese law.

- Definition of site types

While the general idea of differentiating monitoring site types is present in all the monitoring strategies, the site type definitions and descriptions are different. The USA approach starts with different scales of representativity which are then linked to specific monitoring aims. The European Directive [17] approaches site definitions from the other direction, by defining the monitoring aim and describing the site types needed; distances to sources are stipulated and specific siting criteria have to be fulfilled. It is also important to note that sampling and siting is more stringently defined in the European Directive compared with the USA requirements.

- Microscale siting requirements and pollutants

Even though the major air pollutants and limit values appear very similar, significant differences may arise from different siting requirements. CFR 40 [13] states that near road measurements, which are only required for NO_2 and CO, are to be conducted within 50 m of the target road segment. No near-road measurements are required for, e.g. particulate matter $PM_{2.5}$ and PM_{10} in the USA. By contrast, the European Directive [17] requires near-road measurements in a maximum distance of 10 m from the road edge for a larger variety of pollutants (SO_2, NO, NO_2, PM_{10}, $PM_{2.5}$, lead, benzene and CO). Various studies have shown the rapid decrease in primary pollutant concentrations with distance from roads; concentrations can be 20–40% lower for $PM_{2.5}$, PM_{10} and particle number (both <100 nm and 0.5–20 μm in diameter) at 50 m away from a road compared with concentrations at 10 m [19, 20]. This means that the US limit value of 100 ppb for NO_2 is not equivalent to its European counterpart, even though 100 ppb ≈ 200 μg/m^3. It is actually about 40% higher (140 ppb) if concentrations in comparable locations are considered.

- The concept of population exposure

The European Directive [17] is alone in having the concept of regulating population exposure via an average exposure indicator (the AEI for $PM_{2.5}$). This means that an average urban background concentration for a country is calculated and regulated. The concentrations determined at this site type are considered to be the most exposure-relevant for the population. This approach is based on cost–benefit analyses comparing the benefits of focussing on hot spots (in response to a limit value approach) with those from efforts affecting larger areas [21]. Given the absence of evidence for a threshold for PM health effects, it is more beneficial for the general health of the population, and more cost effective, to decrease the average pollution exposure for the whole population than it is to decrease the concentration at highlighted locations in a populated area. The "hot-spot approach" with attainment of limit values at all relevant measurement sites is still pursued, as in the other countries, in parallel.

3 How Do Air Quality Monitoring Technologies Influence Strategies?

Robust air quality monitoring is based on the quantification of pollutant concentrations with a high degree of accuracy, precision, comparability and long-term stability. Accuracy and precision of measurements are essential for the assessment of limit value attainment and law enforcement to improve ambient air quality.

Early techniques used in regular monitoring were bulky and often relied on air sampling with subsequent laboratory analysis. Volz and Kley [22] describe early methods to measure O_3, initially employed in 1876. Here a defined volume of air was bubbled through a solution of AsO_3^{3-} which reacted with O_3 to form AsO_4^{3-}. The amount of AsO_3^{3-} was then titrated with I_2 to determine the O_3 concentration. Other early air quality monitoring techniques focused on the measurement of airborne particles. The use of fibrous filters for sampling airborne particles was invented around 1920 (see [23]) and became standard for ambient air quality monitoring after 1940. The method of manually weighing a filter prior to and after sampling a defined volume of air is still in use, and it is currently the basis of reference methods for the determination of PM_{10} and $PM_{2.5}$ in the EU and US.

Measurements using sampling techniques and subsequent laboratory analysis are labour intensive, costly and do not meet the need for timely dissemination of air pollution information to the public.

Developments in the 1970s and 1980s allowed the first measurements of gases to be made using automated measurement techniques. Good examples include the techniques developed for the measurement of oxides of nitrogen, O_3, CO_2, particle mass and soot (black carbon).

The oxides of nitrogen (NO and NO_2) can be determined with a method based on chemiluminescence, compound-specific light emission after excitation. The principle of this method is based on the following chemical reaction:

$$NO + O_3 \rightarrow NO_2 * + O_2 \qquad (1)$$

$$NO_2 * \rightarrow NO_2 + h\nu \qquad (2)$$

The addition of O_3 and measurement of the light intensity at the specific wavelength (2) in a reaction chamber can be used for continuous NO measurements [24] and for NO_2 if this is catalytically reduced to NO as part of the process.

Light absorption at specific wavelengths is another principle often employed for the online measurement of gaseous pollutants. For example, the absorption of O_3 in the ultraviolet at 254 nm can be used, though this experiences cross sensitivity to SO_2 and PAHs, and CO_2 in the infrared at 4.26 μm (2,350 cm^{-1}), which has little cross sensitivity to other gases. The most common C–H absorption bands are in the range of 3.33–3.57 μm (2,800–3,000 cm^{-1}).

While these methods can readily be employed in "real time", light absorption of soot particles was first developed for laboratory measurements of particles collected on filters and then adapted for online deployment [25]. Sampling on filter tapes allows quasi-continuous measurement by recording the decrease of transmitted light and moving the tape when the exposed area becomes too dark.

Parallel developments in health research, especially empirical studies, have gone hand in hand with new air pollution measurement technologies and have led to a substantial body of evidence on the adverse effects of air pollution. Put simply, new

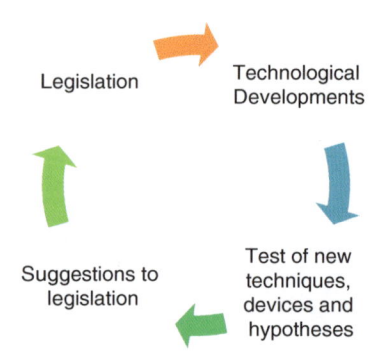

Fig. 1 Interlinkage between new measurement techniques and findings with the development of legislation

measurement technologies and new hypotheses in the environmental health arena have triggered research and taken forward our understanding of air pollution and health, which has in turn fed into legislation and emission controls.

Automated, quasi-online systems with highly time-resolved measurements, and the improvement of data storage and telecommunication technologies, have led to the current standardised system of air quality monitoring. This is based on fixed monitoring sites at different locations representing different site types. The sites are equipped with standardised monitoring devices for the regulated air pollutants and other optional measurements. The standardisation of the measurement methods is undertaken by the standardising and regulation bodies of the corresponding countries or legal jurisdictions, e.g. the US EPA for the USA, the VDI/DIN (Verein deutscher Ingenieure/deutsche Industrienorm) for Germany, CEN (Comité Européen de Normalisation) for the European Union, JIS (Japanese Industrial Standards) for Japan, and by ISO (International Standard Organisation) for worldwide harmonisation of air quality measurements.

Online data exchange is now possible with the standardised automated measurement methods being employed combined with telecommunication links. This data exchange allows for timely public air quality data to be sent via video text, the Internet, telephone and smartphones, as requested in the legislation.

The historical development of air quality measurements over the last 100 years has clearly influenced the monitoring strategies. These changes have not come spontaneously, but in a cycle, or rather in a spiral of related developments (Fig. 1).

Hence air quality monitoring today is a vibrantly developing area, employing new technologies and improving the understanding of human and environmental health effects, to work towards quality-of-life improvements for the population.

4 Recent Technological Developments

Emerging measurement technologies look to enable new monitoring paradigms. An analysis of available and recent developments in air quality monitoring technologies within the European Project "AirMonTech[1]" identified several major areas of development:

(a) Multi-component analysis,

Most current automated measurement techniques focus on one or possibly two air pollutants. A survey of recent developments shows major advances in multi-component analysis. Examples include:

Online liquid chromatography systems (e.g. Marga) which sample airborne particles into a liquid which is subsequently analysed in the field for (soluble) particle components such as SO_4^{2-}, NO_3^-, Cl^-, Ca^{2+}, Mg^{2+}, K^+ and Na^+ [26].

The development of aerosol mass spectrometers (AMS), starting with work by Allen and Gould [27] and Sinha et al. [28], which focusses mainly on compounds ionisable at temperatures below 1,300°C such as sulphates and organic matter. Recently Park et al. [29] have developed a laser-induced ionisation method also allowing the detection of some metals and metal oxides. AMS without particle size separation are now small enough in size and power consumption to be used in long-term monitoring networks [30].

Multi-elemental analysis by non-destructive X-ray fluorescence (XRF) is possible on filter tape samples similar to those collected for particle light absorption measurements. Up to 23 elements with the atomic number of potassium and above can be analysed by this method simultaneously, with a time resolution down to 15 min.

(b) Small, low cost, portable devices with low power consumption,

Existing technologies are being miniaturised and new small devices are being developed. Good examples are handheld condensation particle counters (CPCs) (e.g. TSI 3007) measuring submicrometer particle number concentrations, hand-held and portable devices for submicrometer particle surface area concentrations (sometimes along with particle number concentrations) as summarised and compared in Asbach et al. [31], and small personal black carbon (particle light absorption) monitors (e.g. the microAeth™). The development of sensors based on micro-electro-mechanical systems (MEMS) for PM mass concentrations, and electrochemical and semiconductor systems for gases, which are small enough to be implemented in smartphones also offers new and interesting possibilities (e.g. [32]).

An emerging trend towards combining different sensors and measurement technologies for urban air quality monitoring along with device miniaturisation is evident from new commercially available equipment such as ETL 200 by Casella (Italy), simultaneously measuring NO, NO_2, CO, O_3, benzene and noise, or the

---

[1] www.AirMonTech.eu

airpointer™, simultaneously measuring four compounds using EU reference methods (MLU, Austria).

These small, flexible devices open up two new monitoring paradigms: (1) personal exposure and mobile monitoring and (2) measurements at a multitude of fixed sites which can also be flexibly located in living areas where current monitoring "containers" cannot be placed.

(c) Alternative parameters for particulate air pollution,

The link between exposure to air pollutants and adverse health effects is well established, but the causal biological mechanisms are not clear and this is especially the case for particulate matter health effects. Airborne particulate matter is extremely variable in chemical composition, size and morphology; all parameters of possible health relevance. This and the different health endpoints affected by exposure to ambient PM make the situation very complex. It may well be that more than one particle characteristic is needed to effectively describe the harmful outcomes of exposure. Possible parameters under discussion are particle number concentration, which is dominated by particles below 100 nm in size; the so-called ultrafines [33], particle surface area concentration, which is dominated by particles around 200–800 nm in diameter [34, 35], black carbon or black smoke [36], or the reactivity of particles with respect to redox reactions, or their potential to form radical oxidative species (ROS) [37]. These and some other alternative particulate indicators are currently discussed [38] and investigated in several large European and US studies such as ESCAPE and Transphorm[2].

(d) New methods for data retrieval and analysis.

Next to the development of new measurement technologies, the most important developments are in data retrieval and statistical data analysis. Air pollution data are now regularly remotely retrieved, automatically stored in a database, checked for consistency and made publicly available via the Internet, for example.

Resch et al. [39] summarise and discuss the whole chain from data retrieval, processing, analysis and visualisation. The general design and structure of such a chain is depicted in Fig. 2, starting with generalised sensors. These sensors can be traditional fixed monitors, but can also be mobile sensors installed on cars, ships or at short-term locations, for example on lampposts. These mobile or moveable sensors can be equipped with a geo-positioning system (GPS) to be employed in geo-information systems (GIS) (e.g. [40]). These applications have been made possible by miniaturisation of the GPS as well as the transfer of data by mobile phone systems.

Gross [41] very nicely described the development of such a GIS-based information system:

> In the next century, planet earth will don an electronic skin. It will use the Internet as a scaffold to support and transmit its sensations. This skin is already being stitched together.

---

[2] www.escape.eu, www.transphorm.eu

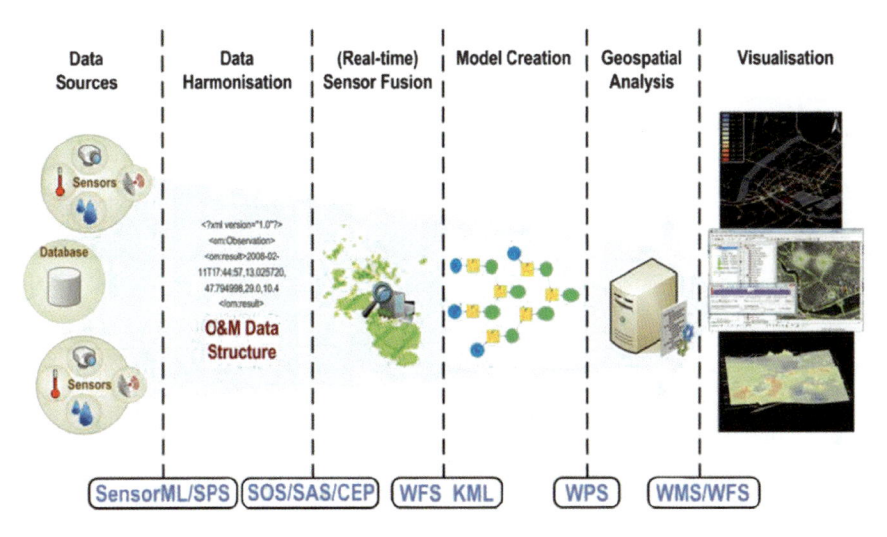

Fig. 2 Flow of data and analyses in modern GIS-based sensor networks (from [39])

It consists of millions of embedded electronic measuring devices: thermostats, pressure gauges, pollution detectors, cameras, microphones, glucose sensors, EKGs, electroencephalographs. These will probe and monitor cities and endangered species, the atmosphere, our ships, highways and fleets of trucks, our conversations, our bodies – even our dreams. (citation from [41])

This envisages the collection of all different types of information, including air quality. The next and very important step towards making such an effort worthwhile is the harmonisation of the collected data, for subsequent modelling, as well as geospatial and logical data analysis. At this point visualisation and heterogeneous data interpretation along with the conditional triggering of actions become possible. This latter analysis is dependent on a very good understanding of the measurement devices, the data and the logical environmental processes, together with an evaluation mechanism or standard for the integrated data.

In summary, significant advances have been made during the last decade and recent and foreseeable developments in air quality monitoring technologies now allow new air quality monitoring concepts to be created and explored.

5 Possible Future Monitoring Strategies

Current monitoring strategies are mainly aimed at fulfilling legal obligations and assessing limit value compliance. These strategies, however, are built around the assessment of background concentrations as a surrogate for population exposure. Future monitoring strategies may harness new technologies to bring the measurements as close

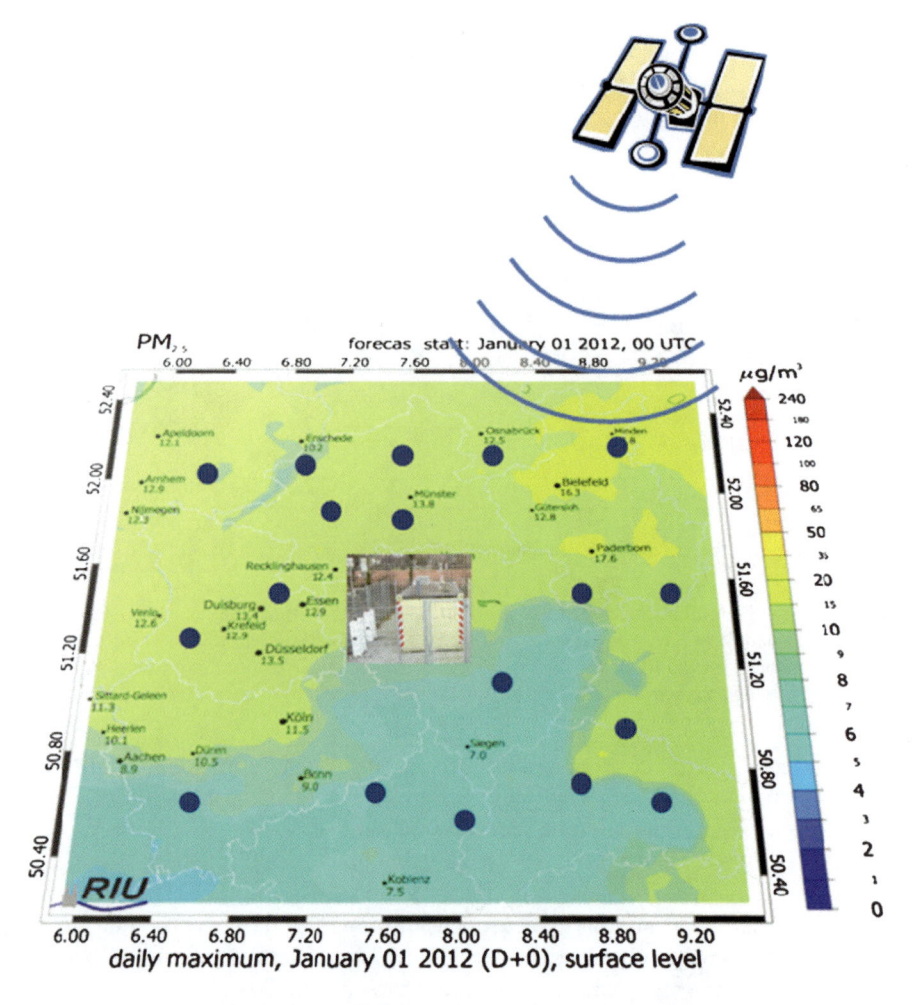

Fig. 3 Visualisation of a future monitoring concept linking all available monitoring tools, fixed site measurements, mobile and flexible measurements, modelling and satellite observations

to exposure assessments as possible. This cannot be pursued based solely on the use of fixed monitoring sites.

A possible future concept is depicted in Fig. 3. This shows the four basic possibilities for ambient air quality monitoring:

– High quality and high time-resolution measurements of air pollutants at fixed locations,
– Mobile and flexibly installable monitoring devices with low power consumption which might still need a protective container, delivering data of relatively high quality, or alternatively could be low cost sensors used very flexibly, e.g. at lampposts or on buses,

- Modelling of the spatial and temporal variation of air pollutants in urban areas, using improved emissions inventories and
- Satellite observations of air pollutants and meteorology to derive information on parameters influencing urban air quality.

Each of the four possibilities has shortcomings with the most significant ones being the

- Limitation of the spatial representativity for fixed monitoring sites,
- Significantly lower precision, comparability and time resolution for the small mobile or flexibly installable devices,
- Relatively high uncertainty of models, which need verification and validation inputs from monitoring data and
- Poor vertical resolution, in the case of satellite measurements, to be of relevance for ground-level human and environmental exposure.

The advantages and limitations of the different monitoring possibilities illustrate the significant challenges that need to be addressed to allow new air quality monitoring tools to be combined to maximum advantage.

Taking the example in Fig. 3, we can envisage that the fixed monitoring sites with high quality measurements act as reference sites for the collocation and ongoing comparison with small, flexible monitoring devices. This "online" reference comparison would allow for dynamic correction and hence decreased measurement uncertainty for the dispersed monitors in an urban area. The combined spatially and temporally resolved data from the fixed and flexible monitors could then be used to validate the model results, along with satellite observations, meteorological data and emissions information, which generally need some improvement [42] (Fig. 4).

The functions of future urban air quality networks will not change dramatically but new facilities will be enabled, new developments can be tested and monitoring can be extended to routinely link air quality to its effects.

The current focus of air quality monitoring on limit value compliance and the calculation of population exposure can be continued on the basis of the fixed monitoring sites. As new technologies are installed and utilised, the use of modelled data based on fewer fixed monitoring sites, networks of mobile, flexible monitors and atmospheric models can be explored. Validated model results may be seen as being more reliable to derive a long-term population exposure value than those based on fixed monitoring sites. This approach allows linkages between the higher data quality obtained by measurements with spatial information obtained from model results to be developed.

The new network design will allow and facilitate new developments through

- New monitoring devices, either for the fixed or the mobile, flexible measurement locations,
- Testing new data collection, analysis and visualisation tools,
- Improved exposure assessments for cohort as well as population-based health effect studies,

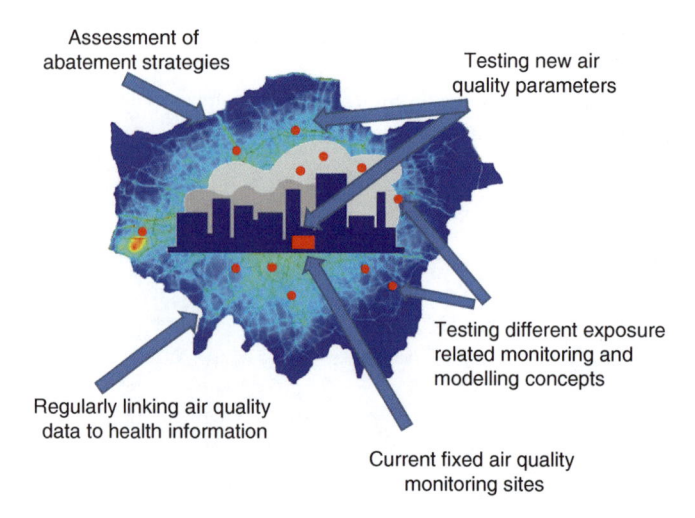

Fig. 4 Conceptual purposes and tasks of the future urban air quality networks

- Routine linkages between public health data and environmental stressors, specifically urban air quality, assess the effectiveness of abatement strategies,
- The assessment of new or alternative air quality parameters to test their use in the context of urban air quality and health.

It is important that we enable and facilitate such developments in Europe to continue the improvement of its air quality. The new directions will improve our understanding of air quality in urban environments, allowing us to bring forward the best and most cost-effective abatement measures for minimising the significant impacts of urban air pollution.

Acknowledgement This chapter was in part sponsored by the EU Project AirMonTech (Project Number) with TK, PQ, UQ, MV and KK being members.

References

1. US EPA (2009) Integrated science assessment for particulate matter (Final Report). http://cfpub.epa.gov/ncea/cfm/recordisplay.cfm?deid=216546
2. Johnson DW, Cresser MS, Nilsson SI, Turner J, Ulrich B, Binkley D, Cole DW (1991) Soil changes in forest ecosystems: evidence for and probable causes. Int J Biol Sci 97B:81–116
3. Kennedy IR (1992) Acid soil and acid rain, 2nd edn. Research Studies Press/Wiley, Taunton, Somerset, UK, 254 pages, ISBN 0 471 93404 6
4. Hessen DO, Henriksen A, Hindar A, Mulder J, Torseth K, Vagstad N (1997) Human impacts on the nitrogen cycle: a global problem judged from a local perspective. Ambio 26:321–325
5. Cowling JE, Roberts ME (1954) Paints, varnishes, enamels, and lacquers. In: Greathouse GA, Wessel CJ (eds) Deterioration of materials: causes and preventive techniques. Reinhold Publishing Corporation, New York, pp 596–645

6. IPCC (2007) In: Solomon S, Qin D, Qin D, Manning M, Chen Z, Marquis M, Averyt KB, Tignor M, Miller HL (eds) Contribution of Working Group I to the Fourth Assessment Report of the Intergovernmental Panel on Climate Change. Cambridge University Press, Cambridge, p 996, http://www.ipcc.ch/publications_and_data/publications_ipcc_fourth_assessment_report_wg1_report_the_physical_science_basis.htm
7. Swap R, Garstang M, Greco S, Talbot R, Kallberg P (1992) Saharan dust in the Amazon Basin. Tellus 44B:133–149
8. EMEP. http://www.emep.int/
9. Bell ML, Bell ML, Davis DL, Fletcher T (2004) A retrospective assessment of mortality from the London smog episode of 1952: the role of influenza and pollution. Environ Health Perspect 112(1):6–8
10. Air-Quality (2011) http://www.air-quality.org.uk/02.php
11. Chen C-H, Liu W-L, Chen C-H (2006) Development of a multiple objective planning theory and system for sustainable air quality monitoring network. Sci Total Environ 354:1–19
12. US EPA, http://www.epa.gov/air/criteria.html, 2011 and reference methods for CO: 76 FR 54294, Aug 31, 2011, Lead: 73 FR 66964, Nov 12, 2008, NO_2: 75 FR 6474, Feb 9, 2010, 61 FR 52852, Oct 8, 1996, O_3: 73 FR 16436, Mar 27, 2008, PM: 71 FR 61144, Oct 17, 2006, SO_2: 75 75 FR 35520, Jun 22, 2010
13. CFR 40 (2011) Part 58, Ambient Air Quality Surveillance, Appendix D. http://ecfr.gpoaccess.gov/cgi/t/text/text-idx?c=ecfr&sid=0228ef3a08fb2915366f10fa0123de5a&rgn=div9&view=text&node=40:5.0.1.1.6.7.1.3.37&idno=40
14. WAQL (Webpages Air Quality Legislation) (2011) Taiwan: http://law.epa.gov.tw/en/laws/atmosph/, India: http://envfor.nic.in/legis/air.htm, China: http://www.chinafaqs.org/library/chinas-new-regional-air-quality-regulation-translated, Clean Air Portal Asia: http://cleanairinitiative.org/portal/knowledgebase/policies
15. AQ Japan (2011) http://www.env.go.jp/en/air/aq/aq.html
16. JAPC, Japanese Air Pollution Control Law (2011) Latest amendment by law No. 32 of 1996, tentative translation. Japanese Ministry of the Environment. http://www.env.go.jp/en/laws/air/air/index.html
17. EC Directive 2008/50/EC, Official Journal of the European Union L 152/1 – L152/44, 2008
18. EC Directive 2004/107/EC, Official Journal of the European Union L 23/3 – 23/16, 2005
19. Hickman AJ, McCrae IS, Cloke J, Davies GJ (2002) Measurement of roadside air pollution dispersion. Project report PR SE/445/02 TRL Ltd, Crowthorne
20. Hitchins J, Morawska L, Wolff R, Gilbert D (2000) Concentrations of submicrometre particles from vehicle emissions near a major road. Atmos Environ 34:51–59
21. CAFE (2004) Second position paper on particulate matter, CAFE working group on particulate matter. http://ec.europa.eu/environment/archives/cafe/pdf/working_groups/2nd_position_paper_pm.pdf. 20 Dec 2004
22. Volz A, Kley D (1988) Evaluation of the Montsouris series of ozone measurements made in the 19th century. Nature 332:240–242
23. Spurny KR (1998) Methods of aerosol measurement before the1960s. Aerosol Sci Technol 29(4):329–349
24. Fontijn A, Sabadell AJ, Ronco RJ (1970) Homogeneous chemiluminescent measurement of nitric oxide with ozone. Anal Chem 42(6):575–579
25. Hansen ADA, Rosen H, Novakov T (1982) Real-time measurement of the absorption coefficient of aerosol particles. Appl Opt 21:3060–3062
26. Orsini DA, Ma Y, Sullivan A, Sierau B, Baumann K, Weber RJ (2003) Refinements to the Particle-Into-Liquid-Sampler (PILS) for ground and airborne measurements of water soluble, aerosol composition. Atmos Environ 37:1243–1259
27. Allen J, Gould RK (1981) Mass-spectrometric analyzer for individual aerosol-particles. Rev Sci Instrum 52(6):804–809
28. Sinha MP, Giffin CE, Norris DD, Estes TJ, Vilker VL, Friedlander SK (1982) Particle analysis by mass spectrometry. J Colloid Interface Sci 87:140–153

29. Park K, Cho G, Kwak J-h (2009) Development of an aerosol focusing-laser induced breakdown spectroscopy (aerosol focusing-LIBS) for determination of fine and ultrafine metal aerosols. Aerosol Sci Technol 43:375–386
30. Ng NL, Herndon SC, Trimborn A et al (2011) An Aerosol Chemical Speciation Monitor (ACSM) for routine monitoring of the composition and mass concentrations of ambient aerosol. Aerosol Sci Technol 45:780–794
31. Asbach C, Kaminski H, von Barany D, Kuhlbusch TAJ, Monz C, Dziurowitz N, Pelzer J, Vossen K, Berlin K, Diertrich S, Götz U, Kiesling H-J, Schierl R, Dahmann D (2012) Comparability of portable nanoparticle exposure monitors. Ann Occup Hyg 56(5):606–621
32. Paprotny I, Doering F, White RM (2010) MEMS particulate matter (PM) monitor for cellular deployment, Proc. IEEE Sensors 2010, pp 2435–2440
33. Hoek G, Boogard H, Knol A, de Hartog J et al (2010) Concentration response functions for ultrafine particles and all-cause mortality and hospital admissions: results of a European expert panel elicitation. Environ Sci Technol 44:476–482
34. Oberdörster G (2001) Pulmonary effects of inhaled ultrafine particles. Int Arch Occup Environ Health 74:1–8
35. Stoeger T, Reinhard C, Takenaka S, Schroeppel A, Karg E, Ritter B, Heyder J, Schulz H (2006) Instillation of six different ultrafine carbon particles indicates a surface area threshold dose for acute lung inflammation in mice. Environ Health Perspect 114:328–333
36. Beelen R, Hoek G, van den Brandt PA, Goldbohm RA et al (2008) Long-term effects of traffic-related air pollution on mortality in a Dutch cohort (NLCS-AIR Study). Environ Health Perspect 116(2):196–202
37. Mudway IS, Stenfors N, Duggan ST, Roxborough H, Zielinski H, Marklund SL, Blomberg A, Frew AJ, Sandstrom T, Kelly FJ (2004) An in vitro and in vivo investigation of the effects of diesel exhaust on human airway lining fluid antioxidants. Arch Biochem Biophys 423 (1):200–212
38. Gu et al (2012) Selection of key ambient particulate variables for epidemiological studies — applying cluster and heatmap analyses as tools for data reduction. Sci Total Environ 435–436:541–550
39. Resch B, Britter R, Outram C, Xiaoji R, Ratti C (2011) Standardised geo-sensor webs for integrated urban air quality monitoring. In: Ekundayo EO (ed) Environmental monitoring, InTech, ISBN 978-953-307-724-6, pp 513–528
40. Matejicek L (2005) Spatial modeling of air pollution in urban areas with GIS: a case study on integrated database development. Adv Geosci 4:63–68
41. Gross N (1999) The earth will don an electronic skin. http://www.businessweek.com, BusinessWeek Online, 30 August 1999. (6 Jan 2012)
42. Winiwarter W, Kuhlbusch TAJ, Viana M, Hitzenberger R (2009) Quality considerations of European PM emission inventories. Atmos Environ 43(25):3819–3828

Number Size Distributions of Submicron Particles in Europe

Ari Asmi

Abstract The aerosol particle number size distribution is a key component in aerosol indirect climate effects, and is also a key factor on potential nanoparticle health effects. This chapter will give background on particle number size distributions, their monitoring and on potential climate and health effects of submicron aerosol particles. The main interest is on the current variability and concentration levels in European background air.

The submicron particle number size distribution controls many of the main climate effects of submicron aerosol populations. The data from harmonized particle number size distribution measurements from European field monitoring stations are presented and discussed. The results give a comprehensive overview of the European near surface aerosol particle number concentrations and number size distributions between 30 and 500 nm of dry particle diameter. Spatial and temporal distributions of aerosols in the particle sizes most important for climate applications are presented. Annual, weekly, and diurnal cycles of the aerosol number concentrations are shown and discussed. Emphasis is placed on the usability of results within the aerosol modeling community and several key points of model-measurement comparison of submicron aerosol particles are discussed along with typical concentration levels around European background.

Keywords Aerosol number concentration, Aerosol number size distribution, Atmospheric aerosols, CCN

A. Asmi (✉)
University of Helsinki, Helsinki, Finland
e-mail: Ari.Asmi@helsinki.fi

Contents

1 Introduction 298
   1.1 Properties of Aerosol Number Concentrations and Size Distributions 299
   1.2 Key Differences Between Number and Mass Measures of Aerosols 301
2 Measurement and Monitoring in Europe 302
   2.1 Measuring the Aerosol Particle Number Size Distributions 302
   2.2 Networks of Measurements in Europe 303
3 Levels and Variability of Aerosol Number Concentrations 304
   3.1 General Properties of Number Size Distributions 304
   3.2 Spatial Differences and Similarities of Aerosol Number Size Distributions 305
   3.3 Correlations Between Aitken and Accumulation Mode 312
   3.4 Short Scale Temporal Variability 314
   3.5 Comparing Measurements of Number Size Distribution with Model Output 316
4 Conclusions and Outlook for the Future 317
References 318

1 Introduction

Air quality effects of aerosols have traditionally been studied in accumulation and coarse mode, concentrating largely in the supermicron particle range, usually by using particle mass as the main property of interest (e.g., PM_x). However, submicron particles have also many important effects to both air quality and the climate system.

The health effects of atmospheric aerosol particles are usually characterized by aerosol particle mass concentration in either PM_{10} or $PM_{2.5}$ [1]. Nanoparticles (commonly defined as particles with $D_p < 100$ nm) have been widely acknowledged to have potential for adverse health effects, although the knowledge which particle property is most important for health effects has not yet been generally agreed on [2]. However, the particle deposition to alveolar region of lungs seems to be especially efficient for particles in the size range between 10 and 50 nm [3]. The research on nanoparticle health effects has been recently increased due to the current wide interest in the industrial nanoparticle processes, with obvious occupational health risks. The health effects of these nanoparticles are potentially important also for the public, as concentrations of sub-50 nm particles are known to be high in many areas. For such particles, a major source of urban air concentrations is traffic, usually diesel soot [4]. As an indication of the potential dangers associated with such ultra-fine particles, recent controls for automotive emissions in the European Union include a number emission limit for particles larger than 23 nm in diameter [5].

The climate effects of atmospheric aerosol particles are a matter of continuous interest in the research community. The aerosol-climate effects are divided into two groups: The direct effect represents the ability of the particle population to absorb and scatter short-wave radiation – directly affecting the radiation balance. These direct effects depend primarily on the aerosol optical properties and particle number size distribution, as the particle size significantly affects the scattering efficiency of shortwave radiation [6]. The indirect effects affect the climate system through the clouds and mostly with the cloud albedo effect and the cloud lifetime effect. The cloud albedo effect (Twomey effect) is the resulting change in cloud radiative properties due to changes in cloud droplet number concentration (CDNC) [7]; the lifetime effect (Albrecht effect) is connected to the changes in cloud properties and in drizzle and precipitation [8]. The aerosol indirect effects are controlled by the ability of particles to activate to cloud droplets (i.e., to be cloud condensational nuclei, CCN) within a cloud [9]. This ability is a strong function of particle size, water supersaturation, and particle hygroscopicity (i.e., chemical composition). An extensive overview of the processes affecting the ability of particles to act as CCN is provided in literature [10], with the conclusion that the particle size is the dominant aerosol property on cloud droplet activation, making particle number size distribution the key factor in aerosol-climate interactions.

1.1 Properties of Aerosol Number Concentrations and Size Distributions

In the submicron range, the aerosol population typically consists of several subpopulations, so-called modes, which are indicative of different loss and formation processes in the atmosphere. The smallest aerosol particles are formed from gas phase vapors via nucleation, forming the nucleation mode. In slightly larger sizes, the Aitken mode particles are produced via growth from nucleation mode and a wide variety of combustion and natural sources. The largest submicron particles are in the accumulation mode, with a major source being growth from Aitken mode, e.g. via cloud processing and condensation. Only the smallest tail of the coarse mode can be seen in the submicron range, and the overall effect of coarse particles in the submicron particle range number concentrations is generally very small. Often as a rough estimate the particles with diameters between 30 and 100 nm are considered to belong to Aitken mode and particles between 100 nm and 1 μm to the accumulation mode. The particle modes can move in the size axis, and they are not always visible in the measured size spectra, especially nucleation mode is often absent during nighttime.

As particle number size distributions can be complex, and the instruments used generate large amount of size distribution data, which can be hard to effectively describe, a common method is to calculate integrated particle number concentrations for specific aerosol particle diameter ranges, depending on which part of the particle number size spectrum is needed for the application. In this chapter, three different ranges are used (Fig. 1a):

- Number concentration of particles between 30 and 50 nm of dry diameter (N_{30-50}), given in particles per cubic centimeter (cm^{-3}). These particles represent smaller end of the Aitken mode particles and are representative of recently emitted or formed nanoparticles, which have high probability to end up in

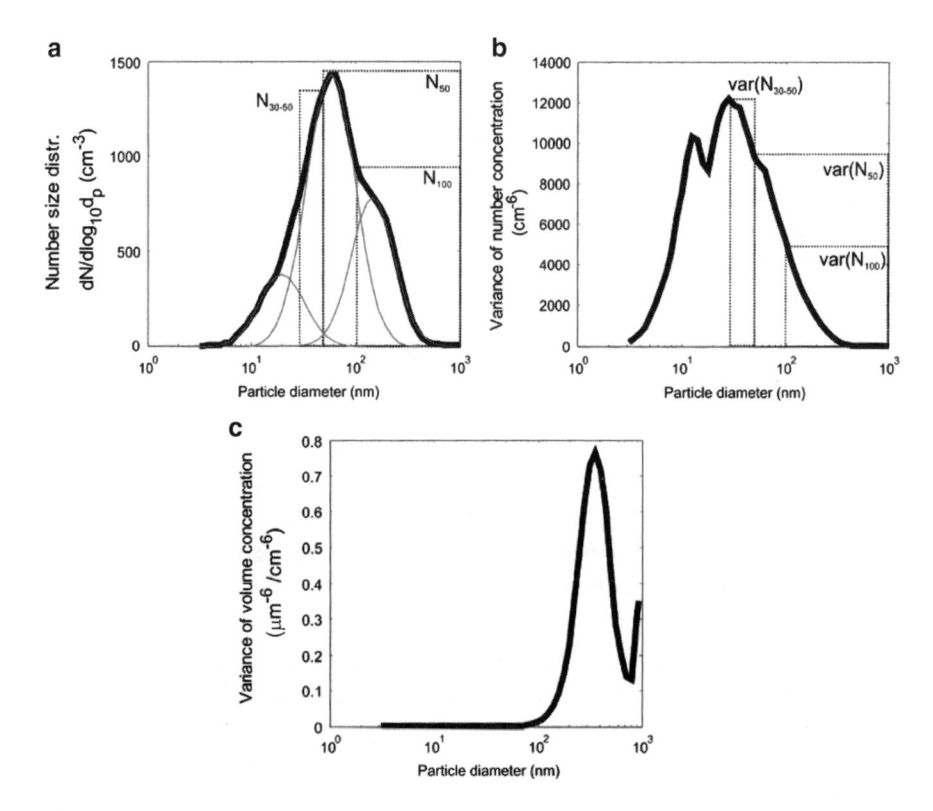

Fig. 1 (a) Typical clean Northern European median number size distribution (measured at SMEAR II station in Hyytiälä, Finland). The approximate modal locations and size ranges of different integral properties of the aerosol number size distribution used are shown; (b) variance of number concentration as a function of particle diameter; (c) variance of (computed) volume concentration in the same station (adapted from Asmi (2012), [11])

alveolar regions of lungs. The smaller diameter limit of 30 nm comes mostly from trying to keep the instrumental datasets comparable. These particles have generally relatively low lifetime (in the order of one or two days) and thus are more representative of the local emissions.

- Particle number concentration between 50 and 500 nm of diameter (N_{50}), which represents particles from larger diameter end (i.e., more aged) Aitken mode, together with the entire accumulation mode. The 500 nm upper limit is due to instrumental limitations. These particles are representative of CCNs for high updraft velocity clouds, as typical aged particles ($\kappa = 0.18$) with 50 nm diameter are activated with water supersaturation of 0.8%. The upper limit of 500 nm is chosen to have comparable values for different instruments. As the particle number size distribution over 500 nm of diameter is usually insignificant in terms of number concentration in comparison with sub-500 nm size ranges (Fig. 1a), making the N_{50} a good representative parameter of the total potential CCN concentration.

- Similar to N_{50}, we also show the number concentrations between 100 and 500 nm in diameter (N_{100}), representing CCNs for lower updraft velocity clouds, and a rough estimate of the number concentration in accumulation mode as the Hoppel minimum between the Aitken and the accumulation modes is often around 100 nm of diameter.

These concentrations in different size ranges give then an approximation of aerosol concentration levels for both air quality (N_{30-50} and lesser extent N_{50}) and for climate applications (N_{50} and N_{100}). The properties are in a sense similar to the mass-based PM_{10} and $PM_{2.5}$ measurements, providing easy way to compare individual concentrations.

1.2 Key Differences Between Number and Mass Measures of Aerosols

The particle number and mass concentrations can sometimes correlate in some timescales, although such behavior should not be automatically assumed. Studies have shown that seasonal correlations between particle number concentrations and mass concentrations in several sites over Europe are high [12]. Thus, there is some indication that if the site is relatively polluted (i.e., PM concentrations are high), the accumulation mode particles are generally high as well. However, in smaller timescales, such as days and hours, these correlations can disappear [11]. The main reasons behind the lack of correlation are the complex dynamics of the particle number size distributions, as larger particles in micrometer range can effectively hinder many particle formation and growth processes, which are crucial to the particle number concentrations. Mass and number are thus different aspects of the aerosol population and should not be considered necessarily similar in behavior.

Number concentrations are dominated by submicron particles, whereas the mass concentrations are strongly influenced by particle concentrations in 0.1–10 μm diameter range [13]. Similarly, the variability of the number-based measurements is strongly dominated by variability in smaller diameter ranges, whereas the variability of mass-based properties, such as PM_{10}, are dominated by variability in the accumulation mode (usually around 500 nm of mass mean diameter) and in the coarse mode. This means the variabilities of these properties are not necessarily similar in shorter timescales, due to sensitivity of variance from very different air masses and thus aerosol types. This is demonstrated in Fig. 1b, where the variance of the each size class of particle number concentrations between 3 and 1,000 nm is shown for SMEAR II station in Hyytiälä, Finland. The variance has similarities to the particle number size distribution (Fig. 1a), but there are also significant differences, especially on smaller particles sizes. Even though in the median particle number size distribution the nucleation mode is visible only weakly, it is a major contributor to submicron particle number concentration variability.

Figure 1c shows the variability of the submicron aerosol volume distribution (calculated from number size distribution using assumption of spherical particles) as the function of size. As the particle volume concentration is closely connected with particle mass concentration, two key findings come apparent: (1) as the aerosol volume variability is dominated by larger accumulation particles in the diameter range from 300 to 700 nm, the variabilities of most mass-based aerosol measurements (especially PM_1, as it is not dependent on supermicron coarse mode, but in lesser extent also $PM_{2.5}$ and PM_{10}) are very sensitive to changes in these particles sizes, and in comparison (2) the variability of particle number concentration is rapidly diminishing above 100 nm in diameter, show that the variability of that aerosol number concentrations are not strongly influenced by changes in accumulation mode particle number, and that the used integral particle number concentrations (N_{30-50}, N_{50} and N_{100}) are mostly sensitive to changes in the smallest particle sizes in each diameter range.

In short, the mass-based measurements are often more sensitive to aerosol concentration changes near the maximum diameter of the size range measured, whereas the number-based measurements are more sensitive on aerosol concentration near the minimum diameter measured.

2 Measurement and Monitoring in Europe

2.1 Measuring the Aerosol Particle Number Size Distributions

Two types of particle mobility particle size spectrometers for measuring submicrometer particle number size distributions are generally used: the differential mobility particle sizer (or DMPS), and the scanning mobility particle sizer (SMPS) [14]. Both are very similar instruments in their operation: they aspirate dried air, use ionizing radiation to establish an equilibrium bipolar charge-distribution in the sampled aerosol, use a cylindrical differential mobility analyzer to select particles based on their electrical mobility, and use a condensation particle counter to measure the resulting concentrations in each of the selected electrical mobility. The main difference between these instruments is the mode of operation, as the DMPS keeps the differential mobility analyzer voltage constant during measurement of a single size interval and the SMPS scans with continuously differing voltages. The size range and time resolution of a DMPS or an SMPS system depends on the system architecture (e.g., the physical dimensions of the instrument and the flow rates used) and on user choice (more size channels vs. time resolution). One key factor is that the measured particle properties are measured for *dry conditions*. The sample air is dried to relative humidity RH < 40% before size selection, meaning that most of the water is evaporated from the particles, and thus they are most likely smaller than in ambient conditions.

Notably, the instruments used in the European Supersites for Atmospheric Aerosol Research (EUSAAR)/Aerosols, Clouds, and Trace gases Research Infra-Structure Network (ACTRIS) and German Ultrafine Aerosol Network (GUAN) measurements used in this chapter are from intercalibrated measurements, where the abilities of the instruments were determined in common intercalibration workshops [15]. Overall, the instruments agree well on particle sizes between 20 and 200 nm, with the differences above 200 nm still relatively minor for number concentrations. In smallest particles sizes the instrument deviation is large, and for this reason we only consider particles larger than 30 nm in diameter in this chapter.

2.2 Networks of Measurements in Europe

Several European intensive short-term ("campaign-type") projects have provided important information on the atmospheric aerosol properties in Europe, usually by concentrating on specific aerosol properties or interactions. However, these kinds of campaign-type measurements do not necessarily represent the seasonal or annual variations of the aerosol concentrations and can overestimate some properties of the aerosol populations. Long-term measurements, especially with intercalibrated instruments and common data handling and calibration protocols make the data comparison between stations much more reliable and provide the end users (e.g., atmospheric modelers) good datasets to compare with.

The EUSAAR project of the Sixth Framework Programme of the European Commission is one of the steps towards a reliable and quality-controlled network of measurements [16]. The EUSAAR project improved and homogenized 20 European sites for measuring aerosol chemical, physical, and optical properties following a standardized protocol of instrument maintenance, measurement procedures, and data delivery in common format to a common database. EUSAAR also provided intercomparison and calibration workshops as well as training for the station operators. The work started in the EUSAAR is continued in ACTRIS infrastructure of the Seventh Framework Programme of the European Commission.

GUAN is a network of multiple German institutes with an interest in submicron aerosol properties, which was established in 2008 [17]. The methodologies of particle number size distribution measurements and data handling procedures in both GUAN and EUSAAR networks are very similar, and the size distribution measurement results are comparable between the two networks. The EUSAAR measurements were available (with some station-to-station variability) for the year 2008–2009 and the GUAN measurements were mostly from 2009. The locations of the stations are shown in Fig. 2.

A key feature of the EUSAAR and GUAN stations is that they are regional background stations. This means in this context that the stations are expected to represent the (possibly polluted) background air and long-range transport, not the direct influence of any obvious local sources. This means that the levels of aerosol

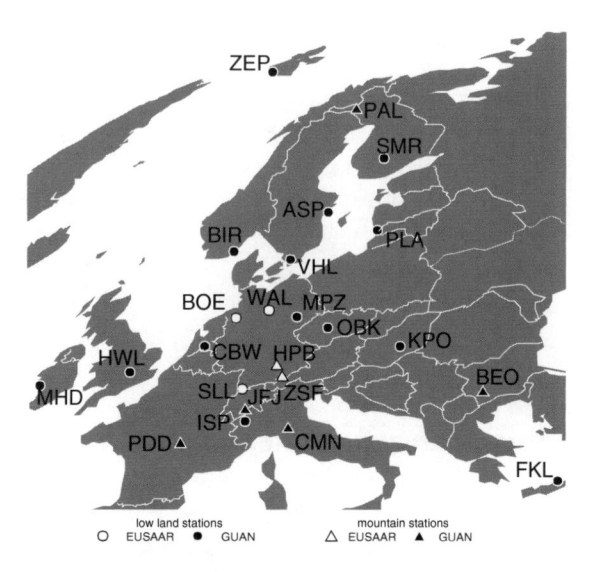

Fig. 2 Stations used in measurements of aerosol number size distributions. *Black symbols* are EUSAAR stations, *white* GUAN (MPZ was in both networks). *Triangles* denote high-altitude mountain stations (over 1,000 m from mean sea level). Figure adapted from figure published in [18], which also has more details on the locations and types of the stations

concentrations measured at these stations should represent the overall regional background concentrations and should be thus representable of a large footprint area around them [19].

3 Levels and Variability of Aerosol Number Concentrations

3.1 General Properties of Number Size Distributions

The particle number size distributions can be, as previously noted, often considered to be a combination of several aerosol subpopulations or modes. These modes are often log-normally distributed in the diameter space. Interestingly, the concentration histograms (i.e., frequency of detecting specific concentration) are typically also log-normal. Thus, using linear measures of concentration or size, such as arithmetic mean, can be strongly influenced by outlier values. This is the reason why either order-based metrics (such as percentiles and medians) or geometric properties (e.g., geometric mean or geometric standard deviation) of distributions are better mean properties to study if one is interested in typical concentration levels or mean particle diameters. The logarithmic occurrence spectrum is not only a feature of particle number concentrations. There are many reasons to expect similar behavior from the particle mass-based metrics [12], even though arithmetic means are the traditional method of averaging in the air quality contexts.

The log-normality of both size distributions (concentration vs. diameter) and in concentration histograms (occurrence vs. concentration) is a product of almost stochastic processes affecting the growth of aerosol populations. Distribution, which has a stochastic growth behavior dependent on the free variable, has tendency to approach log-normal shape [20]. In the case of particle number size distribution, the main growth process is condensation. In the case of particle number concentration, the increasing (or growth) process is both condensational growth from smaller size ranges and emission events. These essentially random growth incidents tend to derive the modes towards log-normal shape both in size and in concentration space. In both cases, the shape of the size distribution is further affected by additional processes, such as coagulation and emissions, which affect the widths and shapes of the distributions, bending and changing the distribution shapes away from log-normality.

Also, the particle number concentration histograms can have multiple "concentration modes." These can be interpreted as different air masses or emission periods affecting the number and mass concentrations measured at the stations. Such behavior is often seen in locations with time-dependent influence of either very clean air (resulting in a log-normal mode of smaller concentrations) or very polluted air (resulting in a log-normal mode of higher concentrations). Similar multi-modal histograms can also be detected in some cases at stations with high seasonality, where each mode corresponds to a seasonal concentration distribution.

3.2 Spatial Differences and Similarities of Aerosol Number Size Distributions

Combining the physical aerosol measurements from a high number of European background stations shows that there are clear similarities between particle number size distributions and concentration levels measured at different locations over wide geographical regions. These similarities are connected to similar emissions, particle loss processes, and meteorological patterns. The main aim of this section is just to provide key factors of each station categorization, more details and complete analysis of individual stations are available in [18] and references therein.

3.2.1 Central European Background (See Fig. 3)

The particle number size distributions and concentration levels measured at the Central European stations were remarkably similar. The median particle number size distributions did not change significantly from season to season, and the differences between the stations were not very large (N_{50} from 2,500 to 3,100 cm^{-3}). The variability of nucleation and small Aitken particles was elevated

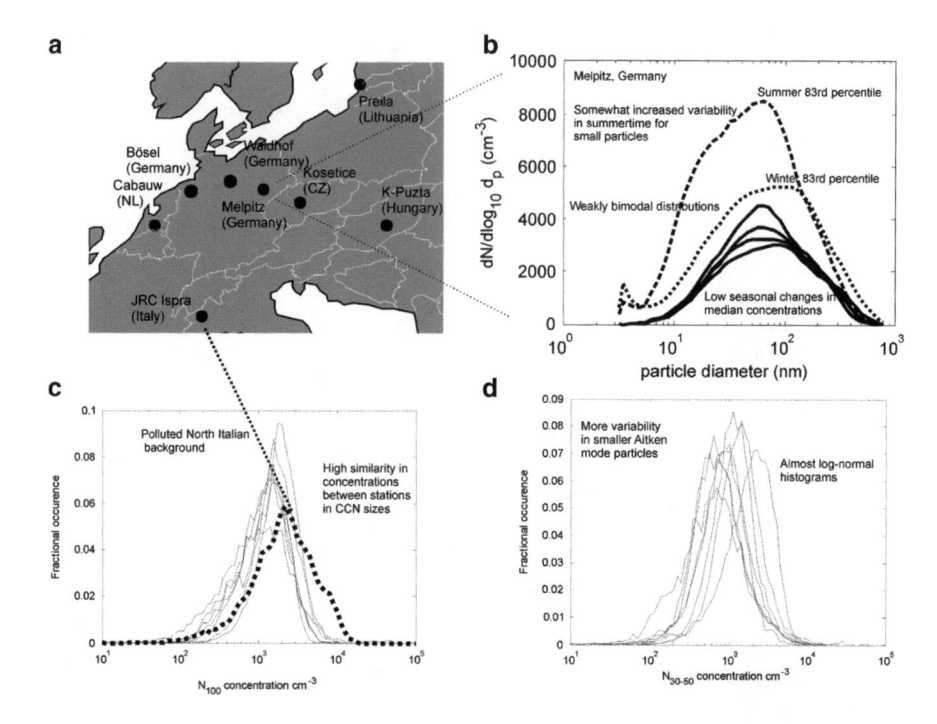

Fig. 3 Key features of submicron aerosol measured at Central European and North Italian EUSAAR and GUAN stations. (**a**) location of the stations, (**b**) key features of size distributions, (**c**) key factors of N_{100} concentration histograms, and (**d**) in N_{30-50} concentration histograms

on summer and autumn months in some stations, leading to short periods of higher concentrations in number concentrations of particle sizes smaller than 100 nm in diameter.

In the histogram (occurrence) plots, the N_{30-50} concentrations showed in some stations weak seasonality, with wintertime concentrations more likely to have decreased concentrations compared to summertime values. However, in the N_{50} and N_{100} concentrations, no seasonality was detected. The histograms are also unimodal and relatively narrow, which suggests very homogenous environment for all of the integrated number concentrations studied. There is also very little evidence of a strong diurnal cycle in the aerosol sizes studied. The station annual median concentrations of N_{30-50} varied between 410 and 1,120 cm^{-3}, N_{50} between 1,330 and 3,387 cm^{-3} with most stations over 2,300 cm^{-3}, and N_{100} between 739 and 1,863 cm^{-3}. The lack of seasonality may have two major reasons. First of all, the relative contribution of the natural aerosol particle sources is smaller than on more remote stations, and secondly, the overall meteorological situation in Central Europe has less strong influence on aerosol concentrations than, e.g., Northern Italy.

Overall, the results suggest that the particle number size distributions in Central Europe are very similar over very large region, and even though the mean concentrations somewhat vary from station to station, the background air in Europe is homogenous from the aerosol point of view.

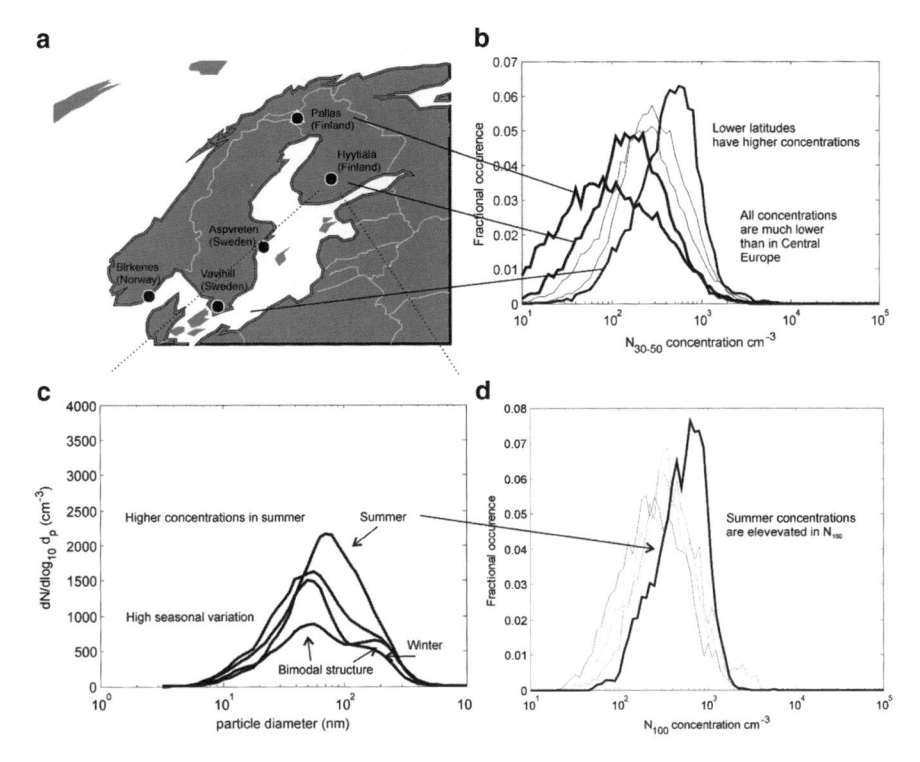

Fig. 4 Key features of submicron aerosol measured at North European EUSAAR stations. (a) location of the stations, (b) histograms of the annual N_{30-50} concentrations, (c) seasonal median size distributions, and (d) seasonal N_{100} variation at Hyytiälä, Finland

3.2.2 Polluted Northern Italian Background (See Fig. 3)

An extreme case of polluted background station was the Joint Research Centre station near Ispra in Northern Italy. Although the N_{30-50} concentrations were similar to Central European stations, the concentration levels in the larger submicron particle sizes were significantly higher, with median N_{50} concentration of 4,448 cm^{-3}. The seasonality of N_{50} and N_{100} particles was also very high, with wintertime median N_{100} concentrations approaching 10,000 cm^{-3}. The station was characterized by a strong anthropogenic influence and wintertime inversions, which trapped the pollution near the surface. These effects lead to extreme episodes with high concentrations in accumulation mode.

3.2.3 North European Background (See Fig. 4)

Northern Europe is characterized by strong seasonal variation in particle number size distributions and much lower overall concentrations than Central Europe. The high seasonality is from a far larger fraction of non-anthropogenic sources

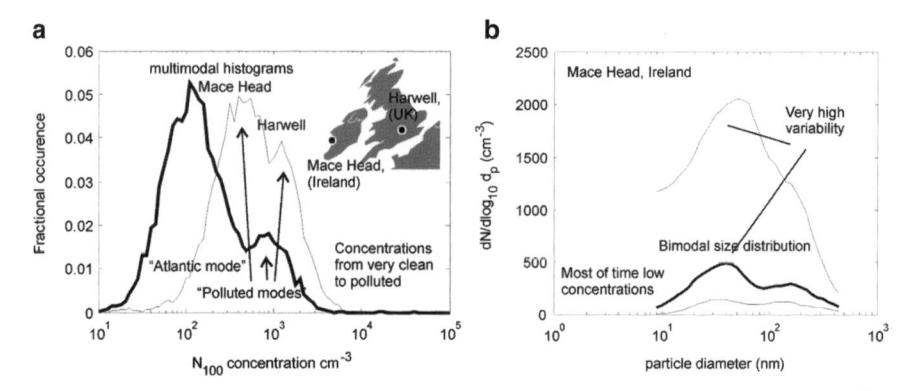

Fig. 5 Britain and Ireland aerosol overview (location shown in the *inset*). (**a**) Histograms of N_{100} concentration in both stations, (**b**) Mace Head size distributions, showing 16th, median (*thick line*), and 83rd percentiles of size distribution functions

influencing the size distributions. Although the similarities between the stations are not as obvious as in Central Europe, the datasets have uniformity in regard to seasonal and size distribution behavior. The winter number concentrations are usually lowest, especially for smaller (diameter less than 100 nm) particle sizes. The seasonality also affects the observed differences of day- and night-time values due to differing length of day in the northern latitudes. All of the stations showed bimodal median particle number size distributions, with clear Aitken and accumulation modes. The concentration distributions on some stations show multiple modes, suggesting a combination of more polluted air masses and cleaner air from the Arctic or Atlantic oceans.

The CCN-sized particle number concentration histograms of the Nordic stations are similar for both N_{50} and N_{100}. The stations in general had greater concentrations in summertime, although the concentrations were also elevated during spring in some stations. Northernmost Pallas station concentrations have a very wide histogram, especially in wintertime, suggesting a wide range of sources affecting the concentrations observed at the station. The seasonal variations in CCN-sized particles are large at almost all Nordic stations.

3.2.4 Britain and Ireland Background (See Fig. 5)

The two EUSAAR/ACTRIS stations in the Britain and Irish stations show not only a significant inter-station variability, but also many similarities. The stations have high seasonal variation and large variance in intra-seasonal concentrations. The station data show a prominent spring–summer maximum in all sizes from 30 to 70 nm. The variability was probably due to occurrences of both clean Atlantic and polluted local air masses, and the maximum value at Mace Head during summer months can be attributed to enchanted marine biota activity, which increases the sub-micron particle mass concentration of non-seasalt sulfate and organic aerosol [21].

The histograms show a clear influence of multiple aerosol sources (clean and polluted) and with a high seasonal cycle for smaller particle sizes. Both stations have a very widely spread histogram in all seasons suggesting high variability in concentrations and multiple types of air masses. Most of the Mace Head N_{100} histogram is located at low concentrations with approximately 200 particles cm^{-3}, but with a second mode at about order on magnitude greater concentrations. This is well in line with previous studies from the both stations showing the importance of the difference between air masses arriving over the relatively clean Atlantic Ocean contrasted by polluted air masses arriving from Britain and mainland Europe [22, 23]. Neither of the stations have strong seasonal signals in CCN-sized concentrations.

3.2.5 Mountain Stations (See Fig. 6)

At high-altitude sites (defined as measurement height approximately 1,000 m above mean sea level, note that the categorization is slightly changed from [18] to include some of the intermediate-height stations), particle number size distributions are similar, even though the stations are located in different parts of the continent. The particle number concentrations were low compared to nearby lowland sites, as most of the aerosol sources are on surface. The particle number size distributions generally show bimodal behavior, although the modes are overlapping at some of the stations. The seasonal cycle is similar at all sites, with greater concentrations during summer, especially for particles over 70 nm in diameter. The variability is considerable especially in summertime, suggesting a range of different types of air masses – most likely boundary layer air during daytime and clear tropospheric air during nighttime. Overall, the concentrations are smaller the higher the station is located.

All of the N_{30-50} histograms at mountain sites have similarities, with almost lognormal shapes, with clear concentration tails towards greater concentrations and similar seasonal cycles. The winter conditions, probably more representative of the free troposphere, were characterized by lower concentrations. The summertime histograms show greatest concentrations, probably due to the planetary boundary layer and/or valley winds influence. The greatest concentrations were generally observed during daytime. The median concentrations of N_{30-50} varied between 79 cm^{-3} for Jungfraujoch station in Switzerland (JFJ, altitude 3,580 m) and 418 cm^{-3} for Schauisland (SSL, altitude 1,210 m).

For the N_{50} and N_{100} particles, in most mountain sites the greatest concentrations were observed during daytime summer and autumn. This daytime effect could be connected with air masses arriving from lower altitudes bringing more polluted air from below [24]. The concentrations are all strongly skewed towards lower concentrations. The lowest concentration tails of the distributions are probably indicative of concentrations of the free tropospheric air. The concentrations of the stations with highest altitude had a more pronounced clean mode with N100 concentrations below 100 cm^{-3}.

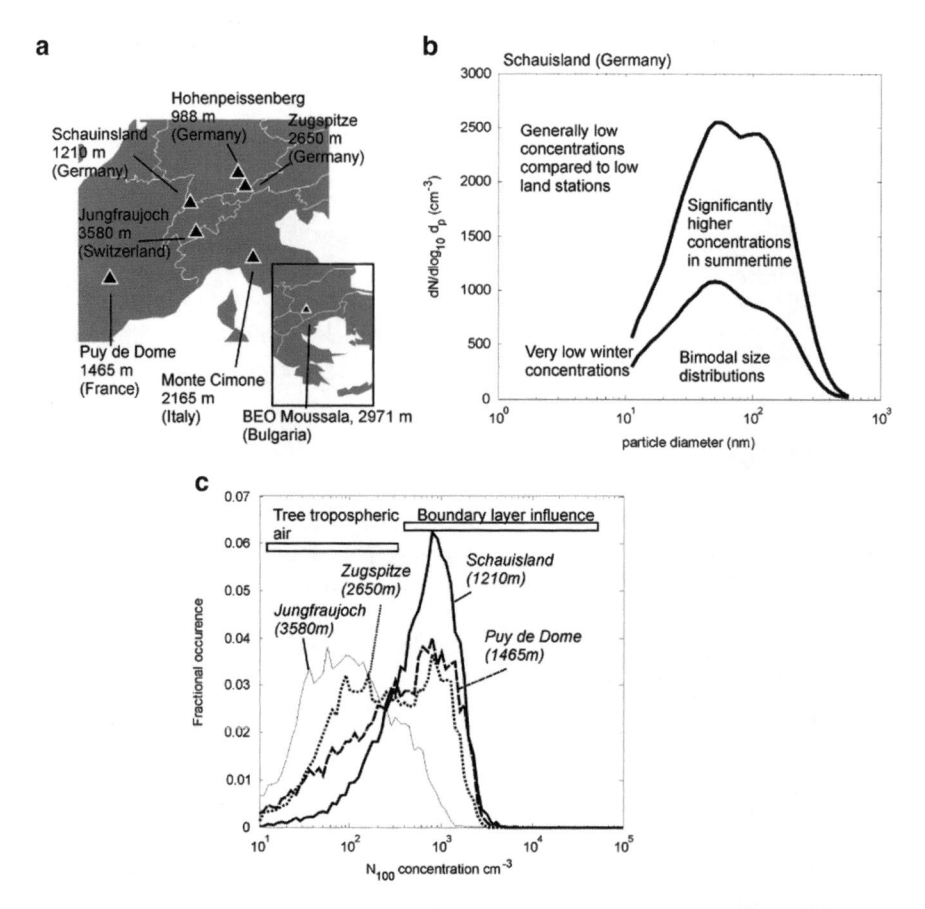

Fig. 6 Features of mountain stations. (**a**) Location of the stations with their height from the mean sea level, BEO Moussala in Bulgaria is shown in an *inset*, (**b**) median summer and winter size distributions in Schauisland showing typical features of mountain size distributions, (**c**) histograms of N_{100} concentrations in four mountain stations, showing the prevalence of cleaner "free tropospheric" air in the stations located in high locations

3.2.6 Specifics of Arctic Size Distributions (See Fig. 7)

One of the stations, Zeppelin (ZEP), is located far north of the European mainland, on the Svalbard archipelago, 78°N. This far northern position creates many environmental drivers for the aerosol size distribution, almost never seen at the more southern stations. As the station is located north of the northern polar circle, the station is good part of the year in complete daylight (midnight sun), and in complete darkness (polar night). Although the data quality was not always optimal, some indication of the aerosol number size distributions can be made.

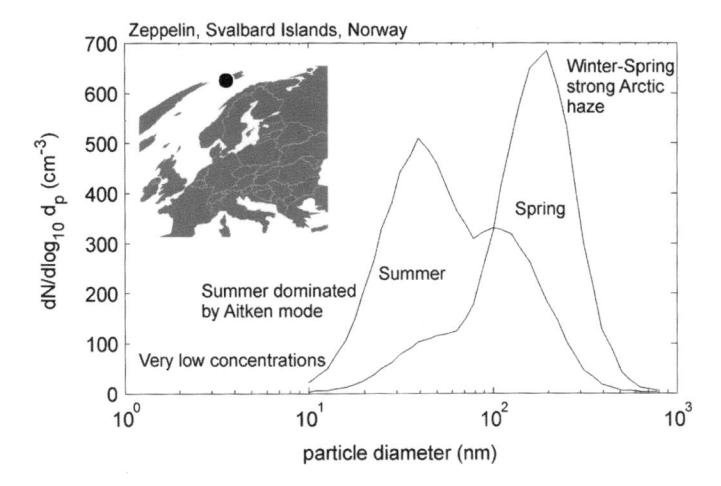

Fig. 7 Features of Arctic station Zeppelin size distributions. Location of the station is shown in the *inset*

In addition to changes in light, the changes in sea-ice and general Arctic circulation make the number size distributions observed at the station very different from southern locations. During winter and autumn, the particle number concentrations were very low with their maxima at around 200 nm. The springtime distributions are dominated by Arctic haze, strongly increasing the concentrations in accumulation mode. In summertime, the distribution changes to very clean marine bimodal distribution, with a strong Aitken mode around 30 nm. This seasonal change is connected to different meteorological situations, daylight as well as changes in ocean ice cover. Concentrations at ZEP were very low compared to European mainland concentrations. The key feature of the Arctic aerosol is the extreme seasonality. During summertime, the particle number size distributions are dominated by a strong Aitken mode, whereas in the winter–spring seasons, a strong accumulation mode is completely dominating the size distributions.

The summertime N_{30-50} concentrations of around 100 cm^{-3} are high, compared to concentrations of around 10 cm^{-3} in other seasons. At CCN-concentrations, a completely different behavior can be seen, as the, e.g., N_{100} concentration is during springtime close to 800 cm^{-3}, in comparison with summertime median of 200 cm^{-3}. The day/night cycle at the ZEP station is very strongly connected to the seasonal cycle.

3.2.7 Eastern Mediterranean Background (See Fig. 8)

At the Mediterranean station Finokalia (FKL) located in the Greek island of Crete, the particle number size distributions were bimodal for winter with an Aitken mode around 50 nm particle diameter and accumulation mode at 150 nm. The spring and summer were dominated by strong accumulation mode at around 100 nm.

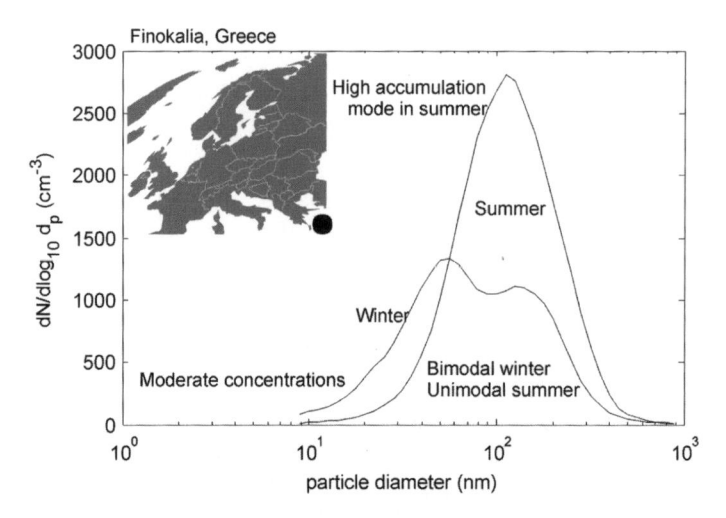

Fig. 8 Features of Eastern Mediterranean Finokalia size distributions. Location of the station is shown in the *inset*

The station had greatest N_{30-50} concentrations during spring and winter with no apparent day/night difference. The concentration levels were relatively low, with a median concentration of 220 cm^{-3}.

The particle histograms show peaks in N_{50} and N_{100} concentrations during spring of over 1,000 particles cm^{-3}. Another smaller mode of low concentrations was also visible around 500 particles cm^{-3} in both N_{50} and N_{100}.

3.3 Correlations Between Aitken and Accumulation Mode

Figure 9 (adapted from [18]) shows some of the typical correlations between particle number concentrations between 30 and 100 nm (here referred to as "Aitken mode," although a more rigorous derivation would require actual modal fitting) and concentrations between 100 and 500 nm ("accumulation mode"). The idea of this kind of plot is to show the possible correlation between the two aerosol modes, to indentify some of the main particle number size distribution types, and whether the particle number concentrations in both modes increase in the same rate.

In general, the Aitken and accumulation mode number concentrations are correlating (on logarithmic scales), but on different environments, the correlation is not always along the 1:1 line. Aerosol at the Nordic, Mediterranean and polluted continental (Central European) conditions has roughly distributions along the diagonal, which shows that in most cases the Aitken and accumulation modes are behaving in the same way. This suggests that the shape of the particle number size distribution is relatively nonsensitive to the overall concentration levels. In contrast, the free tropospheric part of the mountain distributions is almost round in shape, a

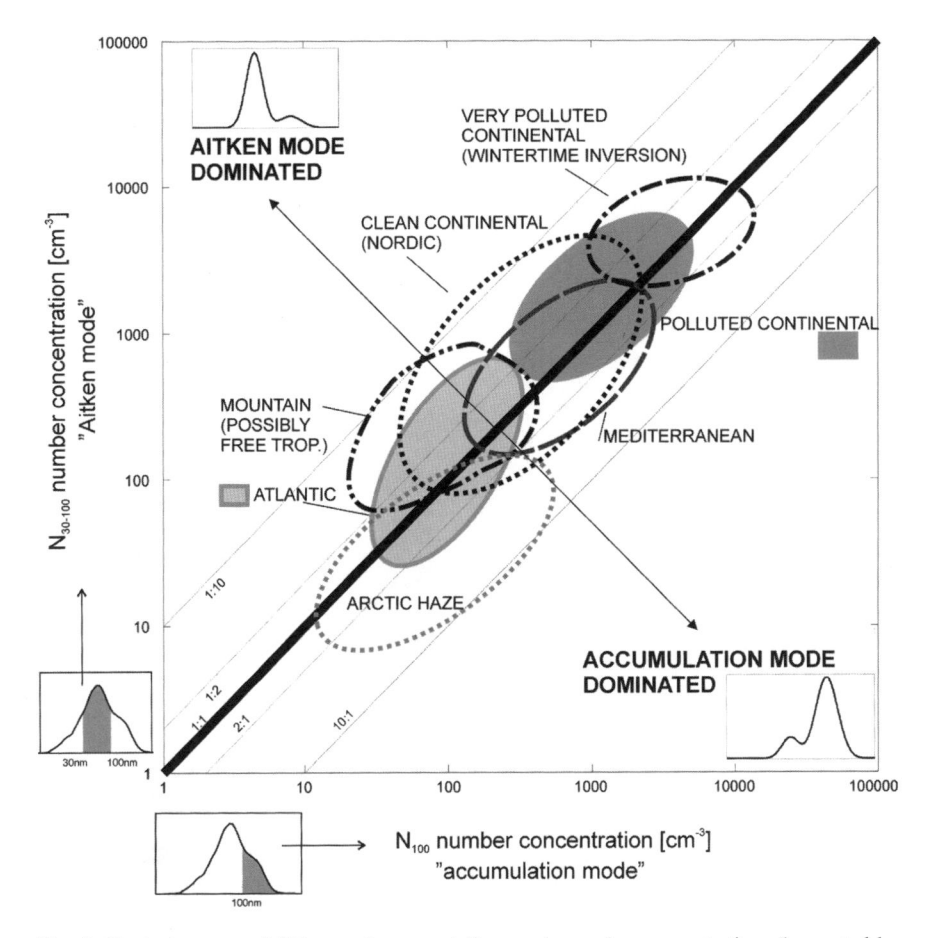

Fig. 9 Typical ranges of Aitken and accumulation mode number concentrations (separated by 100 nm dry diameter) in different European environments. The areas show the densest parts of the different environment concentration scatter (see Asmi et al. [18] for details, adapted from the same article)

sign that the two modes are relatively non-correlated in these air masses. In the case of Atlantic and polluted Italian background, the two modes are correlated, but not in diagonal direction. Atlantic air masses seem to increase in concentration mostly in Aitken mode, suggesting that most of the variability is in sub-100 nm range. The polluted background environments instead have a strong increase in accumulation mode, with relatively low variability in Aitken mode. Only Arctic haze environment has very strong dominance of the accumulation mode, although considerable parts of most other environments sometimes show such behavior.

Overall, in most environments the two modes are correlated, but different environments have very varying concentration levels of the two modes (see Table 1), and their relationship varies strongly from environment to environment and has relatively large scatter.

Table 1 Representational (geometric) mean aerosol number concentrations in European background environments (data from [18])

Environment	Typical particle number concentration between 30 and 100 nm (cm^{-3})	Particle number concentration between 100 and 500 nm (cm^{-3})	Spatial variability
Central European polluted background	1,500 (400–8,000)	1,000 (400–6,000)	Low
Northern European background	800 (80–5,000)	400 (50–2,000)	High
Atlantic air masses on Britain and Ireland	200 (30–700)	100 (30–300)	High
Eastern Mediterranean sea	800 (200–2,000)	700 (100–3,000)	Unknown (1 station)
Mountains (probably free tropospheric)	100 (20–700)	100 (20–400)	Smaller concentrations on higher sites
Arctic Haze period	40 (8–100)	80 (10–500)	Unknown (1 station)
Polluted inversion (Northern Italy)	5,000 (2,000–10,000)	3,000 (800–10,000)	Unknown (1 station)

The values in parentheses give the approximate typical variability range, including the spatial and temporal variability

3.4 Short Scale Temporal Variability

Sub-seasonal differences in particle number size distributions can come from many sources. In very short timescales, the spatial variability of local sources, atmospheric turbulence, cloud processing, precipitation, and other short-time variabilities can create significant differences in particle number size distribution. Diurnal cycles can also be important for the particle number size distributions, as the differences in emissions or mixing layer heights can lead to changes in particle mass and number concentrations in the surface layer. In Central European stations, the differences between day and night concentrations were not large, but some indication of the highest concentrations occurring more likely in the nighttime could be detected, especially for N_{100}. This behavior was much stronger in polluted Northern Italian station, further suggesting the meteorological background of the high concentrations in larger particle sizes. For the Northern European (excluding Northern Lapland Pallas) mountain and Britain and Ireland sites, no such strong diurnal effect could be found. The strongest day/night differences were observed in Pallas station (PAL in Fig. 2) in Northern Finland, and (with some reservations in data quality) in Arctic Zeppelin (ZEP) stations, although in these stations, the observed differences do not come from actual diurnal cycles, but from the fact that the stations are located north of the Polar Circle, meaning the strong difference between daytime and nighttime concentrations is actually an artifact of the seasonal variation.

Of particular interest are the differences between days of the week. The so-called "weekend effect" cycle in aerosol–weather interactions is based on observations of day of the week-related variations of meteorological data [25]. These effects have been connected to similar variations in particle mass concentrations and optical thickness detected at measurements sites in urban and suburban and remote locations [26]. The key point of the "weekend effect" is that the weekday-related changes in anthropogenic emissions of aerosol particles could change the regional meteorology in such an extent that the precipitation and air temperature could be significantly affected. This could then have implications on, e.g., weather prediction. The main mechanism behind this proposed weekend effect could be (semi)-direct or indirect aerosol effects. As this could be a direct anthropogenic influence to the short-term weather systems, such existence of such phenomena would influence many atmospheric fields.

As discussed earlier, the indirect effects of aerosols are controlled by the number of CCN, not by aerosol particle mass or optical properties used in many weekend-effect publications. The datasets of CCN-sized aerosol number concentrations in EUSAAR/ACTRIS and GUAN stations do not support indirect effects as a major contributor to the (possible) weekend effect. Figure 10 shows distributions (25–50–75th percentiles) of each weekday in a long time series of two stations. The statistical tests and wavelet frequency analysis could not detect any consistent statistically ($p < 0.05$) significant differences in CCN-sized number concentrations in annual or seasonal datasets from the stations. This means that the concentrations are not generally different in different weekdays, making a strong continent-wide weekend effect unlikely to occur from aerosol indirect effects. The main reason for the differences between number and mass-based weekday variation is probably a combination of different sources and the fact that mass-based measurements measure particles with much lower lifetime [11]. As lower lifetime particles can be removed efficiently from the atmosphere during the lower emission period, the weekday signal for such aerosol properties is larger than for most CCN-sized particles.

Even though there is no significant change in the CCN number concentrations between days of the week, this is not yet the complete picture of potential aerosol-cloud weekend effects. There are many other potential aerosol processes, such as semi-direct effect, which could have significant effect on the local meteorology. Also, the CCN number concentrations have a weak weekday variation within the cities, which could then lead to meteorological weekend effect directly above the urban environment. However, as the urban or semi-urban areas cover spatially quite small area of the Europe (around 5%) [27], the weekend effect is then of much more a local effect, if it exists at all.

The difference between mass and number-based metrics in weekly variation is a good example on the short-time scale differences between the properties. One should be careful on generalizing correlations from long timescales to small timescales, especially near the lifetime of the particles in question.

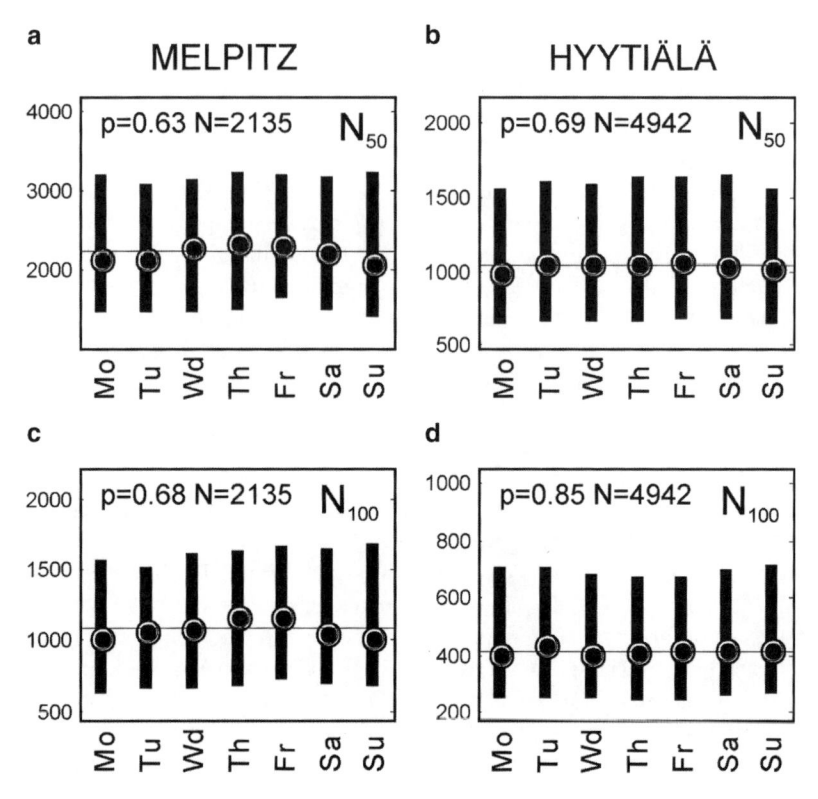

Fig. 10 (**a–b**) Weekday variation of the N_{50} concentration in Central European background (**a**, Melpitz, Germany) and in Northern European background (**b**, Hyytiälä, Finland). *Dots* indicate median concentrations (cm^{-3}) and the *bars* show the variation between 25th and 75th percentile. The p values represent the possibility that all of the weekday distributions are from the same distribution as a result of a U-test. The period of measurements in days is given as N. (**c** and **d**) same for N_{100} concentrations. Adapted from [11]

3.5 Comparing Measurements of Number Size Distribution with Model Output

One of the main uses of comparable size distribution dataset is model-measurement comparison. Without comparable data, the air quality and climate models do not have a way to reliably validate their results. In model-to-measurement comparison, many application-based conditions must be taken into account, and especially for aerosol number size distribution, some of the key points are [18]:

- The comparison must be done with similar properties (STP conditions ($T = 293.15$ K, $P = 101,300$ Pa), dry aerosol).
- Similar timescales for comparison. Short-term simulations can provide easily different or similar particle number size distributions than measured, just from overall model internal variability.

- Longer datasets give possibility to compare distributions of concentrations (distribution statistics, histograms), which give better view of concentrations, processes, and variability than short-term comparisons.
- Multiple peaked concentration histograms are usually a sign of multiple types of airmasses arriving to the station. The model ability to produce all of the peaks is also dependent on capturing correct advection.
- Particle number concentrations in different sizes in the submicron range are strongly interrelated. Comparing just one size range can give too optimistic view of the model performance and multiple size range comparison is useful in pinpointing the processes needing improvement.
- It is worthwhile to consider the special environments of some of the stations. In Arctic, the Arctic haze, in mountains, the transport from lower altitudes and the common inversion situations in Northern Italy can lead to very hard to reproduce size distributions, at least in large spatial scale models.

The EUSAAR/ACTRIS and GUAN data together with some of the analyses is available freely for model-to-measurement comparison uses at http://www.atm.helsinki.fi/eusaar/.

4 Conclusions and Outlook for the Future

The submicron aerosol populations in the European background air are variable from location to location. The concentrations and variability of aerosol distributions do, however, show similarities over large geographical areas (Fig. 11). The particle number concentrations are generally lower in more northern and higher mountain locations, naturally as they are generally located farther from the emission areas.

Standardized long-term measurements provide reliable information on statistical behavior of atmospheric aerosols, far beyond what could be obtained in short-term campaign-wise measurements. Although data from a period of only two years is shown, the results already provide a previously unavailable variety of information on the sub-micron aerosol physical properties and variability in Europe. Such information would also be hard to achieve based on information collected from separately managed stations, especially if the instrumentation and data handling are not harmonized.

The similarities within the regions give a good chance of useful air quality model-measurement comparisons. The actual choice of what should be used for the models to compare with depends on the application and complexity needed. The most straightforward way is just to compare one or more mean parameters, such as median concentrations. This approach is simple to do, but can easily lose many features of the data, and, in cases of strongly bimodal histograms, can even be misleading. Comparing modeled histograms to results should pay attention to the histogram mode location (mean or median concentration), width, and relative abundance (height) of each mode in the histograms.

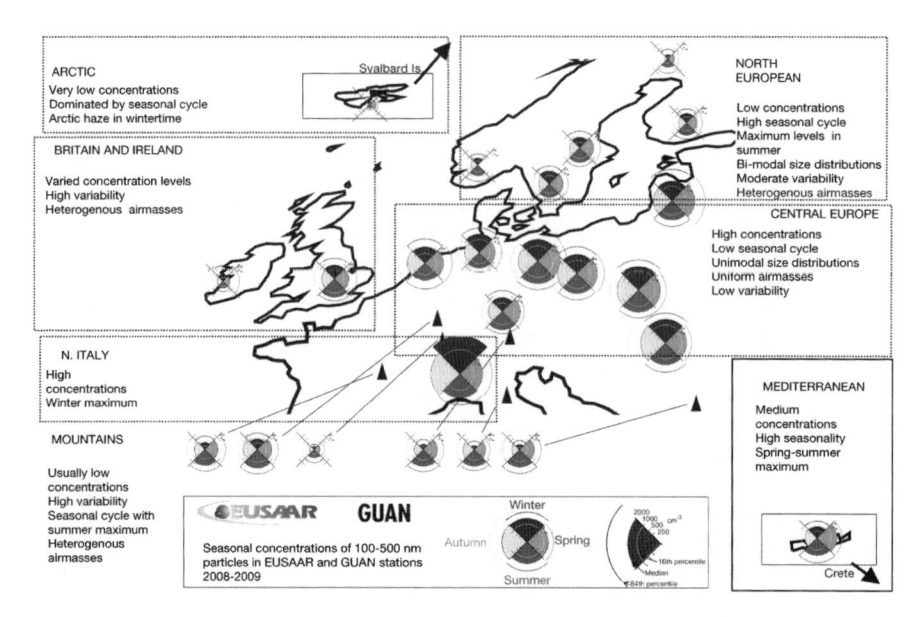

Fig. 11 Overview of submicron aerosol number concentrations. The *symbols* indicate the typical concentrations, variability, and seasonality of N_{100} particles. Adapted from [18]

The EUSAAR/ACTRIS and GUAN networks are globally unique both in data quality and relatively dense network. Building a similar global network is a major undertaking, but would enable the community to efficiently characterize the aerosol number distribution, and thus to improve the potential of characterizing the climate impacts of aerosols in the global boundary layer.

Acknowledgments The author wishes to thank Dr. A. Wiedensohler for the useful comments in the review. The extensive work of all scientists and technical staff maintaining and operating the stations and the instruments is gratefully acknowledged.

References

1. Dockery DW, Pope C (1994) Acute respiratory effects of particulate air pollution. Annu Rev Public Health 15:107–132
2. Wittmaack K (2007) In search of the most relevant parameter for quantifying lung inflammatory response to nanoparticle exposure: particle number, surface area, or what? Environ Health Perspect 115:187–194
3. Oberdörster G, Oberdörster E, Oberdörster J (2005) Nanotoxicology: an emerging discipline evolving from studies of ultrafine particles. Environ Health Perspect 113:829–839
4. Beddows DCS, Dall'osto M, Harrison RM (2009) Cluster analysis of rural, urban, and curbside atmospheric particle size data. Environ Sci Technol 43:4694–4700
5. Commission regulation (EC) No 682/2008 of 18 July 2008 (2008) Off J Eur Union: Legis, L199, 1–136

6. Seinfeld JH, Pandis SN (2006) Atmospheric chemistry and physics - from air pollution to climate change. Wiley, New Jersey, USA
7. Twomey S (1977) The influence of pollution on the shortwave albedo of clouds. J Atmos Sci 34:1149–1152
8. Albrecht BA (1989) Aerosol, cloud microphysics, and fractional cloudiness. Science 245:1227–1230
9. Andreae M, Rosenfeld D (2008) Aerosol–cloud–precipitation interactions. Part 1. The nature and sources of cloud-active aerosols. Earth Sci Rev 89:13–41
10. McFiggans G et al (2006) The effect of physical and chemical aerosol properties on warm cloud droplet activation. Atmos Chem Phys 6:2593–2649
11. Asmi A (2012) Weakness of the weekend effect in aerosol number concentrations. Atmos Environ 51:100–107. doi:10.1016/j.atmosenv.2012.01.060
12. van Dingenen R et al (2004) European aerosol phenomenology – I: physical characteristics of particulate matter at kerbside, urban, rural and background sites in Europe. Atmos Environ 38:2561–2577
13. Heintzenberg J et al (1998) Mass-related aerosol properties over the Leipzig basin. J Geophys Res D 103:13125–13135
14. Laj P et al (2009) Measuring atmospheric composition change. Atmos Environ 4:5351–5414
15. Wiedensohler A et al (2012) Particle mobility size spectrometers: harmonization of technical standards and data structure to facilitate high quality long-term observations of atmospheric particle number size distributions. Atmos Meas Tech 5:657–685
16. Philippin S et al (2009) EUSAAR an unprecedented network of aerosol observation in Europe. Earozoru Kenkyu 24:78–83
17. Birmili W et al (2009) Atmospheric aerosol measurements in the German Ultrafine Aerosol Network (GUAN), – Part 1: soot and particle number distributions. Gefahrstoffe Reinhalt Luft 69:137–145
18. Asmi A et al (2011) Number size distributions and seasonality of submicron particles in Europe 2008–2009. Atmos Chem Phys 11:5505–5538
19. Henne S et al (2010) Assessment of parameters describing representativeness of air quality in-situ measurement sites. Atmos Chem Phys 10:3561–3581
20. Grönholm T, Annila A (2007) Natural distribution. Math Biosci 210:659–667
21. Yoon YJ et al (2007) Seasonal characteristics of the physicochemical properties of North Atlantic marine atmospheric aerosols. J Geophys Res, Vol. 112, 14 p., doi:10.1029/2005JD007044
22. McGovern FM et al (1996) Aerosol and trace gas measurements during the Mace Head experiment. Atmos Environ 30:3891–3902
23. Charron A, Birmili W, Harrison RM (2008) Fingerprinting particle origins according to their size distribution at a UK rural site. J Geophys Res 113:07202
24. Weingartner E, Nyeki S, Baltensberger U (2000) Seasonal and diurnal variation of aerosol size distributions ($10 < D < 750$ nm) at high-alpine site. J Geophys Res 104:26809–26820
25. Forster PM, Solomon S (2003) Observations of a "weekend effect" in diurnal temperature range. Proc Natl Acad Sci USA 100:11225–11230
26. Bäumer D, Vogel B (2007) An unexpected pattern of distinct weekly periodicities in climatological variables in Germany. Geophys Res Lett, Vol. 34, 4 p., doi:10.1029/2006GL028559
27. European Environmental Agency (2007) Land-use scenarios for Europe: qualitative and quantitative analysis on a European scale (PRELUDE), EAA Technical report No 9/2007, ISBN: 987-92-9167-927-0, http://www.eea.europa.eu/publications/technical_report_2007_9

Indoor–Outdoor Relationships of Particle Number and Mass in European Cities

Gerard Hoek, Otto Hänninen, and Josef Cyrys

Abstract Human exposure to air pollutants is often characterized by measured or modeled outdoor concentrations. In Western societies, subjects spend about 90% of their time indoors, of which a large fraction in their own home. Hence indoor air quality is an important determinant of the true personal exposure for many components. Indoor air quality is affected both by infiltration of outdoor air in buildings and indoor sources such as smoking, gas cooking, and use of consumer products. In this chapter we separately describe the impact of indoor sources and outdoor air on indoor pollution. We first illustrate differences in outdoor and personal exposure using data on real-time particle number concentrations from a recent study in Augsburg, Germany. We then present a model of indoor PM concentrations, illustrating the factors that affect indoor air quality. We summarize empirical studies that have assessed indoor–outdoor relationships for particle mass, particle number, and specific components of particulate matter.

Outdoor air pollution significantly infiltrates in buildings. Combined with the large fraction of time that people typically spend indoors, a major fraction of human exposure to outdoor pollutants occurs indoors. Understanding the factors affecting infiltration is therefore important. Infiltration factors have been shown to vary substantially across seasons, individual homes and particle size and components. Important factors contributing to these variations include air exchange rate, characteristics of the building envelop (e.g., geometry of cracks), type of ventilation, and use of filtration. Penetration and decay losses are particle size dependent with the lowest losses for submicrometer particles and higher losses for ultrafine

---

G. Hoek (✉)
IRAS, Institute for Risk Assessment Sciences, Utrecht University, PO Box 80178, 3508 TD Utrecht, The Netherlands
e-mail: g.hoek@uu.nl

O. Hänninen
Unit of Environmental Epidemiology, THL-National Public Health Institute, Kuopio, Finland

J. Cyrys
HMGU Institute of Epidemiology II, Neuherberg, Germany

and especially coarse particles. The largest infiltration factors are consistently found for sulfate and black carbon. Volatilization and chemical decay may also result in losses of specific components, including nitrates and organic components. The large variability of $PM_{2.5}$ infiltration factors reported may further be due to different composition of PM across locations. In locations with relatively high sulfate and EC contributions, higher infiltration factors can be anticipated than in locations with high nitrate and OC concentrations.

Keywords Indoor air, Infiltration, Outdoor, Particle size, Particles, Penetration, Ultrafine

Contents

1 Introduction	322
2 Example of Outdoor and Indoor Exposure: Augsburg Study	323
3 Model for Indoor Air Quality	324
4 Empirical Studies on Indoor–Outdoor Relationships of Particles	328
4.1 Methodological Issues	328
4.2 Indoor–Outdoor Relationships of $PM_{2.5}$	330
4.3 Indoor–Outdoor Relationships of Coarse (PM_{10}–$PM_{2.5}$) Particles	331
4.4 Indoor–Outdoor Relationships of Ultrafine Particles	331
4.5 Indoor–Outdoor Relationships of Specific Particle Components	332
5 Conclusions	334
References	335

1 Introduction

Particulate matter (PM) has been identified in many studies as an important component of the complex air pollution mixture responsible for various health effects [1, 2]. In many epidemiological studies, PM is characterized as the mass of particles smaller than 2.5 or 10 μm ($PM_{2.5}$ and PM_{10}). This is largely because of the availability of routinely measured concentrations of these particle metrics, related to the promulgation of air quality guidelines in many countries, including the European Union. It has, however, been hypothesized that the number of ultrafine particles (UFP) may be another health relevant particle metric [3] Nevertheless, there remains considerable uncertainty on the health effects of UFP observed in epidemiological studies, largely related to measurement error associated with characterizing exposure to UFP by central site outdoor measurements [4, 5]. A major component of measurement error is how well central site outdoor concentrations are reflected in indoor air at the thousands of different locations where the population actually is exposed for the majority of time. In Western societies, subjects spend about 90% of their time indoors, of which a large fraction in their own home. Hence indoor air quality is an important determinant of personal exposure for many components.

Indoor air quality has been a public health concern for several decades now. Indoor air quality is affected both by infiltration of outdoor air in buildings and indoor sources such as smoking, gas cooking, and use of consumer products [6]. Penetration of particles into indoor environments depends on particle size, air exchange rates, and other factors. Consideration of indoor sources is important because they may be associated with significant health effects, e.g., environmental tobacco smoke. Presence of indoor sources may further complicate assessment of the impact of outdoor air on indoor air. In this chapter we separately describe the impact of indoor sources and outdoor air on indoor pollution, because health effects of outdoor and indoor generated particles may differ as their composition differs [7].

We focus in this chapter on particles from ambient origin. We first illustrate differences in outdoor and personal exposure using data on real-time particle number concentrations (PNC) from a recent study in Augsburg, Germany. We then present a model of indoor PM concentrations, illustrating the factors that affect indoor air quality. We summarize empirical studies that have assessed indoor–outdoor relationships for particle mass, particle number, and specific components of particulate matter. The focus is on European studies, but we included key studies from outside Europe as well. We conclude by comparing the strength of indoor–outdoor relationships of various particle fractions and components.

2 Example of Outdoor and Indoor Exposure: Augsburg Study

Within the framework of a study conducted in Augsburg, Germany, personal exposure to UFP has been measured during 2011. The study participants followed three different scenarios: (A) commute by car to their home, spend the morning there, and commute back by car; (B) commute by public transport to their home, spend the morning there, and commute back by public transport; and (C) walk to the city center, spend about 4–5 h there with at least 2 h outside, and walk back. All trips started and ended at the study center, located near the central railway station in the center of Augsburg. The measurements were done by each of the participants on the same day of the week over 3 consecutive weeks in winter, spring, and summer. In all three scenarios, the subjects kept a detailed diary and their whereabouts were recorded with geographic positioning systems (GPS).

In Fig. 1 the average time series of personal and ambient PNC for scenarios A and B are presented. Between 7:00 and 8:30 as well as after 13.15 the majority of participants were outdoors traveling between the study center and their home. During this time the difference between the personal and ambient levels of PNC was rather small. In contrast large differences between personal and ambient levels of PNC were observed when participants were in their homes or other indoor microenvironments. In indoor microenvironments the personal exposure to PNC exceeded significantly ambient PNC levels. Domestic activities such as cooking or lightning candles were found to greatly increase personal indoor exposure. Especially cooking (between 12:00 and 13:00) strongly contributed to increases in personal PNC.

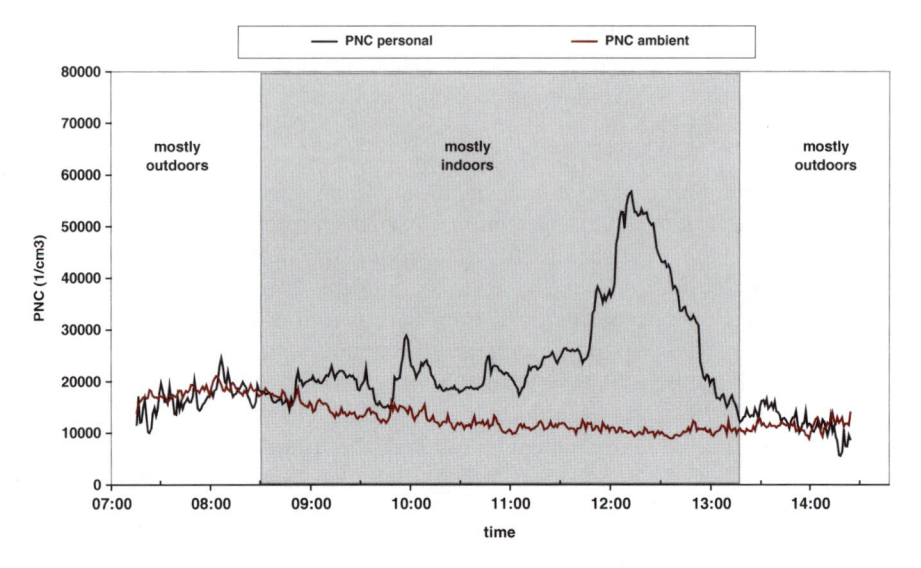

Fig. 1 Time series of personal and ambient PNC averaged for all measurements scenarios A and B ($n = 58$)

Individual patterns differed substantially from each other. In Fig. 2 two examples of individual patterns are presented. In Fig. 2a the personal exposure to PNC measured indoors was much lower as the ambient PNC. The study participant spent almost the whole time indoors in an office. The windows were closed during the whole day. Activities in the office contributed less to PNC as the typically domestic activities such as cooking or cleaning. The small peak of PNC around 12:30 was recorded during a coffee break in the entrance hall outside of the office.

In Fig. 2b the influence of two different indoor microenvironments is clearly visible. Whereas in the office the indoor (personal) PNC levels are lower than the ambient PNC (similar to Fig. 2a), the personal PNC were much higher in a cafeteria. Furthermore, it seems that the air conditioning system which turned on and off automatically together with some cooking activities caused a very pronounced pattern of PNC going down and up.

3 Model for Indoor Air Quality

Mass balance models are used extensively to describe the concentration in indoor air as a function of outdoor air and indoor sources. According to Dockery and Spengler [6] the mass-balance of indoor and outdoor generated pollutants can be expressed as:

$$\overline{C_i} = \frac{Pa}{a+k}\overline{C_a} + \frac{\overline{Q}}{V(a+k)} - \frac{\Delta C_i}{\Delta t(a+k)} \qquad (1)$$

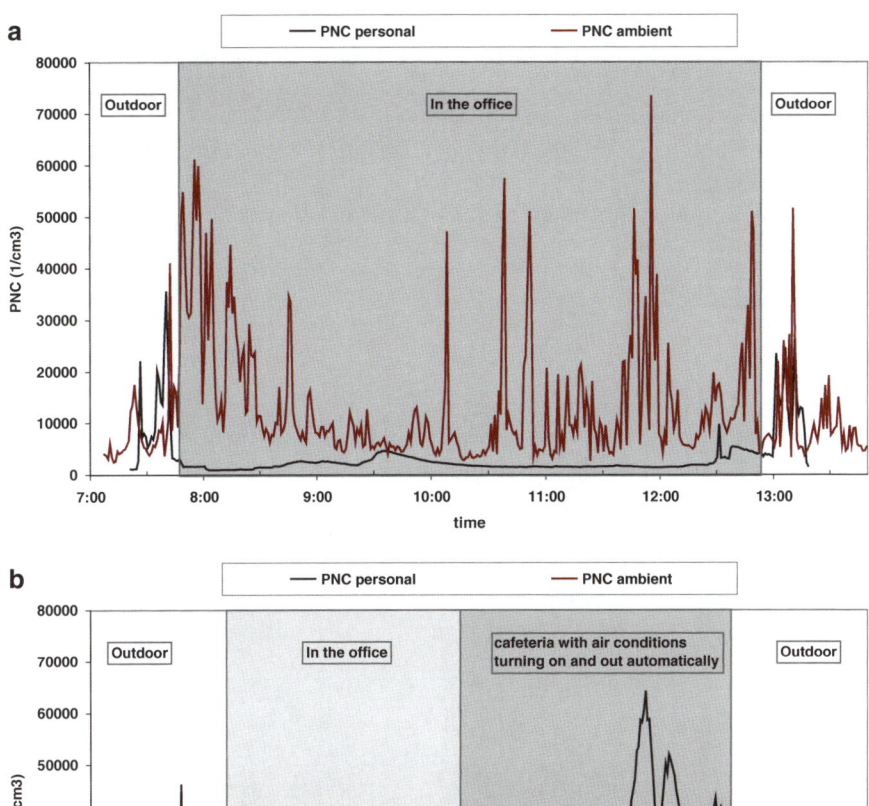

Fig. 2 Two examples of time series of personal and ambient PNC: (**a**) participant spent mostly the whole time indoors in an office and (**b**) participant spent time indoors in an office and a cafeteria

where
- C_i = indoor concentration (µg m^{-3})
- C_a = ambient (outdoor) concentration (µg m^{-3})
- P = penetration efficiency (dimensionless)
- a = air exchange rate (h^{-1})
- k = decay rate indoors (h^{-1})
- Q = source strength (µg h^{-1}) (symbol used by Dockery and Spengler was S)

V = interior volume of the building (m³)
$\Delta C = C_i(t_1) - C_i(t_0)$, indoor concentration change during the sampling period (µg m⁻³)
$\Delta t = t_1 - t_0$, sampling period (h)

The third term represents the transient component, which in steady state and for long-term averages can be set to zero.

Looking separately at outdoor originating and indoor generated components, the equation can be simplified into Wilson et al. [7]

$$C_i = C_{ai} + C_{ig} \qquad (2)$$

where
C_i = total indoor concentration (µg m⁻³)
C_{ai} = indoor concentration originating from outdoors (µg m⁻³)
C_{ig} = indoor concentration from indoor sources (µg m⁻³)

Now, combining the above equations, the fractional indoor concentrations from outdoor and indoor sources can be expressed separately as

$$C_{ai} = \frac{Pa}{a+k} C_a \qquad (3)$$

$$C_{ig} = \frac{Q}{V(a+k)} \qquad (4)$$

Equation (3) is also written as $C_{ai} = F_{inf} C_a$, with F_{inf} is defined as the infiltration factor. The infiltration factor describes the fraction of ambient pollution that penetrates indoors and remains suspended.

The model describes the influence of the key factors affecting indoor air quality, including air exchange rate and aerosol properties dependent on the particle size distribution, namely penetration efficiency (P) and decay rate (k). The model has been successfully used for estimation of indoor source strengths from observed concentration data [8]. Due to seasonal trends in meteorology, air exchange rates and consequently also infiltration vary with season [9]. However, data on actual air exchange rates and seasonal patterns is still quite limited due to complexities in air exchange rate measurement techniques.

Penetration efficiency represents the probability of a particle suspended in the air flowing into the building through windows, doors, cracks in the walls, and ventilation channels remaining in the air. The penetration efficiency is thus dependent on several factors including the air velocity, the opening dimensions, and the particle size. Liu and Nazaroff [10] developed a theoretical model for penetration efficiency assuming air flow through cracks of a given dimension. According to this model the penetration efficiency is very close to unity for all particle sizes from 10 nm to 10 µm if the crack width is about 1 mm. When the crack width is reduced to 0.25 mm, the particles effectively penetrating through it are reduced to a range from a few tens of nanometers to a few micrometers [10]. The pressure difference across the cracks has a modest additional impact (Fig. 3).

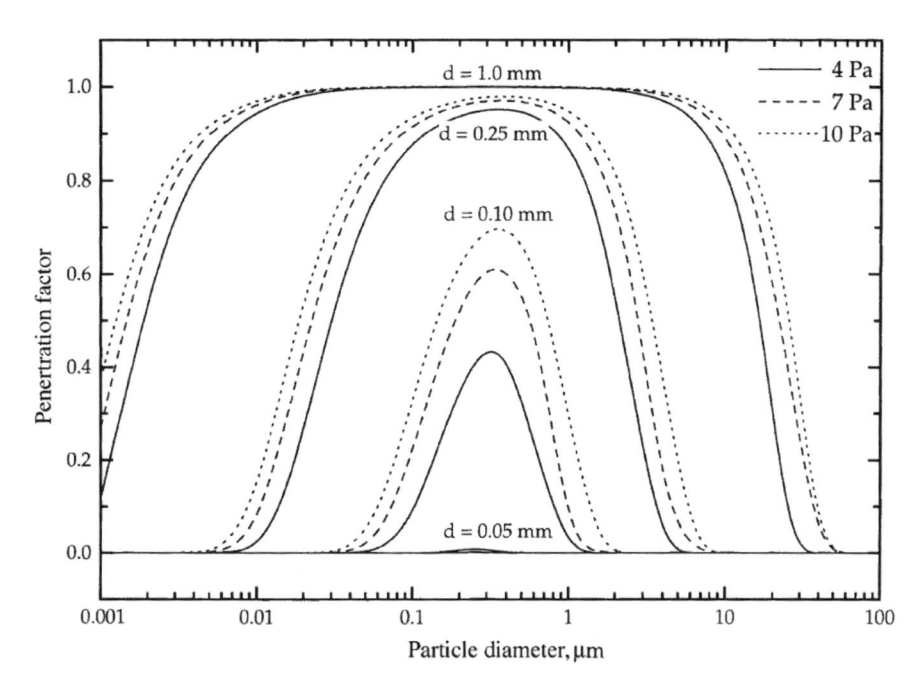

Fig. 3 Penetration efficiency as function of particle size and crack width [10]

The type of ventilation plays an important role as well. Chen and Zhao [11] distinguish mechanical ventilation (artificial systems usually including filters), natural ventilation (through open windows and doors), and infiltration (uncontrolled flow through cracks).

The second particle size-dependent parameter in the mass-balance equation is the decay term (k). Decay of nonvolatile particles indoors is dominated by thermokinetic and gravimetric deposition. Thermokinetic deposition affects UFP in the nanometer size range, while particles larger than few hundreds of nanometers in diameter are strongly affected by the gravimetric deposition. Between these two domains remains a particle size range, which has the largest tendency of remaining suspended in the indoor air. In the thermokinetic domain the deposition rate is strongly affected also by friction velocity, i.e. turbulence of the air in indoor spaces, bringing the particles effectively close to the deposition surfaces (Fig. 4).

The deposition rate depends also on the availability of surfaces. While gravimetric deposition takes place only on horizontal upward facing surfaces, thermokinetic deposition occurs also on downward facing and vertical surfaces. The more such surfaces are available for deposition, the faster the corresponding deposition rate is. Relationship of the deposition surface availability and the deposition velocity in a rectangular space is characterized by Lai and Nazaroff [12] as

$$\beta = \frac{v_{dv}A_v + v_{du}A_u + v_{dd}A_d}{V} \quad (5)$$

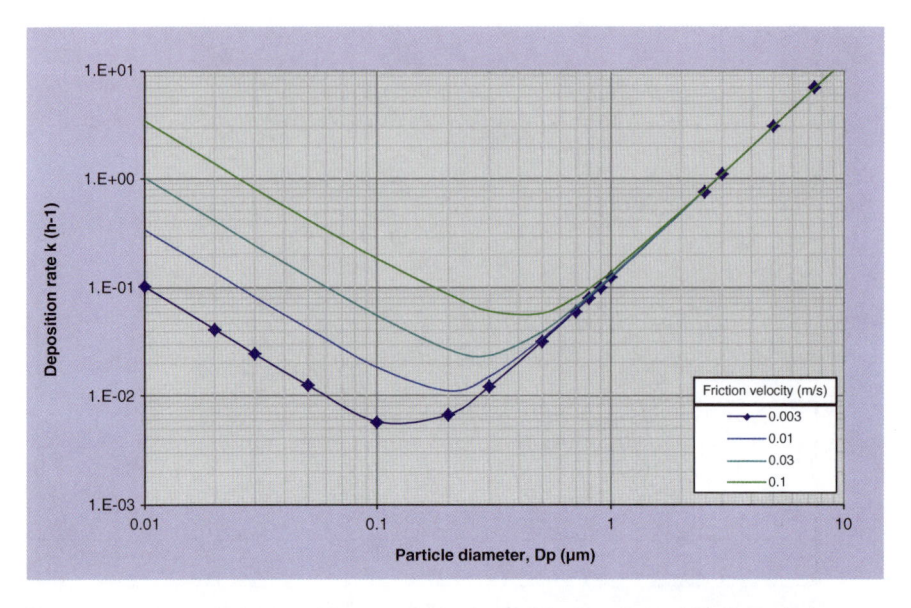

Fig. 4 Dependency of the deposition rate (k) on the particle diameter and friction velocity

where β is the first-order loss coefficient for deposition in a rectangular space V, v_{dv}, v_{du}, and v_{dd}, are the deposition velocities for vertical, upward and downward facing surfaces, respectively, and A_v, A_u, and A_d are the corresponding deposition surface areas.

As a result of the particle size-dependent properties the accumulation mode particles having highest penetration efficiencies and lowest deposition rates tend to enter indoors most efficiently and remain suspended there, thus substantially contributing to indoor exposures. Another implication is that the particle size distribution indoors differs significantly from that outdoors, even in the absence of indoor sources. Finally particle infiltration varies from home to home, resulting in higher variability across homes in indoor particle concentrations compared to outdoor concentrations.

4 Empirical Studies on Indoor–Outdoor Relationships of Particles

4.1 Methodological Issues

For assessment of exposure in epidemiological studies, the *correlation* between indoor and outdoor concentration is of importance in addition to the quantitative contribution of outdoor air to indoor air quality. Figure 5 presents the median of the

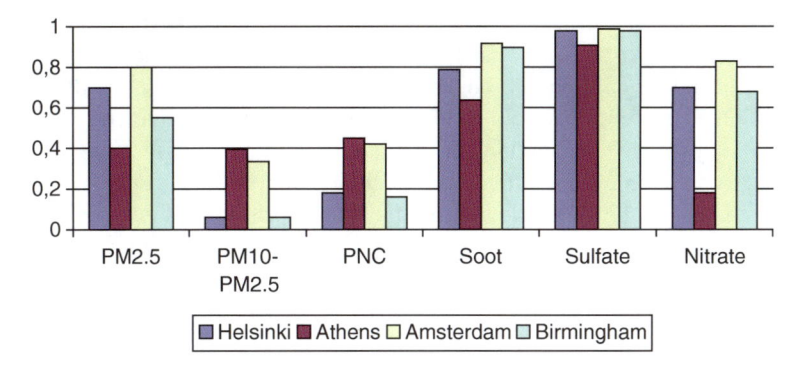

Fig. 5 Median Pearson correlation between central site outdoor and indoor 24-h average particle concentrations in four European cities. The correlation reflects the temporal variation of indoor and outdoor concentrations. Source: Hoek et al. [13]

longitudinal correlation of indoor and outdoor 24-h average particle concentrations in four European cities. These correlations were obtained by first calculating the correlation per home and then calculating the median per city. This correlation measures the temporal correlation and takes care of differences between homes related to, e.g., indoor sources. Correlations were moderately high for $PM_{2.5}$. For PNC and coarse particles correlations were much lower, due to a combination of lower infiltration of ultrafine and coarse particles in the homes compared to fine particles and time-varying indoor sources. Several studies have suggested that indoor sources contribute least to fine particles, defined as particles smaller than 1 μm (Sects. 4.2–4.4). For soot and sulfate, correlations were higher than for $PM_{2.5}$. This reflects fewer indoor sources for these components and that these components are concentrated in the submicrometer fraction that infiltrates most efficiently in homes.

The correlation only measures whether the time pattern of indoor concentrations track outdoor concentration trends, it does not provide information on how much a pollutant infiltrates in the indoor environment. In the RUPIOH study, outdoor and indoor nitrate concentrations correlated well in three of the four cities (Fig. 5). However, the infiltration factor was very low, ranging from 0.05 to 0.13.

Empirical studies have used different methods to describe quantitative indoor–outdoor relationships [11]. Many studies have used the *indoor/outdoor ratio* (I/O ratio). The I/O ratio of occupied homes is difficult to interpret as it is affected by indoor sources and by infiltration of outdoor particles indoors. I/O ratios from different studies have shown a very wide range: from well below 1 to well above 1 [11]. Some studies have attempted to exclude homes with major indoor sources such as smoking and use of gas appliances for cooking to obtain more interpretable values [13, 14, 19]. Because of the variety of indoor sources, this has generally not been successful. In a Swedish study that purposefully selected homes without smokers and gas appliances, indoor $PM_{2.5}$ concentrations were still for 60–90% caused by indoor sources [14]. The authors speculate that resuspension of particles by human movement and particle formation from organic components with ozone may have been significant indoor sources. Regression analyses of indoor and outdoor concentrations and indoor source use data within the EXPOLIS and RIOPA studies also documented that a large fraction of the indoor $PM_{2.5}$ source contribution was unidentified [8, 15]. Another approach is to limit the calculation of I/O ratio to hours with little indoor source activity, typically the nighttime [16, 17, 19]. The influence of this approach was demonstrated in a study of hourly ultrafine particle concentrations in four European cities [13]. When all 24 h were considered, the correlation between simultaneously measured hourly indoor and residential outdoor concentrations was weak (range 0.25–0.62). When only the nighttime period was included, a strong correlation was found in the four cities between indoor and outdoor ultrafine particle concentrations (range 0.64–0.87).

Many studies have used the *infiltration factor* (F_{inf}), defined as the fraction of outdoor particles that penetrate indoors and remain suspended [11]. Compared to the I/O ratio, a major advantage of the use of the infiltration factor is that it is not affected by indoor sources. Equation (3) describes the factors that influence it, which includes both penetration and decay rates.

Finally studies have reported the *penetration factor* which describes penetration and not decay. Penetration factors are not directly observable and they are usually calculated from infiltration rates and decay rates [11].

4.2 Indoor–Outdoor Relationships of $PM_{2.5}$

Hänninen and coworkers [9] have recently evaluated the original data of European studies of indoor–outdoor relationships for $PM_{2.5}$. The overall average infiltration factor was 0.55, illustrating significant infiltration of outdoor fine particles. A recent review including European and North American studies reported infiltration factors from 0.3 to 0.82 for $PM_{2.5}$ [11]. Since western people spend on average about 90% of their time indoors, human exposure to fine particles of outdoor origin largely occurs indoors. Infiltration factors were consistently higher in the summer season than in the winter season (Fig. 6). The implication is that for the same outdoor concentration, actual human exposure of subjects in the summer season is higher than in the winter season [9]. This could be one of the explanations for the often higher health effects reported in the summer months compared to the winter months in epidemiological studies that are based upon outdoor concentrations. Higher infiltration factors in the summer are explained by higher air exchange rates in the summer season compared to the winter. The air exchange rate (AER) was measured in the EXPOLIS and RUPIOH study in different seasons. Mean AER (h^{-1}) was 1.38 (summer) versus 0.56 (winter) in EXPOLIS and 1.07 (summer) versus 0.36 (winter) in RUPIOH [9]. Differences in infiltration factors across Europe were less consistent, with, e.g., little difference between Northern and southern European cities (Fig. 6). Methodological differences across studies may have complicated assessment of regional differences; however, two studies

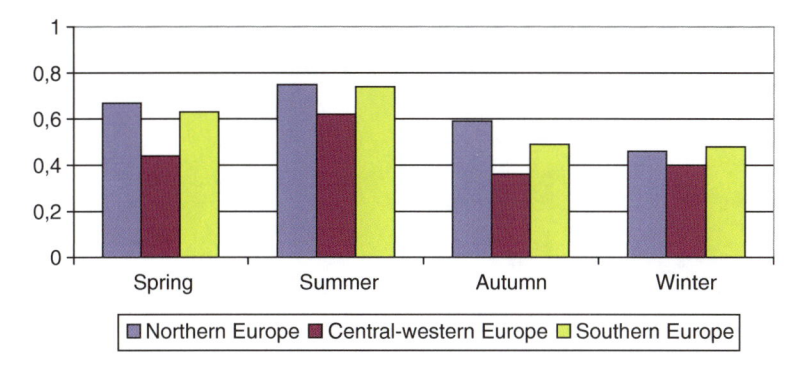

Fig. 6 Mean infiltration factor of $PM_{2.5}$ in relationship to season and region of Europe. Source: Hänninen et al. [9]

including four cities across Europe also did not find higher infiltration in southern compared to northern European cities [8, 13]. Large differences were observed between individual homes/days within regions, likely related to differences in air exchange rates and geometry of the building envelop (e.g., dimension of cracks).

4.3 Indoor–Outdoor Relationships of Coarse (PM_{10}–$PM_{2.5}$) Particles

Much lower infiltration factors were found for coarse particles than for $PM_{2.5}$ in four European cities (Table 1). Infiltration factors were estimated from linear regression analyses of simultaneously measured indoor and outdoor coarse particle concentrations. The lower infiltration of coarse particle is consistent with lower penetration and higher decay rates of coarse particles due to gravitational settling and impaction, see Sect. 3. Studies conducted in the USA have reported substantially lower infiltration rates for coarse particles as well [17, 18].

4.4 Indoor–Outdoor Relationships of Ultrafine Particles

Infiltration factors for UFP assessed by total particle number counts were somewhat lower than for $PM_{2.5}$ in the four European cities included in the RUPIOH study (Table 1), but higher than for coarse particles. A large Canadian study in Windsor, Ontario in which total particle number counts were measured with PTraks reported infiltration factors of 0.16, 0.26, and 0.21 in the first summer, winter, and second summer, respectively, with a large variability for individual homes [19]. The lower infiltration of ultrafine particles is consistent with lower penetration and higher decay rates due to diffusion losses compared to accumulation mode particles.

Table 1 Infiltration factors estimated as regression slope for the relationships between indoor and outdoor 24-h average concentrations of different particle metrics from RUPIOH study [13]

	Helsinki	Athens	Amsterdam	Birmingham
$PM_{2.5}$	0.48	0.42	0.39	0.34
PM_{10}–$PM_{2.5}$	0.14	0.16	0.11	0.13
PNC	0.42	0.42	0.19	0.22
Soot	0.63	0.84	0.78	0.71
Sulfate	0.59	0.61	0.78	0.61

Several studies in the USA that measured particle size distributions have also found lower infiltration factors for particles in the ultrafine range [17, 18, 20, 21]. In a study in nine Boston homes without smokers, the infiltration factor for particles between 0.02 and 0.1 μm was between 0.5 and 0.7, whereas it was 0.28 for coarse particles and 0.74 for $PM_{2.5}$ [16]. In a study of 17 Los Angeles homes, the infiltration factor was 0.75 for particles between 0.08 and 0.3 μm; 0.50 for particles between 0.02 and 0.03 μm and 0.17 for particles between 5 and 10 μm [18]. A study conducted in a Helsinki office found that indoor PNC tracked outdoor concentrations well but were only at 10% of the outdoor concentrations [22]. A study in two empty hospital rooms in Erfurt, Germany reported a high correlation and an I/O ratio of 0.42 for total PNC, compared to 0.79 for $PM_{2.5}$ [23]. There is thus a large range in reported infiltration factors, related to differences in air exchange rates, building characteristics and likely also differences in measurement methods across studies.

A significant fraction of UFP may be volatile, particularly in particles smaller than 50 nm [24, 25]. Volatility increased with decreasing distance to a major freeway [24]. In a study in four apartments near a US freeway, I/O ratios were highest for particles of about 100 nm, decreased up to about 20 nm and then increased again for particles smaller than 20 nm [25]. The pattern below 20 nm did not agree with previous theory (Sect. 3). A potential explanation is that upon infiltration losses may occur of volatile components, resulting in a shift of the particle size distribution towards particles less than 20 nm [25].

4.5 Indoor–Outdoor Relationships of Specific Particle Components

$PM_{2.5}$ is a complex mixture of components derived from multiple sources. Major components of $PM_{2.5}$ include secondary inorganic components (sulfate, nitrate), elemental carbon and organic carbon. Infiltration factors and indoor sources differ for these components, as illustrated in Table 1.

4.5.1 Sulfate and Nitrate

Sulfate has no major indoor sources and has therefore been used as a tracer for outdoor particles [7, 8, 17, 26, 27]. Infiltration factors obtained from linear regression for sulfate were substantially higher than for $PM_{2.5}$ in the RUPIOH study (Table 1). Other studies in Europe [8, 28] and the USA [15, 29] have found the same trend. In the EXPOLIS study the indoor–outdoor regression slope ranged between 0.75 and 0.88 for sulfur and between 0.64 and 0.69 for $PM_{2.5}$ in the four cities [8]. The main explanation for this difference is that sulfate is concentrated in submicron particles that have the lowest penetration and decay losses. The second explanation is that $PM_{2.5}$ in many locations includes a significant nitrate fraction that poorly infiltrates (below). A study in retirement homes in California found that using the infiltration factor for sulfur overestimated the $PM_{2.5}$ infiltration significantly [29]. The authors attributed this difference with studies in the eastern USA to the high nitrate concentrations in fine particles in California.

Studies in the USA have documented that the cation associated with sulfate may change upon infiltration indoors [30]. While outdoor sulfates were partly acidic, indoor particle acidity was largely neutralized by the high ammonia concentration indoors. This is an illustration of a chemical decay process in addition to the mainly physical processes discussed in Sect. 3.

In contrast, for *nitrates* low infiltration factors were found in the RUPIOH study (range 0.05–0.13) and a study in California homes (infiltration factor 0.2) [18]. The low infiltration is due to the volatility of ammonium nitrate, resulting in a shift towards the gaseous components ammonia and nitric acid upon infiltration because of temperature and relative humidity changes [31]. Nitric acid has high deposition on walls, resulting in a further shift of the equilibrium [31].

4.5.2 Elemental or Black Carbon

Infiltration factors for EC exceeded those for $PM_{2.5}$ significantly (Fig. 7). EC is concentrated in submicrometer particles and is nonvolatile. Though smoking affects EC levels, the impact is less than on $PM_{2.5}$ concentrations [28, 32]. Furthermore, there are few other indoor sources of EC [15]. A detailed analysis of the RIOPA study showed that 92% of the indoor EC concentration was due to outdoor EC, whereas the corresponding contribution for $PM_{2.5}$ was 53% [15]. Tobacco smoking, incense burning, and sweeping were marginally significant sources for EC and the intercept of the regression model was insignificant suggesting no unidentified sources of EC. In contrast, for $PM_{2.5}$ a highly significant intercept was found suggesting that most of the indoor sources were unidentified [15].

Fig. 7 Infiltration factors for PM$_{2.5}$ and soot (EC, BC) measured in the same study. Numbers for Hoek 2008 are averages for four cities

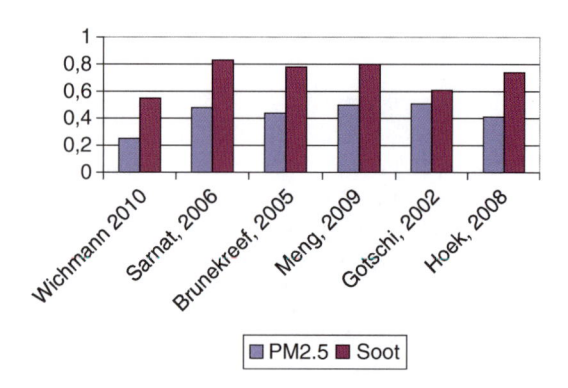

4.5.3 Organic Carbon

There is much more limited information available on infiltration of organic carbon (OC), but as many components are volatile significant changes occur upon infiltration indoors. The OC measurement reflects a complex mixture of components, including polycyclic aromatic hydrocarbons (PAH), which differ widely in volatility. In ambient air a distinction is made in primary and secondary organic components. Organic PM components can also be formed indoors, e.g. by reactions of organic gases with ozone. Measurement of particulate OC is complicated as gas phase artifacts may occur, resulting both in positive and negative biases [33]. In a study in an unoccupied home in California, significant losses of OC were found due to volatilization upon infiltration in the home [33]. In the RIOPA study, the indoor–outdoor regression slope was 0.4 for OC, about half the slope for EC [15]. The contribution of outdoor OC to indoor OC was estimated at 22% only. High indoor OC in PM$_{10}$ concentrations in schools in Aveiro, Portugal was found which were explained for 26% by ambient OC and further by small paper, skin, and cloth particles [34]. A study in Cracow, Poland showed that personal exposure of pregnant women to PAH was explained for typically 90% by the outdoor concentration [35].

5 Conclusions

Outdoor air pollution significantly infiltrates in buildings. Combined with the large fraction of time that people typically spend indoors, a major fraction of human exposure to outdoor pollutants occurs indoors. Understanding the factors affecting infiltration is therefore important. Infiltration factors have been shown to vary substantially across seasons, individual homes, and particle size and components. Important factors contributing to these variations include air exchange rate, characteristics of the building envelop (e.g., geometry of cracks), type of ventilation, and use of filtration. Penetration and decay losses are particle size dependent with the lowest losses for submicrometer particles and higher losses for ultrafine and especially coarse particles. The largest infiltration factors are consistently found for sulfate and black carbon. Volatilization and chemical decay may also result in losses of specific components, including nitrates and organic components.

The difference in infiltration between components and particle sizes documents that indoor air quality differs from outdoor air quality, even in the absence of indoor sources. This may result in measurement error in exposure estimates that exclusively focus on outdoor concentrations. Furthermore heterogeneity in the size of estimated health effects may occur in epidemiological studies using exposure estimates based upon outdoor concentrations performed in different locations that differ in factors affecting infiltration.

The large variability of $PM_{2.5}$ infiltration factors reported may further be due to different composition of PM across locations. In locations with relatively high sulfate and EC contributions, higher infiltration factors can be anticipated than in locations with high nitrate and OC concentrations.

References

1. Brunekreef B, Holgate S (2002) Air pollution and health. Lancet 360:1233–1242
2. World Health Organization (2006) Systematic review of air pollution, a global update
3. Seaton A, MacNee W, Donaldson K, Godden D (1995) Particulate air pollution and acute health effects. Lancet 345:176–178
4. Pekkanen J, Kulmala M (2004) Exposure assessment of ultrafine particles in epidemiologic time-series studies. Scand J Work Environ Health 30(Suppl 2):9–18
5. Sioutas C, Delfino RJ, Singh M (2005) Exposure assessment for atmospheric ultrafine particles (UFPs) and implications in epidemiological research. Environ Health Perspect 113:947–955
6. Dockery DW, Spengler JD (1981) Personal exposure to respirable particulates and sulfates. J Air Pollut Control Assoc 31:153–159
7. Wilson W, Mage DT, Grant LD (2000) Estimating separately personal exposure to ambient and nonambient particulate matter for epidemiology and risk assessment: why and how. J Air Waste Manag Assoc 50:1167–1183
8. Hänninen OO, Lebret E, Ilacqua V, Katsouyanni K, Künzli N, Srám RJ, Jantunen MJ (2004) Infiltration of ambient $PM_{2.5}$ and levels of indoor generated non-ETS $PM_{2.5}$ in residences of four European cities. Atmos Environ 38(37):6411–6423
9. Hänninen O, Hoek G, Mallone S, Chellini E, Katsouyanni K, Kuenzli N, Gariazzo C, Cattani G, Marconi A, Molnár P, Bellander T, Jantunen M (2011) Seasonal patterns in ventilation and PM infiltration in European cities: review, modelling and meta-analysis of available studies from different climatological zones. Air Qual Atmos Health 4(3–4):221–233
10. Liu D-L, Nazaroff W (2001) Modeling pollutant penetration across building envelopes. Atmos Environ 35:4451–4462
11. Chen C, Zhao B (2011) Review of relationship between indoor and outdoor particles: I/O ratio, infiltration factor and penetration factor. Atmos Environ 45(2):275–288
12. Lai A, Nazaroff W (2000) Modeling indoor particle deposition from turbulent flow onto smooth surfaces. J Aerosol Sci 31:463–476
13. Hoek G, Kos G, Harrison R, de Hartog J, Meliefste K, ten Brink H, Katsouyanni K, Karakatsani A, Lianou M, Kotronarou A, Kavouras I, Pekkanen J, Vallius M, Kulmala M, Puustinen A, Thomas S, Meddings C, Ayres J, van Wijnen J, Hameri K (2008) Indoor-outdoor relationships of particle number and mass in four European cities. Atmos Environ 42(1):156–169

14. Wichmann J, Lind T, Nilsson MA-M, Bellander T (2010) PM 2.5, soot and NO 2 indoor-outdoor relationships at homes, pre-schools and schools in Stockholm, Sweden. Atmos Environ 44(36):4536–4544
15. Meng QY, Spector D, Colome S, Turpin B (2009) Determinants of indoor and personal exposure to PM2.5 of indoor and outdoor origin during the RIOPA study. Atmos Environ 43 (36):5750–5758
16. Long CM, Suh HH, Koutrakis P (2000) Characterization of indoor particle sources using continuous mass and size monitors. J Air Waste Manag Assoc 507:1236–1250
17. Long CM, Suh HH, Catalano PJ, Koutrakis P (2001) Using time- and size-resolved particulate data to quantify indoor penetration and deposition behavior. Environ Sci Technol 3510:2089–2099
18. Sarnat SE, Coull BA, Ruiz PA, Koutrakis P, Suh HH (2006) The influences of ambient particle composition and size on particle infiltration in Los Angeles, CA, residences. J Air Waste Manag Assoc 56(2):186–196
19. Kearney J, Wallace L, MacNeill M, Xu X, VanRyswyk K, You H, Kulka R, Wheeler AJ (2011) Residential indoor and outdoor ultrafine particles in Windsor, Ontario. Atmos Environ 45:7583–7593
20. Abt E, S H, Allen G, Koutrakis P (2000) Characterization of indoor particle sources: a study conducted in the metropolitan Boston area. Environ Health Perspect 108:35–44
21. Abt E, Suh HH, Catalano P, Koutrakis P (2000) Relative contribution of outdoor and indoor particle sources to indoor concentrations. Environ Sci Technol 34:3579–3587
22. Koponen IK, Asmi A, Keronen P, Puhto K, Kulmala M (2001) Indoor air measurement campaign in Helsinki, Finland 1999 – the effect of outdoor air pollution on indoor air. Atmos Environ 35:1465–1477
23. Cyrys J, Pitz M et al (2004) Relationship between indoor and outdoor levels of fine particle mass, particle number concentrations and black smoke under different ventilation conditions. J Expo Anal Environ Epidemiol 144:275–283
24. Kuhn T, Krudysz M, Zhu Y, Fine PM, Hinds WC, Froines J, Sioutas C (2005) Volatility of indoor and outdoor ultrafine particulate matter near a freeway. J Aerosol Sci 36(3):291–302
25. Zhu Y, Hinds WC, Krudysz M, Kuhn T, Froines J, Sioutas C (2005) Penetration of freeway ultrafine particles into indoor environments. J Aerosol Sci 36(3):303–322
26. Molnár P, Bellander T, Sällsten G, Boman J (2007) Indoor and outdoor concentrations of PM2.5 trace elements at homes, preschools and schools in Stockholm, Sweden. J Environ Monit 9(4):348–357
27. Sarnat JA, Long CM, Koutrakis P, Coull BA, Schwartz J, Suh HH (2002) Using sulfur as a tracer of outdoor fine particulate matter. Environ Sci Technol 36:5305–5314
28. Brunekreef B, Janssen NA, de Hartog JJ, Oldenwening M, Meliefste K, Hoek G, Lanki T, Timonen KL, Vallius M, Pekkanen J, Van Grieken R (2005) Personal, indoor, and outdoor exposures to PM2.5 and its components for groups of cardiovascular patients in Amsterdam and Helsinki. Research Reports Health Effects Institute 127
29. Polidori A, Cheung KL, Arhami M, Delfino RJ, Schauer JJ, Sioutas C (2009) Relationships between size-fractionated indoor and outdoor trace elements at four retirement communities in southern California. Atmos Chem Phys 9(14):4521–4536
30. Suh HH, Koutrakis P, Spengler JD (1994) The relationship between airborne acidity and ammonia in indoor environments. J Expo Anal Environ Epidemiol 4(1):1–22
31. Lunden MM, Revzan KL, Fischer ML, Thatcher TL, Littlejohn D, Hering SV, Brown NJ (2003) The transformation of outdoor ammonium nitrate aerosols in the indoor environment. Atmos Environ 37(39–40):5633–5644
32. Gotschi T, Oglesby L, Mathys P, Monn C, Manalis N, Koistinen K, Jantunen M, Hanninen O, Polanska L, Kunzli N (2002) Comparison of black smoke and PM2.5 levels in indoor and outdoor environments of four European cities. Environ Sci Technol 36(6):1191–1197

33. Lunden MM, Kirchstetter TW, Thatcher TL, Hering SV, Brown NJ (2008) Factors affecting the indoor concentrations of carbonaceous aerosols of outdoor origin. Atmos Environ 42(22):5660–5671
34. Pegas PN, Nunes T, Alves CA, Silva JR, Vieira SLA, Caseiro A, Pio CA (2012) Indoor and outdoor characterisation of organic and inorganic compounds in city centre and suburban elementary schools of Aveiro, Portugal. Atmos Environ 55:80–89
35. Choi H, Perera F, Pac A, Wang L, Flak E, Mroz E, Jacek R, Chai-Onn T, Jedrychowski W, Masters E, Camann D, Spengler J (2008) Estimating individual-level exposure to airborne polycyclic aromatic hydrocarbons throughout the gestational period based on personal, indoor, and outdoor monitoring. Environ Health Perspect 116(11):1509–1518

Nanoparticles in European Cities and Associated Health Impacts

Prashant Kumar, Lidia Morawska, and Roy M. Harrison

Abstract Atmospheric nanoparticles are a pollutant currently unregulated through ambient air quality standards. The aim of this chapter is to assess the environmental and health impacts of atmospheric nanoparticles in European environments. This chapter begins with the conventional information on the origin of atmospheric nanoparticles, followed by their physical and chemical characteristics. A brief overview of recently published review articles on this topic is then presented to guide those readers interested in exploring any specific aspect of nanoparticles in greater detail. A further section reports a summary of recently published studies on atmospheric nanoparticles in European cities. This covers a total of about 45 sampling locations in 30 different cities within 15 European countries for quantifying levels of roadside and urban background particle number concentrations (PNCs). Average PNCs at the reviewed roadside and urban background sites were found to be $3.82 \pm 3.25 \times 10^4$ and $1.63 \pm 0.82 \times 10^4$ cm^{-3}, respectively, giving a roadside to background PNC ratio of ~2.4. Engineered nanoparticles are one of the key emerging categories of airborne nanoparticles,

P. Kumar (✉)
Civil Engineering (C5); Office 03AA03, Faculty of Engineering and Physical Sciences, University of Surrey, Guildford GU2 7XH, United Kingdom

Faculty of Engineering and Physical Sciences (FEPS), Environmental Flow (EnFlo) Research Centre, University of Surrey, Guildford GU2 7XH, UK
e-mail: P.Kumar@surrey.ac.uk; Prashant.Kumar@cantab.net

L. Morawska
International Laboratory for Air Quality and Health, Queensland University of Technology, 2 George Street, Brisbane, QLD 4001, Australia

R.M. Harrison
Division of Environmental Health and Risk Management, School of Geography, Earth and Environmental Sciences, University of Birmingham, Edgbaston, Birmingham B15 2TT, UK

Department of Environmental Sciences/ Center of Excellence in Environmental Studies, King Abdulaziz University, PO Box 80203, Jeddah 21589, Saudi Arabia especially for the indoor environments. Their ambient concentrations may increase in future due to widespread use of nanotechnology integrated products. Evaluation of their sources and probable impacts on air quality and human health are briefly discussed in the following section. Respiratory deposition doses received by the public exposed to roadside PNCs in numerous European locations are then estimated. These were found to be in the $1.17–7.56 \times 10^{10}$ h^{-1} range over the studied roadside European locations. The following section discusses the potential framework for airborne nanoparticle regulations in Europe and, in addition, the existing control measures to limit nanoparticle emissions at source. The chapter finally concludes with a synthesis of the topic areas covered and highlights important areas for further work.

Keywords Aerosol number and size distributions, Engineered nanoparticles, European environment, Exposure–response doses, Ultrafine particles

Contents

1 Introduction	340
2 State-of-the-Art Summary of Recent Review Articles	341
3 Physico-Chemical Characteristics of Airborne Nanoparticles	343
4 Origin of Atmospheric Nanoparticles	344
4.1 Natural Sources	344
4.2 Anthropogenic Sources	345
5 Airborne Nanoparticle Concentrations in European Cities	349
6 Exposure Assessment: Respiratory Deposition Doses	355
7 Regulatory Measures for Atmospheric Nanoparticles	356
8 Synthesis and Future Work	358
References	360

1 Introduction

Atmospheric particles in the nano-size range contribute significantly to particle number concentrations (PNCs). These also have the potential to adversely affect human health due to their ability for deep penetration into the lungs [1] and beyond. Hence, this chapter only focuses on particle "numbers." In what follows, the terms "ultrafine particles" and "nanoparticles" are used interchangeably according to the context for representing the "total PNCs." In the recent past, there has been a significant increase in number of studies related to characterisation, monitoring, modelling and human exposure assessment of airborne nanoparticles, though this progress has not been sufficient to inform any regulatory framework or to establish ambient air quality standards [2]. There still remain a number of inconclusive results on various aspects such as exposure–response relationships and standardised sampling methodology [3]. This is further complicated by the lack of information on physico-chemical characteristics and behaviour of emerging sources such as biofuel-derived nanoparticles or synthesised, manufactured or engineered nanomaterials or nanoparticles; hereafter referred to as ENPs [4, 5].

Numerous review articles have been published in the recent past covering different aspects of atmospheric nanoparticles such as measurements, instrumentation, characteristics, health-exposure assessment and dispersion modelling. To our knowledge, there is currently no review available which has specifically been dedicated to assess the ambient concentrations of nanoparticles in European cities. The aim of this chapter is to synthesise the existing knowledge on numerous aspects (i.e. origin, characteristics and regulatory control) of atmospheric nanoparticles and to highlight research gaps and future research priorities. An intensive review of published studies is also conducted for making preliminary estimates of both ambient and roadside concentrations of airborne PNCs at about different 45 locations in European cities, and assessing associated health impacts.

This chapter begins by summarising the recent review articles on this topic. Other topics such as sources and physico-chemical characteristics of ambient and emerging nanoparticles (i.e. ENPs) are then covered briefly for the completeness of the article. This is then followed by the assessment of nanoparticles in numerous European cities, estimation of respiratory deposition doses and a brief discussion on current and future prospects of their regulatory control. In what follows, the terms "airborne nanoparticle" and "ENP" refer to total particles, currently mainly produced by vehicles, and nanomaterials-derived products, respectively.

2 State-of-the-Art Summary of Recent Review Articles

This section provides a brief summary of recently published review articles on this topic providing readers an opportunity to explore any individual topic in detail. Vardoulakis et al. [6–8] summarised street scale modelling of gaseous and particulate matter and discussed the challenges, sensitivity and uncertainty associated with them. Buseck and Adachi [9] discussed the nature of airborne nanoparticles, instruments and techniques for characterising their physical and chemical properties and their significance from the health and climate change perspective. Measurements of airborne nanoparticles are generally made by using aerodynamic and optical detection techniques. Detailed description of the operating principles of optical, aerodynamic and electrical mobility analysers can be found in Flagan [10], McMurry [11, 12], and Simonet and Valcárcel [13]. Biswas and Wu [14] presented the life history of airborne ENPs from their formation to potential use and their eventual fate in the environment. Nowack and Bucheli [15] presented a comprehensive review covering the occurrence, behaviour and effects of nanoparticles in the environment. Ju-Nam and Lead [16] presented a widespread overview on the chemistry, interactions and potential environmental implications associated with ENPs. Later, Seigneur [17] focused on measurement techniques and ambient measurements of the physico-chemical characteristics of ultrafine particles. They also summarised some of the model studies on the evolution of particles in vehicle exhaust plumes. Pey et al. [18] were one of the first ones to apply receptor modelling tools to PNCs for apportioning the PNC sources and processes in the urban background of Barcelona, Spain. Recently, Harrison et al. [19] applied the similar technique on PNC data collected at roadside environment (Marylebone Road) in London to identify contribution from various sources.

Morawska and co-workers have produced a number of review articles on this topic. For example, Holmes and Morawska [20] reviewed several simple and complex models covering a wide range of urban scales for the dispersion of particulate matter. Morawska et al. [21] focused on vehicle produced ultrafine particles and discussed limitations of measurement methods, sources, characteristics, transport and exposure of these particles in urban environments. Their further review focused on indoor and outdoor monitoring of airborne nanoparticles [3]. Morawska [22] discussed the importance of airborne ENPs from the health perspective. Regulations and policy measures related to the reduction of ambient particulate matter were discussed in their follow-up article [23]. Their recent review article discussed the commuters' exposure to ultrafine particles and associated health effects [24].

There has been around half a dozen review articles published by Kumar and co-researchers during the last 2 years on various aspects of urban atmospheric nanoparticles. Their first review synthesised the existing knowledge on characteristics, measurements and currently available instruments for measuring atmospheric nanoparticles in the urban environment. In addition, they discussed the potential prospects of regulatory control for atmospheric nanoparticles, recent advances on this topic and future research priorities [2]. A further article presented the comparison of the behaviour of vehicle derived airborne nanoparticles and the ENPs produced from nanomaterials integrated products, besides discussing the consequences for prioritising research and regulation activities [4]. Later, they reviewed the characteristics and impacts of biofuelled vehicles derived nanoparticles on the number-based regulations [5]. Their further article in this series focused on the dispersion modelling of nanoparticles in the wake of moving vehicles, and presented a critical analysis of the information on dispersion models and techniques (numerical and computational) used for dispersion modelling at this fine scale [25]. A recent review by Kumar et al. [26] illustrated dynamics and dispersion modelling of nanoparticles at five spatial scales (vehicle wake, street scale, neighbourhood, city and tunnel). Also were presented comprehensive discussions on the importance of sinks and transformation processes in nanoparticle dispersion models at various spatial scales. Their recent article discussed the technical challenges in tackling regulatory concerns over atmospheric nanoparticles [27] and the follow-up work summarised their impacts on urban air quality and public health [28]. Their most recent work discussed the importance of nanoparticles produced by building activities such as construction, demolition or recycling, besides highlighting the need for developing risk assessment and management strategies [29].

The ever-increasing number of published studies on airborne nanoparticles, which is also evident from the brief summary of various reviews presented above, clearly demonstrates a mounting importance of this research topic among the air quality science and management communities.

3 Physico-Chemical Characteristics of Airborne Nanoparticles

Various mathematical distribution fits such as Rosin–Rammler, Nukiyama–Tanasawa, power law, exponential and Khrgian–Mazin find limited application in aerosol science [30]. The lognormal distributions are generally used to fit empirically the wide range and skewed shape of most particle number distributions (PNDs) in the urban environment (see Fig. 1). The PNDs are often expressed in terms of the logarithmic function of the particle diameter (log D_p) on the x-axis and the measured differential concentration on the y-axis i.e. $dN/d\log D_p$; the number of particles per cm^{-3} of air having diameters in the size range from $\log(D_p + dD_p)$. If $dN/d\log D_p$ is plotted on a linear scale, the number of particles between D_p and $D_p + dD_p$ is proportional to the area under the curve between $dN/d\log D_p$ and $\log D_p$. The same is true if distributions are plotted in similar plots on the basis of mass, volume or surface area.

The most common way in aerosol science to represent PND data is in terms of various modes. Generally, these modes are nucleation (typically in the 1–30 nm range), Aitken (typically in the 20–100 nm range), accumulation (typically in the 30–300 nm range) and coarse (typically over 300 nm size range). Each mode contains different sources, size range, formation mechanisms, and chemical compositions [30].

Particles in nucleation mode are generally formed due to condensation of the vapour present in the exhaust gases and nucleation (gas-to-particle conversion) in the atmosphere after rapid cooling and dilution of exhaust emissions [31, 32]. These particles originate mainly from unburned fuel and lubricating oil consisting of sulphates, nitrates and organic compounds [33].

Aitken mode particles arise from the growth or coagulation of nucleation mode particles, and are also produced in high numbers by primary combustion sources such as vehicle exhausts [34]. These are mainly composed of a soot/ash core with a readily absorbed layer of readily volatilisable material [35]. Accumulation mode includes combustion particles, smog particles and nucleation mode particles that have coagulated with accumulation mode particles [30]. These particles are formed in the combustion chamber (or shortly thereafter) with associated condensed organic matter [36]. They are composed of carbonaceous agglomerates (soot and/or ash) coming mainly from the combustion of engine fuel and lube oil by diesel-engined or direct injection gasoline-engined vehicles [37].

Particles in the coarse mode are deposited on cylinder and exhaust system surfaces and later re-entrained [38]. These are also produced by brake wear, tyre wear, windblown dust (also called advected mineral particles), large salt particles from sea spray, re-suspension of particles by traffic and wind produced turbulence, and anthropogenic activities such as agriculture and surface mining [39–41]. These particles are notable for their significant mass and volume concentrations. Their lifetime in the atmosphere is relatively shorter (a few hours or days) because of the effect of gravitational settling due to their large size or due to their washout [39].

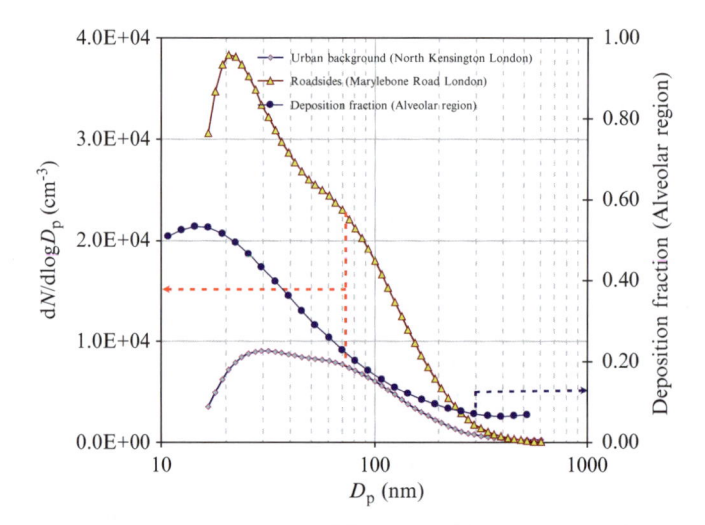

Fig. 1 Typical example of particle size distributions at roadside (Marylebone Road, London) and urban background site (North Kensington, London), both taken in 2007. Also are shown size dependent deposition in alveolar region [102]

4 Origin of Atmospheric Nanoparticles

Nanoparticles originate from both natural and anthropogenic activities. This section provides a brief overview of the anthropogenically produced nanoparticles from gasoline and diesel-fuelled road vehicles, with a special focus on the emerging class of nanoparticles (i.e. ENPs). A brief overview of natural and anthropogenic sources is also presented for the completeness of the chapter. Wherever felt necessary, readers are directed to the relevant literature for detailed information.

4.1 Natural Sources

Nanoparticles in the natural environment originate from atmospheric (organic acids, sea salts), biogenic (organic colloids, organisms, magnetite and metals), geogenic or pyrogenic (soot such as fullerenes, Al- and Fe-oxides and Allophane) formation [15]. Besides sea salt and mineral matter, atmospheric formations of organic acids through secondary aerosol formation are of particular importance. Sea salt and mineral dust are primary constituents with rather small nanoparticle components. Secondary nanoparticles generally predominate in less polluted environments and are formed in the atmosphere by the condensation of semi-volatile vapours, usually generated through atmospheric photochemistry [42–44]. The most common formation mechanism involves sulphuric acid nucleation, followed by condensational growth, with oxidised organic compounds playing a major role in the latter process [45].

The literature suggests two other nucleation mechanisms, one involving oxidation of terpenes or other organic compounds released from trees [46] and the other based upon iodine oxides [47]. Heintzenberg et al. [48] summarised that new particle formation events in the boundary layer generally show following distinct features. Firstly, a rapid increase in number concentrations of particles below 20 nm due to nucleation and subsequent particle growth into detectable sizes. Secondly, further particle growth within the size range of the instrumentation, as a result of continuing condensation of vapours. Finally, a gradual decrease in PNCs within several hours due to coagulation and mixing with other air masses. Such a particle formation and successive growth is informally called as "banana event" because of their appearance in two-dimensional contour plots of PNCs as a function of time and particle diameter [48]. However, the rate of nanoparticle formation varies in different environments. For example, Kulmala et al. [34] concluded that the formation rate of 3 nm size particles within the boundary layer, urban areas, coastal areas and industrial plumes can be in the range 0.01–10, up to 100, 10^4–10^5 and 10^4–10^5 # cm^{-3} s^{-1}, respectively. They also reported that particle growth rate varies between 1 and 20 nm h^{-1}. This growth rate is mainly driven by the ambient temperature, clumping, availability of condensable vapours, condensation of other materials onto particle surfaces, deliquescence or hygroscopic properties of particles (e.g. sea salt particles grow by collecting water) [49]. Ketzel et al. [50] found that during periods with low primary particle emissions in Copenhagen, particle formation events are due to nucleation in background air. They observed the particle growth rate between 1 and 6 nm h^{-1}. These are similar to those found at another European suburban background location in Prague where the average value of particle growth rate was found to be about 5.4 nm h^{-1} [51]. These are well within the range of 1–10 nm h^{-1} given for urban locations by Kulmala et al. [34]. Ketzel et al. [50] also reported that total number of concentrations of particles in Copenhagen increased by up to 5–10 times within a few hours in connection with clean air and high solar radiation. Generally, new particle formation events occur in defined conditions such as low wind speed, low relative humidity and high global radiation. Other factors accompanying the new particle formation events are lowered condensation sinks preceding the particle burst or low concentrations of SO_2 and NOx [51]. Detailed information on the formation and growth of particles in different environments can be found in Holmes [43], Kulmala et al. [34] and the references therein.

4.2 Anthropogenic Sources

Nanoparticles produced from anthropogenic sources can either be formed inadvertently as a by-product of combustion activities (i.e. emissions from road vehicles and industries) or produced intentionally (e.g. ENPs) due to their particular characteristics [15]. The following section describes the ENPs in detail and provides only a brief overview of other anthropogenic sources.

4.2.1 ENPs Produced from Emerging Sources

Before going into the details of the ENPs, this is important to define the use of this term in the context of this article. The International Standard Organisation (ISO/TC 146/SC 2/WG1 N 320) defines these as "a particle with a nominal diameter smaller than about 100 nm." The ENP is a widely accepted term to represent those nanoparticles originating from various manufacturing or engineering processes. This is used to make these nanoparticles distinct from those originating from fossil fuel combustion sources such as vehicles or industries [4].

Due to ever-growing use of nanomaterials in consumer products, unintentional release of airborne ENPs can occur into the atmospheric environment during their use or production in commercial and research units. A review by Aitken et al. [52] summarised that there are four main production processes (gas-phase, vapour deposition, colloidal and attrition), and these all can result in exposure by inhalation, dermal or ingestion routes. Studies report that more than 15% of all products globally will incorporate nanotechnology in them during their manufacture, with a value of about $1 trillion per year by 2015 [53, 54]. Similarly, a recent review by Peralta-Videa et al. [55] reported that due to ever-increasing demand for consumer electronics and household cleaning products, revenues for nanotechnology and nanomaterials in consumer products were approximately US $1,545 million in 2009 which is expected to increase to $5,335 million by 2015.

Use of nanotechnology has surely improved the capabilities of current products in many areas such as biotechnology, biomedical, energy, catalytic, electronics, packing and logistics and cosmetics, etc. [56]. As a by-product of their use, it has also been found that nano-sized materials can escape into the environment during manufacture, use and disposal of ENP-integrated products [57, 58]. Their physico-chemical characteristics differ in an adverse manner to the nanoparticles produced by vehicles [4]. For instance, ENPs are generally solid, non-volatile and can be in different shapes and sizes (round or having large aspect ratios, e.g. nanotubes, nanopillars, nanowires, nanospheres, nanocubes). They can remain suspended for a relatively longer time in air and are available for inhalation, hence increasing the chances of human exposure and health risks [59]. This is mainly because of their small sizes contributing to negligible gravitational settling. Once separated, properties of individual ENPs can be substantially different from their bulk counterpart in terms of having high surface/volume ratio and changes in optical, electrical, reactive and physico-chemical properties [55, 60]. Consequently, their exposure can cause adverse effect on human health, resulting in demands to develop health and safety regulations associated with nanotechnology integrated products. A thorough hazard assessment of all ENP-integrated products may take 10s of years, e.g. 30–50 years for USA only [61], and this time will increase as the new nanomaterials emerge [55].

Production of ENPs can be categorised by two approaches: top-down and bottom-up [62]. The first category includes the processes which start with a bulk material and then break into smaller pieces due to the influence of some sort of energy. These processes are high-energy ball milling, mechanochemical processing, etching, electro-explosion, sonication, sputtering and laser ablation [14]. The second category includes synthesis of material from the atomic or molecular level by growth and assembly to form the desired ENPs through sol–gel, aerosol routes, chemical vapour deposition, plasma or flame spraying, laser pyrolysis, atomic or molecular condensation, supercritical fluid, spinning and self-assembly processes [14]. The major classes of these particles produced by above processes fall in the category of fullerenes, carbon nanotubes, metal oxides (e.g. oxides of iron and zinc, titania, ceria) and metal nanoparticles. BSI [63] provides categorisation of nanomaterials into four main following groups:

- Fibrous (i.e. insoluble nanomaterials with high aspect ratio)
- CMAR (i.e. any nanomaterial which is already classified in its larger particle form as carcinogenetic (C), mutagenic (M), asthmagenic (A) or a reproductive toxin (R))
- Insoluble (i.e. insoluble or poorly soluble nanomaterials not in the fibrous or CMAR category)
- Soluble (i.e. nanomaterials not in fibrous or CMAR category)

This document [63] also suggests exposure control measures in the following hierarchical form: *eliminate* (i.e. avoid using hazardous substance), substitute (i.e. replace with less risky material), *enclose* (i.e. perform operations in enclosed cabins), *engineering control* (i.e. carrying out potential dust formation processes with extract ventilation), *procedural control* (i.e. limiting the number of personnel exposed, their time and the process to specified areas), and *personal protective equipment* (i.e. use of protection measures to limit inhalation and dermal exposure).

One of the questions to address on this topic is: what are concentrations of airborne ENPs in indoor (i.e. at work places) and outdoor (i.e. ambient) environments? Few studies can be identified that have monitored the ambient concentrations of airborne ENPs in workplaces and indicated their increased concentrations near the production site, especially during their handling [58, 64, 65]. As highlighted by Morawska [22], even a slight increase in ENP concentrations is important from a health perspective but currently available literature does not provide an opportunity to generalise their results and form conclusive answers about the level of exposure. Furthermore, there is scarcely any study available which addresses the latter part of the above question. This is because once the ENPs are dispersed into the outdoor environments, the capability of currently available instrumentation for nanoparticle measurements [2] is not sufficient to apportion them from other nanoparticles in real time. However, ENPs can be apportioned from other nanoparticles through indirect methods such as modelling [66] or through physical and chemical characterisation in the laboratory [16]. A rare study on this topic by Muller and Nowack [66] derived information on ENP (silver, titanium oxide and carbon nanotubes) mass concentrations by differentiating from other nanoparticles in the ambient environment of Switzerland in the range of about 10^{-2}–10^{-3} µg m^{-3}. To date, no study has reported ambient concentrations of ENPs on a *number* basis. There is certainly a need for better instrumentation which can measure the ENPs, as distinct from other nanoparticles, and an increased number of studies measuring the number and size distributions in both indoor and outdoor environments are needed for reporting accurate ambient concentrations.

Future Research Directions

Understanding of the safe design and use of airborne ENPs and their health and environmental related impacts is still in its infancy. There are still numerous questions to be answered. Few of them were highlighted by Kumar et al. [4] as: (1) do the characteristics of the ENPs differ from those of other airborne nanoparticles? (2) should the ENPs be regarded as an ENPs, especially in outdoor environments? (3) what should be an appropriate measurement metric to represent their health impacts? (4) can the same instruments be applied to measure airborne ENPs and other nanoparticles? (5) are the dispersion characteristics of ENPs and other nanoparticles similar? and (6) is exposure of ENPs a major concern? Further, Morawska [22] extended the list of questions related to the exposure of ENPs to: (7) are the particles in the nano-size range more toxic than larger particles of the same material? (8) does the surface chemistry of the lung alter the toxicity of inhaled particles? (9) do nano-fibres pose the same risk as toxic fibres such as quartz and asbestos? and (10) do the currently deployed methods assess the health risk appropriately? Currently, most of the above questions cannot be answered precisely until more comprehensive information on these topics becomes available. Description of the various aspects of the ENPs is discussed in a nutshell here due to brevity reasons but further details related to sources, characteristics, toxicity, physical and chemical interactions of ENPs can be found in recent reviews by Handy et al. [67, 68], Valant et al. [69], Bystrzejewska-Piotrowska et al. [70], Ju-Nam and Lead [16], Kumar et al. [4] and Peralta-Videa et al. [55].

4.2.2 Road Vehicles and Other Anthropogenic Sources

Numerous studies based on the mass metric show that vehicular sources can comprise up to 80% of total PM_{10} and/or $PM_{2.5}$ in urban areas [71–77]. Similarly, road vehicles dominate with their contribution towards the total PNCs in urban environments. These can contribute up to 86% of total PNCs in the polluted urban environment depending on the measurement location, meteorological and traffic conditions [2, 18, 19, 78]. A majority of these PNCs generally fall below 300 nm diameter [79–82]. While most of the PNCs are contributed by particles in the ultrafine size range (i.e. <100 nm), majority of the mass concentrations are contributed by particles over 100 nm in size [27]. For instance, Charron and Harrison [31] observed about 71–95% of total PNCs in central London in the 11–100 nm size range. There can be an appreciable contribution from particles below 10 nm, mainly arising from secondary formation, towards the total PNCs [83]. For example, Shi et al. [84] found this contribution between 36% and 44% at roadsides in Birmingham, UK. This contribution was between 16% and 24% in the 3–10 nm range in Leipzig, Germany [85], and between 4% and 12% in Cambridge, UK for the 5–10 nm size range [44].

Other anthropogenic sources of nanoparticles can include: brake and tyre wear [40], industrial emissions such as from power plants [86], idling, taxiing and take-off from aircraft at airports [87, 88], ship journeys from ports or harbours [89], construction, demolition or recycling of concrete [29, 90, 91] or cooking [92], biomass burning, fuel combustion during gardening, waste incineration, agriculture processes, cigarette smoke and fugitive emissions [23]. We have not covered these sources in detail here as their influence is either more localised or site specific. Their individual contribution is likely to be relatively small in urban areas in comparison with strong nanoparticle sources (e.g. road vehicles). Figure 2 illustrates results of two recent source apportionment studies indicating typical contributions of various sources towards the total PNCs along the roadside at Marylebone road in London [19] and urban background site in Barcelona [18]. Further details on aforementioned sources can be seen elsewhere [2, 21, 23].

5 Airborne Nanoparticle Concentrations in European Cities

A review of recent nanoparticle studies is carried out, covering about 45 different locations within 30 and 15 European cities and countries, respectively. The reviewed PNC data have been classified into two categories: (1) roadside (i.e. measurement locations situated along the freeways, roadsides, curbsides, crossroads in city centres or street canyons) and (2) urban background (i.e. measurement locations situated at 10s of metres away from the roads, highways or at the rooftops of buildings). Table 1 provides a comprehensive summary of the name of country, study location and year, measured size range and instrument used along with the original source of this information. Figures 3 and 4 give a summary of average PNCs in roadside and urban background environments in various European cities, respectively. Birmingham (1.70×10^5 cm^{-3}) and Zurich (8.0×10^4 cm^{-3}) sites in the UK and Switzerland, respectively, appear to demonstrate the most polluted roadside sites; Zurich (3.50×10^4 cm^{-3}) also falls in the same category for an urban background site, although these are now relatively old studies and it is likely that PNC have decreased since that time. Antwerp ($1.24 \pm 0.91 \times 10^4$ cm^{-3}) and Prague (0.72×10^4 cm^{-3}) seem to have least polluted roadside and urban background sites, respectively. Average PNCs at roadside are about 2.4 times higher than the average PNCs at the urban background sites. This ratio is based on a global average over all the European sites and is comparable to those found during individual local studies. For instance, a recent study by Reche et al. [93] reported PNCs at roadside (Marylebone Road) and urban background (North Kensington) sites in London; their roadside to background PNC ratio turns out to be about 1.83.

Average roadside PNCs (i.e. $3.82 \pm 3.25 \times 10^4$ cm^{-3}) show a great range of variability, showing over an order of magnitude (~14 times) difference between the minimum and maximum PNCs (Fig. 4). This difference, however, comes down to ~5 times if we ignore the PNC values measured at the Birmingham site during 1996–1997. This variability can be explained by many dissimilar factors acting on different sites such as: position of sampling head, type of fuel used in the vehicles in

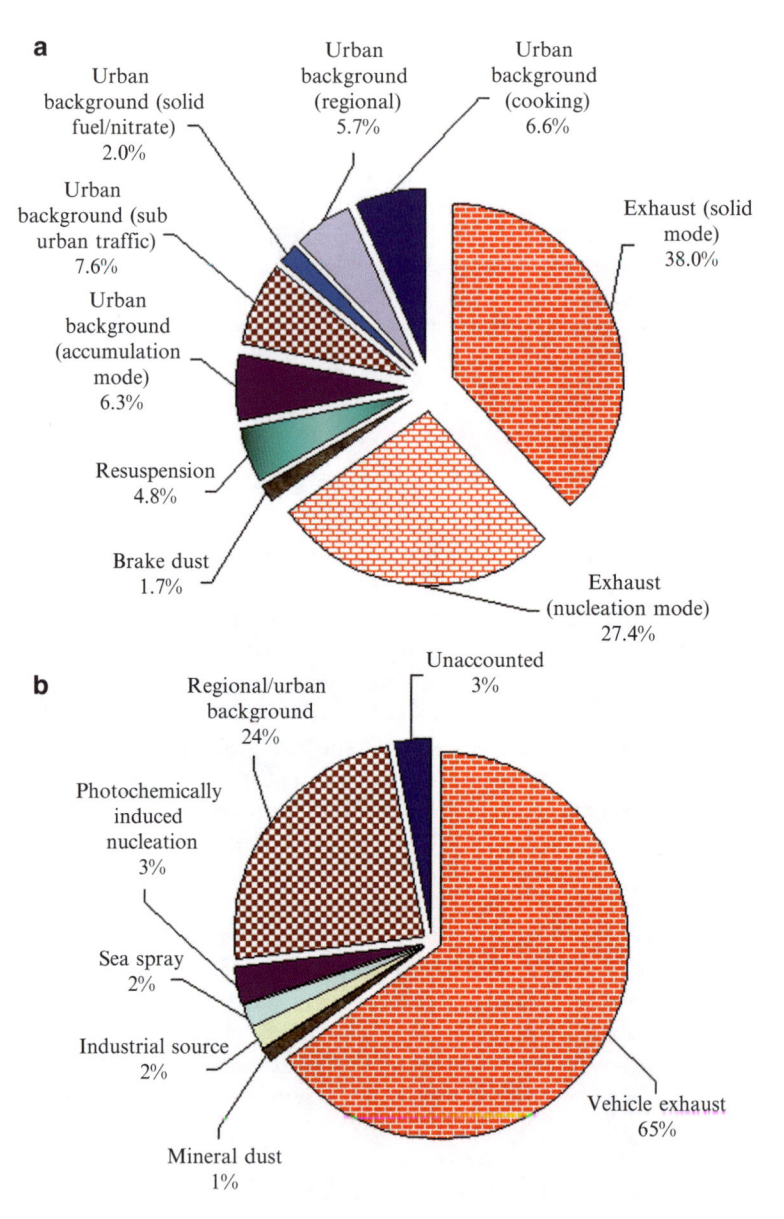

Fig. 2 Typical contribution of various sources towards (**a**) roadside PNCs in the 14.9–10,000 nm size range London, and (**b**) urban background PNCs in the 13–800 nm size range in Barcelona. Roadside and urban background data has been taken from Harrison et al. [19] and Pey et al. [18], respectively

different years, meteorological conditions, traffic density, minimum and maximum cut-off size by instruments used for measurements, sampling duration, averaging time of samples, year of study, and the type and length of sampling tubes used and

Table 1 Summary of sample studies for atmospheric nanoparticles in European countries

Country	Study location	Study year	Size range (nm)	Instrument	Reference
Roadside environments					
UK	Cambridge	2007	10–2,500	DMS500	Kumar et al. [79]
UK	Manchester	2001	4–100	SMPS	Longley et al. [105]
UK	Leicester	2005	5–1,000	DMS500	Agus et al. [106]
UK	Birmingham	1996–1997	10–352	SMPS	Shi et al. [107]
UK	London	2009	7–1,000	CPC	Reche et al. [93]
Belgium	Antwerp	2009	20–500	TSI Monitor 3031	Can et al. [108]
Germany	Essen	2008	20–750	CPC-SMPS	Weber [109]
Germany	Leipzig	2003–2004	3–800	TDMPS	Voigtlander et al. [110]
Finland	Lahti	1995	6–300	UCPC	Vakeva et al. [42]
Finland	Helsinki	2002–2004	7–3,000	CPC	Puustinen et al. [111]
Greece	Athens	2002–2004	7–3,000	CPC	Puustinen et al. [111]
Netherlands	Amsterdam	2002–2004	7–3,000	CPC	Puustinen et al. [111]
Netherlands	Utrecht	2008	10+	CPC	Boogaard et al. [112]
Spain	Barcelona	2001	10+	CPC (3022A, TSI)	Paatero et al. [113]
Italy	Rome	2001	10+	CPC (3022A, TSI)	Paatero et al. [113]
France	Strasbourg	2003	7–10,000	ELPI	Roth et al. [114]
Denmark	Copenhagen	2007	6–700	DMPS	Wåhlin [94]
Czech Republic	Prague	2008	25–2,500	SMPS + APS	Ondráček et al. [115]
Austria	Vienna	2002–2003	7+	CPC	Gomiscek et al. [116]
Austria	Linz	2002–2003	7+	CPC	Gomiscek et al. [116]
Austria	Graz	2002–2003	7+	CPC	Gomiscek et al. [116]
Austria	Salzburg (Rudolfsplatz)	1998	13–830	SMPS	Morawska et al. [117]
Switzerland	Zurich	2001–2002	3+	SMPS + CPC + OPC	Bukowiecki et al. [118]

(continued)

Table 1 (continued)

Country	Study location	Study year	Size range (nm)	Instrument	Reference
Switzerland	Bern	2009	7–1,000	CPC	Reche et al. [93]
Urban background					
UK	London (Bloomsbury)	2003	10–415	DMPS	Rodríguez et al. [119]
UK	Cambridge	2006	5–1,000	DMS500	Kumar et al. [44]
UK	Leicester	2005	5–1,000	DMS500	Agus et al. [106]
UK	Birmingham		10–352	SMPS	Shi et al. [107]
Italy	Milan (via-Messina)	2003–2004	10–800	DMPS	Rodríguez et al. [119]
Spain	Barcelona	2003–2004	10–800	DMPS	Rodríguez et al. [119]
Spain	Madrid	2006–2008	15–600	TSI-SMPS	Gómez-Moreno et al. [120]
Spain	Huelva	2009	2.5–10,000	CPC	Reche et al. [93]
Spain	Barcelona	2009	5–1,000	CPC	Reche et al. [93]
Spain	Santa Cruz de Tenerife	2011	2.5–10,000	CPC	Reche et al. [93]
Germany	Augsburg	2001	10+	CPC (3022A, TSI)	Paatero et al. [113]
Germany	Erfurt		10–500	MAS	Ruuskanen et al. [121]
Germany	Leipzig	2003–2004	3–800	TDMPS	Voigtlander et al. [110]
Finland	Helsinki	2001	10+	CPC (3022A, TSI)	Paatero et al. [113]
Finland	Lahti	1995	10+	CPC	Vakeva et al. [42]
Netherlands	Alkmaar	1996–1997	10–500	SMPS + PMS LAS-X	Ruuskanen et al. [121]
Netherlands	Utrecht	2008	10+	CPC	Boogaard et al. [112]
Sweden	Stockholm	2001	10+	CPC (3022A, TSI)	Paatero et al. [113]
Czech Republic	Prague	2009	10–2,500	SMPS + APS	Ondrácek et al. [115]
Austria	Vienna	2002–2003	7+	CPC	Gomiscek et al. [116]
Switzerland	Zurich	2001–2002	3+	SMPS + CPC + OPC	Bukowiecki et al. [118]
Switzerland	Lugano	2009	7–1,000	CPC	Reche et al. [93]

Note: *DMS* differential mobility spectrometer, *SMPS* scanning mobility particle sizer, *CPC* condensation particle counter, *TDMPS* twin differential mobility particle sizer, *DMPS* differential mobility particle sizer, *OPC* optical particle counter, *APS* aerodynamic particle sizer, *MAS* mass aerosol spectrometer, *LAS-X* optical laser aerosol spectrometer, *ELPI* electrical low pressure impactor

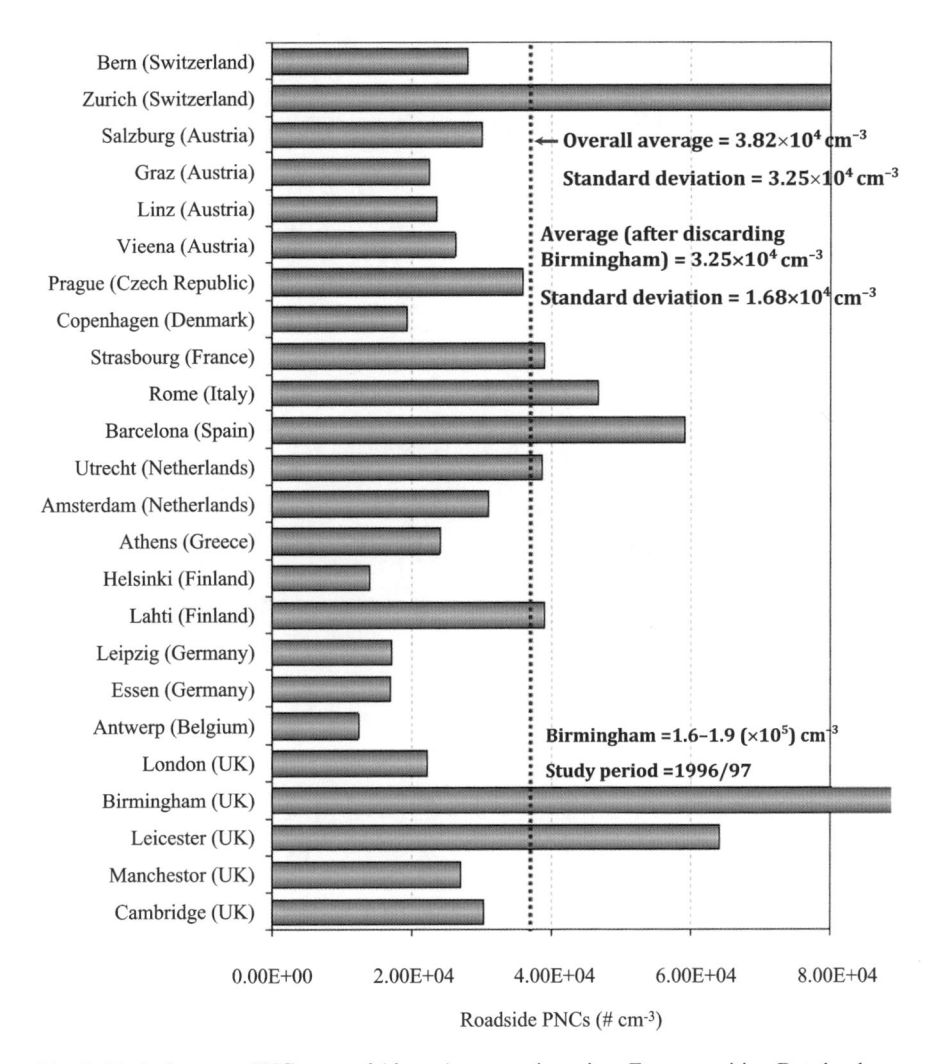

Fig. 3 Typical average PNCs at *roadside environments* in various European cities. Data has been taken from the references listed in Table 1

treatment for particle losses within them (see Table 1). Highest PNCs are found to be on the sampling location adjacent to the roadside having a high traffic volume (i.e. at Birmingham and Zurich sites), where the vehicle sources are close to the sampling location. It is also worth nothing that even though traffic is the main source of PNCs, new particle formation by photochemistry in Mediterranean regions may also be a significant source [93]. This is clearly evident from Fig. 3 that Barcelona shows the fourth highest PNC despite the fact that the sampling station is near traffic, but not such a roadside location as Marylebone Road in London or Zurich. The ultrafine size range of particles contributes about 80% of the PNCs and their concentrations are elevated near the sources, majorly contributing

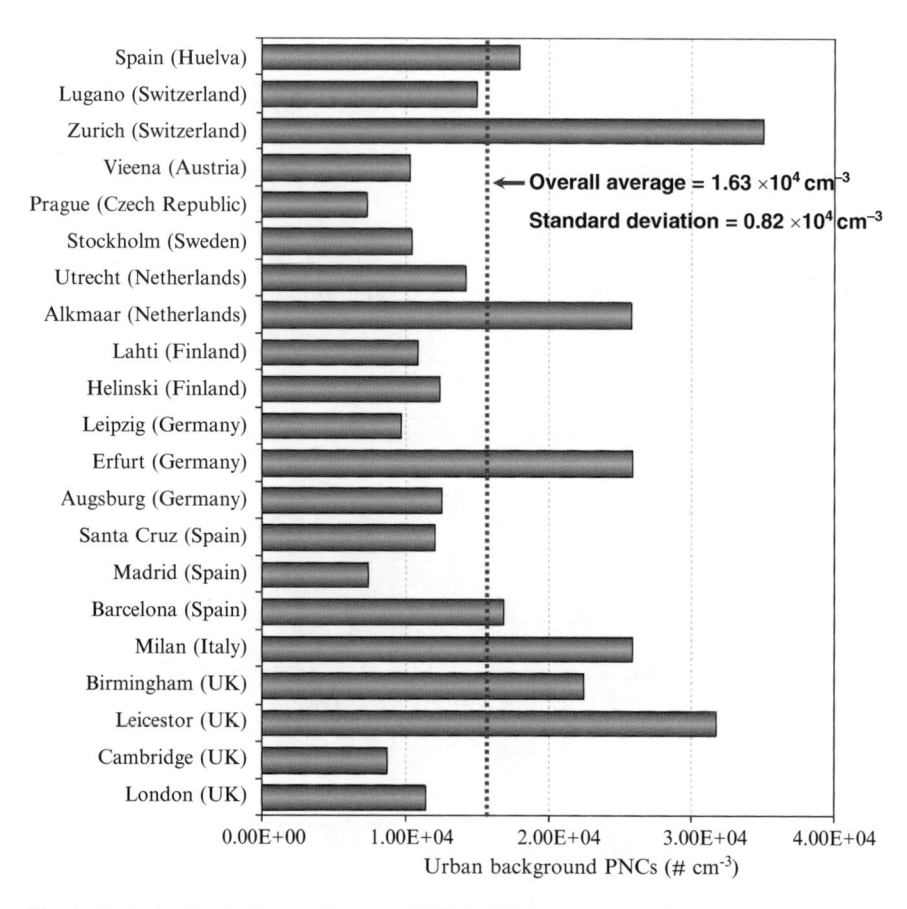

Fig. 4 Typical *urban background* average PNCs in different European cities. Data has been taken from the references listed in Table 1

towards this variability [2]. Furthermore, this difference can be caused by the fuel type in different years. For example, a reduction in ultrafine size range has been experienced in recent years compared with the late 1990s or early 2000s in some of the European cities due to a reduction in sulphur content in fuel and use of emission control technology (e.g. oxidation catalytic convertors, diesel particulate filters) in modern vehicles. Wåhlin [94] reported a 27% decrease in PNCs in the 6–700 nm size range on a busy street in Copenhagen in the period 2005–2007 compared with 2002–2004 levels due to implementation of reduced sulphur contents (from 30–50 to 6–7 ppm) in 2005. If we consider the use of cleaner fuels, stricter emission control technologies and regulation policies (e.g. Euro-5 and Euro-6 emission standards) in recent years, a reduction in airborne nanoparticles is expected with the passage of time. This trend is somewhere reflected by both the roadside and background PNCs plotted in Figs. 3 and 4 if we consider them in conjunction with the study years mentioned in Table 1. For instance, Harrison and Jones [95] measured monthly average roadside concentrations at the Marylebone Road supersite in London during 2001 up to 1.17×10^5 cm^{-3} compared with $2.22 \pm 1.29 \times 10^4$ cm^{-3} reported by Reche et al. [93] for 2009 for the same site.

Urban background PNCs show quite consistent values ($1.63 \pm 0.82 \times 10^4$ cm^{-3}) despite having several uncertainties derived by the measured size range, sampling locations, meteorological and traffic conditions (Fig. 4). Furthermore, it should be noted that all these studies were conducted in different years which could have influenced the nanoparticle emissions from the vehicles continuously experiencing changes in fuel type and strict control measures (see Table 1). All these factors lead to dissimilar conditions for comparison. Despite this, a difference between the minimum (Prague) and maximum (Zurich) PNCs at different locations was found to be within a factor of 5, which is quite plausible considering the dynamic nature of airborne nanoparticles influenced by atmospheric processes [26, 96].

6 Exposure Assessment: Respiratory Deposition Doses

A UK government report entitled Air Quality Strategy 2007 for England, Scotland, Wales and Northern Ireland estimated an average loss of 7–8 months in life expectancy to UK residents, with equivalent health costs of about £20 billion per year due to PM$_{2.5}$ exposure. Such figures are not available separately for the nanoparticulate fraction. The estimate of health costs is based upon the chronic effects of exposure to PM$_{2.5}$ mass. Since comparable epidemiological studies have not been conducted using ultrafine particle concentrations, it is unclear whether such an estimate includes the impact of nanoparticles, or whether such an impact is not captured by the use of the PM$_{2.5}$ particle metric. Atkinson et al. [97] have shown in a time series epidemiological study that PNC and particle mass metrics are predictive of different health outcomes, and Harrison et al. [98] have shown that measurements of particle mass concentrations do not well describe particle exposures to ultrafine particles. Likewise, Morawska et al. [99] reported a lack of relationship between number and mass concentrations for submicrometer airborne particles. Once inhaled, nanoparticles can lead to detrimental health effects such as oxidative stress, pulmonary inflammation and cardiovascular events [9]. Besides the nasal and mouth passages, they can enter the body through the skin and can also penetrate epithelial cells and accumulate in lymph nodes [100]. However, uncertainties still exist on the exact biological mechanism through which nanoparticles cause disease or death. Neither is there adequate knowledge of any safe quantities of ambient PNCs for inhalation nor the levels of harmless deposition doses, if any.

Inhalation dosimetry plays a vital role in determining the links between exposure and human health effects. In the following paragraph, preliminary estimates of respiratory deposition doses are made for indicating exposure–response-doses at various European locations. These estimates are made based on the observed typical roadside PNCs (Fig. 3) for the condition of light exercise; the volume of inhaled air by an adult men is considered as 1.5 m^3 h^{-1}[30]. Information on size-resolved PNDs is required for making accurate deposition estimates but here we have assumed a total respiratory deposition fraction as 0.63 ± 0.03, as is used in recent exposure studies [101]. The reason for adopting this deposited fraction for our estimates is that this comes out to be nearly identical if compute this for particle size distributions with the typical roadside PND data that: (1) assume two main particle modes having mean geometrical diameters as 16 nm (with ~65% of total PNCs) and 65 nm (with 35% of total PNCs) [79], (2) respiratory deposited fraction using ICRP curves as ~0.82 and 0.36 at 16 and 65 nm diameter, respectively [102] and (3) then the resulting weighted average of deposited fraction from (1) and (2) turns out to be 0.66.

Average respiratory deposition doses over the studies presented in Fig. 5 are estimated as 3.61 ± 0.17 × 10^{10} h^{-1}; these come down to 3.07 ± 0.17 × 10^{10} h^{-1} if we ignore values of a pre-2000 study carried out in Birmingham. Since we did not have the PNDs available for each study and the deposition doses are estimated based on the total PNCs, these follow the same expected trend as is seen in Fig. 3 for roadside PNCs. For instance, smallest deposition doses (1.17 ± 0.06 × 10^{10} h^{-1}) were found for Antwerp (Belgium) corresponding to the lowest PNCs (1.24 ± 0.91 × 10^4 cm^{-3}) among all locations. After Birmingham, largest deposition doses were for Zurich showing ~6.5 times larger than those for Antwerp.

It is worth noting that these estimates should be taken as indicative figures since these may change depending on the nature of emission sources that can affect the particle number and size distributions, and the type and condition of population exposed (i.e. children, male, female or elderly adults; inhalation conditions such as at rest, light or heavy exercise) which can affect the values of deposited fraction. However, the computed range (1.17–7.56 × 10^{10} h^{-1}) of deposition estimates seems to cover reasonably well the similar estimates made for road emissions exposure in different urban compartments. For example, we changed the deposition doses estimated by Int Panis et al. [101] in "# m^{-1}" to "# h^{-1}" by multiplying the route length with the deposition doses (# m^{-1}), and then dividing the resulting product by the time taken (h^{-1}) to cover this route. These estimates for a typical car journey in Brussels were found to be 1.54 × 10^{10} h^{-1} which falls within the range estimated by our study for roadside European locations.

7 Regulatory Measures for Atmospheric Nanoparticles

There is currently no air quality regulation in Europe or any other part of the world to control the *ambient* concentrations of particles on a *number* basis. However, particle numbers are currently being regulated at vehicle tailpipe exhausts through Euro-5 and Euro-6 emission standards [103]. These are the first ever limits of this kind, though only applicable in Europe, to control the emissions of particle numbers at source. These standards include a lower size limit of 23 nm for minimising the

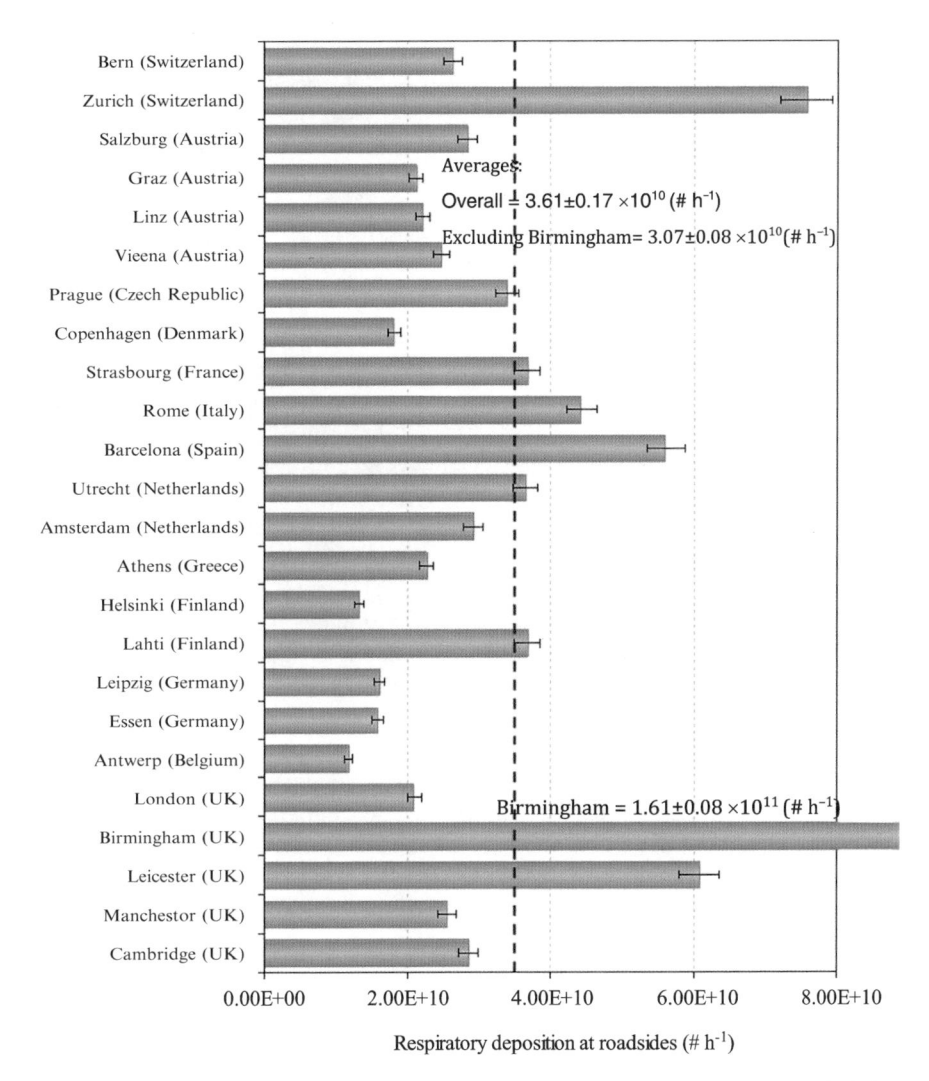

Fig. 5 Typical respiratory deposition doses in different European cities. Standard deviation values are derived used the average PNCs plotted in Fig. 3 and the standard deviation values of deposition fraction (i.e. ±0.03)

Averages:
Overall = $3.61 \pm 0.17 \times 10^{10}$ (# h^{-1})
Excluding Birmingham = $3.07 \pm 0.08 \times 10^{10}$ (# h^{-1})
Birmingham = $1.61 \pm 0.08 \times 10^{11}$ (# h^{-1})

effects of both small volatile particles and diffusion losses during sampling [80], and a requirement to remove volatile particles (through heating and dilution) to avoid large variations in the results. Vehicle emission standards will certainly help in reducing the total PNCs in ambient European environments but similar ambient air quality standards for airborne nanoparticles are desirable to limit public exposure. However, such regulations seem to be some way off because of a number of practical and technical constraints [27]. These include lack of harmonisation of evidence between the toxicological and epidemiological studies favouring particle numbers as a metric over others such as surface area or chemical composition. Further concerns include lack of standardisation of the key measurement parameters, including sampling, which is necessary for robust evaluation of PNCs [27]. What should be the lower cut-off size for particles for ambient regulations is another question. Adopting 23 nm lower cut-off size as in the Euro vehicle emission standards will leave out over one third of total ambient PNCs, as highlighted in above sections. Variations in PNCs can be easily up to two orders of magnitude or more between the background and roadside environments [2, 21], raising a question about choosing an appropriate limit value which can address this remarkable spatial variation seen at different locations (i.e. at road, roadside, street canyons) within an urban area. A further question can be raised on appropriate sampling height that can represent exposure to the entire population living at a particular location (i.e. at ground floor or above); past studies have shown appreciable changes in PNCs near the road level within about the first 2 m and then decreased concentrations as move upward from the road surface [79, 104].

At present, there is no agreed safe threshold limit for exposure to ambient nanoparticles due to the lack of a sufficient knowledge on the exposure–response relationships, making developing any regulatory framework even harder. Vehicle emissions can increase the PNCs up to an order of magnitude higher in urban environments compared with natural environments, meaning that future control and management strategies should target a decrease of PNCs in urban environments by more than one order of magnitude which is not a trivial task [21]. Last but not the least, other challenges can include an adequate treatment of PNC peaks which can arise due to secondary particle formation, enhancing the complexity whether the regulations should be set around the baseline PNCs without taking into account the peak PNCs, or should include the peaks which needs to be defined first [21]. All the aforementioned questions warrant further research and some definite answers before any regulatory framework is proposed.

8 Synthesis and Future Work

A considerable amount of development has happened in the last two decades in the area of measurements, dispersion modelling and exposure assessment studies related to airborne nanoparticles. This is clearly evident from the ever-increasing number of published studies in Europe, and elsewhere in general. This study presented PNCs over 45 sampling locations covering about 30 cities and 15 European countries. While reviewing the literature, it was felt that there are still a number of European countries where nanoparticle-related studies are scarce.

Airborne nanoparticles empirically fit well to log normal distributions and exhibit bimodal distributions in atmospheric urban environments. These arise from both natural and anthropogenic sources. Road vehicles remain a dominant source, contributing up to 90% of total PNCs, in polluted urban environments.

ENPs are emerging class of airborne nanoparticles having a main impact on the air quality of indoor environments; these are unintentionally released into the ambient environment during the manufacture (commercial or research), handling, use or disposal of nanomaterials integrated products. Their physical and chemical characteristics differ from other nanoparticles produced through traffic [4]. The health consequences of their inhalation are not yet well known. A number of studies have reported their number concentrations and size distributions in workplaces but their concentrations in ambient urban environments are largely unknown and warrant further research. Adequate methods have yet to be developed to quantify them in the presence of nanoparticles from other sources.

Closer inspection of the PNC studies at various European locations indicates a factor of ~2.4 differences between the overall average PNCs at roadside and urban background locations. As expected, roadside PNCs exhibit higher concentrations and up to a factor of 14 differences between the PNCs observed at different roadside locations. Such a difference is nearly one-third (i.e. ~5 times) of roadside PNCs for urban background sites. Sites in Birmingham and Zurich appear to show highest PNCs among all the *roadside* sites in Europe while Antwerp site seems to be the cleanest one among all other sites (Fig. 3). However, changes in concentrations over time complicate such cross-sectional inter-comparisons. Among the reviewed European locations, Zurich and Prague seem to show the most and the least polluted *urban background* sites, respectively (Fig. 4).

Deposition doses in the human respiratory system were estimated based on the roadside PNCs in various locations. These showed the similar trend to the roadside PNCs since the dose estimates did not take into account the size distributions of particles at each site. Average deposition doses over all the considered locations were found to be $3.61 \pm 0.17 \times 10^{10}$ h^{-1} for male subjects, with exceptionally high values ($1.61 \pm 0.08 \times 10^{11}$ h^{-1}) for site at Birmingham where the study was carried out along the roadside about 15 years ago (i.e. in 1996–1997).

Nanoparticle emissions are progressively being controlled at source in European countries through the Euro-5 and Euro-6 emission standards. However, there are currently no air quality regulations for airborne nanoparticles available in any part of the world for controlling exposure at the receptor. As discussed in various sections of this review chapter, there are still a number of practical and technical constraints to overcome before any regulatory framework for atmospheric nanoparticles can be put in practice. Some of the key additional challenges with respect to emerging novel pollutants such as ENPs may include: (1) quantification of the mass and number concentrations of ENPs in the presence of nanoparticles from other sources such as traffic and (2) establishing exposure–response functions for human health effects of different types of nanomaterials upon inhalation.

Acknowledgements Prashant Kumar thanks the volume editor, Dr. Mar Viana, for inviting him to write this chapter.

References

1. Donaldson K, Tran L, Albert Jimenez LA, Duffin R, Newby DE, Mills N, MacNee W, Stone V (2005) Combustion-derived nanoparticles: a review of their toxicology following inhalation exposure. Particle Fibre Toxicol 5(6):553–560
2. Kumar P, Robins A, Vardoulakis S, Britter R (2010) A review of the characteristics of nanoparticles in the urban atmosphere and the prospects for developing regulatory controls. Atmos Environ 44:5035–5052
3. Morawska L, Wang H, Ristovski Z, Jayaratne ER, Johnson G, Cheung HC, Ling X, He C (2009) JEM spotlight: environmental monitoring of airborne nanoparticles. J Environ Monit 11:1758–1773
4. Kumar P, Fennell P, Robins A (2010) Comparison of the behaviour of manufactured and other airborne nanoparticles and the consequences for prioritising research and regulation activities. J Nanopart Res 12:1523–1530
5. Kumar P, Robins A, ApSimon H (2010) Nanoparticle emissions from biofuelled vehicles – their characteristics and impact on the number-based regulation of atmospheric particles. Atmos Sci Lett 11:327–331
6. Vardoulakis S, Fisher BEA, Gonzalez-Flesca N, Pericleous K (2002) Model sensitivity and uncertainty analysis using roadside air quality measurements. Atmos Environ 36:2121–2134
7. Vardoulakis S, Fisher BRA, Pericleous K, Gonzalez-Flesca N (2003) Modelling air quality in street canyons: a review. Atmos Environ 37:155–182
8. Vardoulakis S, Valiantis M, Milner J, ApSimon H (2007) Operational air pollution modelling in the UK – street canyon applications and challenges. Atmos Environ 41:4622–4637
9. Buseck PR, Adachi K (2008) Nanoparticles in the atmosphere. Elements 4:389–394
10. Flagan RC (1998) History of electrical aerosol measurements. Aerosol Sci Technol 28:301–380
11. McMurry PH (2000) The history of condensation nucleus counters. Aerosol Sci Technol 33:297–322
12. McMurry PH (2000) A review of atmospheric aerosol measurements. Atmos Environ 34:1959–1999
13. Simonet BM, Valcárcel M (2009) Monitoring nanoparticles in the environment. Anal Bioanal Chem 393:17–21
14. Biswas P, Wu C-Y (2005) Nanoparticle and the environment. J Air Waste Manag Assoc 55:708–746
15. Nowack B, Bucheli TD (2007) Occurrence, behavior and effects of nanoparticles in the environment. Environ Pollut 150:5–22
16. Ju-Nam Y, Lead JR (2008) Manufactured nanoparticles: an overview of their chemistry, interactions and potential environmental implications. Sci Total Environ 400:396–414
17. Seigneur C (2009) Current understanding of ultra fine particulate matter emitted from mobile sources. J Air Waste Manag Assoc 59:3–17
18. Pey J, Querol X, Alastuey A, Rodríguez S, Putaud JP, Van Dingenen R (2009) Source apportionment of urban fine and ultra fine particle number concentration in a Western Mediterranean city. Atmos Environ 43:4407–4415
19. Harrison RM, Beddows DCS, Dall'Osto M (2011) PMF analysis of wide-range particle size spectra collected on a major highway. Environ Sci Technol 45:5522–5528
20. Holmes NS, Morawska L (2006) A review of dispersion modelling and its application to the dispersion of particles: an overview of different dispersion models available. Atmos Environ 40:5902–5928
21. Morawska L, Ristovski Z, Jayaratne ER, Keogh DU, Ling X (2008) Ambient nano and ultrafine particles from motor vehicle emissions: characteristics, ambient processing and implications on human exposure. Atmos Environ 42:8113–8138
22. Morawska L (2010) Airborne engineered nanoparticles: are they a health problem? Air Qual Climate Change 44:18–20

23. Morawska L, Jayaratne ER, Knibbs LD, Megatmokhtar M (2010) Regulations and policy measures related to the reduction of ambient particulate matter. In: Zereini F, Wiseman C (eds) Urban particulate matter: origins, chemistry, fate and health impacts. Springer, Berlin
24. Knibbs LD, Cole-Hunter T, Morawska L (2011) A review of commuter exposure to ultrafine particles and its health effects. Atmos Environ 45:2611–2622
25. Carpentieri M, Kumar P, Robins A (2011) An overview of experimental results and dispersion modelling of nanoparticles in the wake of moving vehicles. Environ Pollut 159:685–693
26. Kumar P, Ketzel M, Vardoulakis S, Pirjola L, Britter R (2011) Dynamics and dispersion modelling of nanoparticles from road traffic in the urban atmospheric environment – a review. J Aerosol Sci 42:580–603
27. Kumar P, Robins A, Vardoulakis S, Quincey P (2011) Technical challenges in tackling regulatory concerns for urban atmospheric nanoparticles. Particuology 9:566–571
28. Kumar P (2011) Footprints of airborne ultrafine particles on urban air quality and public health. J Civ Environ Eng 1:e101. doi:10.4172/jcee.1000e101
29. Kumar P, Mulheron M, Fisher B, Harrison RM (2012) New Directions: Airborne ultrafine particle dust from building activities – a source in need of quantification. Atmos Environ, In Press, doi: 10.1016/j.atmosenv.2012.04.028
30. Hinds WC (1999) Aerosol technology: properties, behaviour and measurement of airborne particles. Wiley, London, p 483
31. Charron A, Harrison RM (2003) Primary particle formation from vehicle emissions during exhaust dilution in the road side atmosphere. Atmos Environ 37:4109–4119
32. Kittelson DB, Watts WF, Johnson JP (2006) On-road and laboratory evaluation of combustion aerosols – Part 1: summary of diesel engine results. J Aerosol Sci 37:913–930
33. Seinfeld JH, Pandis SN (2006) Atmospheric chemistry and physics, from air pollution to climate change. Wiley, New York, p 1203
34. Kulmala M, Vehkamaki H, Petaja T, Dal Maso M, Lauri A, Kerminen V-M, Birmili W, McMurry PH (2004) Formation and growth rates of ultrafine atmospheric particles: a review of observations. J Aerosol Sci 35:143–176
35. Lingard JJN, Agus EL, Young DT, Andrews GE, Tomlin AS (2006) Observations of urban airborne particle number concentrations during rush-hour conditions: analysis of the number based size distributions and modal parameters. J Environ Monit 8:1203–1218
36. Kittelson DB, Watts WF, Johnson JP, Lawson DR (2006) On-road and laboratory evaluation of combustion aerosol – Part 2: summary of spark ignition engine results. J Aerosol Sci 37:931–949
37. Graskow BR, Kittelson DB, Abdul-Khaleek IS, Ahmadi MR, Morris JE (1998) Characterization of exhaust particulate emissions from a spark ignition engine. Society of Automotive Engineers, Warrendale, PA, 980528
38. Kittelson DB (1998) Engines and nano-particles: a review. J Aerosol Sci 29:575–588
39. Anastasio C, Martin ST (2001) Atmospheric nanoparticles. In: Banfield JF, Navrotsky A (eds) Nanoparticles and the environment, vol 44. Mineralogical Society of America, pp 293–349
40. Dahl A, Gharibi A, Swietlicki E, Gudmundsson A, Bohgard M, Ljungman A, Blomqvist G, Gustafsson M (2006) Traffic-generated emissions of ultrafine particles from pavement-tire interface. Atmos Environ 40:1314–1323
41. Patra A, Colvile R, Arnold S, Bowen E, Shallcross D, Martin D, Price C, Robins A (2008) On street observations of particulate matter movement and dispersion due to traffic on an urban road. Atmos Environ 42:3911–3926
42. Vakeva M, Hameri K, Kulmala M, Lahdes R, Ruuskanen J, Laitinen T (1999) Street level versus rooftop concentrations of submicron aerosol particles and gaseous pollutants in an urban street canyon. Atmos Environ 33:1385–1397
43. Holmes N (2007) A review of particle formation events and growth in the atmosphere in the various environments and discussion of mechanistic implications. Atmos Environ 41:2183–2201

44. Kumar P, Fennell P, Hayhurst A, Britter RE (2009) Street versus rooftop level concentrations of fine particles in a Cambridge street canyon. Boundary-Layer Meteorol 131:3–18
45. Alam A, Shi JP, Harrison RM (2003) Observations of new particle formation in urban air. J Geophys Res 108:4093–4107
46. O'Dowd CD, Aalto P, Hameri K, Kulmala M, Hoffmann T (2002) Aerosol formation: atmospheric particles from organic vapours. Nature 416:497–498
47. O'Dowd C, Jimenez JL, Bahreinl R, Flagan RC, Seinfeld JH, Hamerl D, Pirjola L, Kulmala M, Jennings SG, Hoffmann T (2002) Marine aerosol formation from biogenic iodine emissions. Nature 417:632
48. Heintzenberg J, Wehner B, Birmili W (2007) 'How to find bananas in the atmospheric aerosol': new approach for analyzing atmospheric nucleation and growth events. Tellus 59B:273–282
49. Wise ME, Semeniuk TA, Bruintjes R, Martin ST, Russell LM, Buseck PR (2007) Hygroscopic behaviour of NaCl-bearing natural aerosol particles using environmental transmission electron microscopy. J Geophys Res 112:D10224. doi:10210.11029/12006JD007678
50. Ketzel M, Wahlin P, Kristensson A, Swietlicki E, Berkowicz R, Nielsen OJ, Palmgren F (2003) Particle size distribution and particle mass measurements at urban, near-city and rural level in the Copenhagen area and Southern Sweden. Atmos Chem Phys 4:281–292
51. Rimnácová D, Zdímal V, Schwarz J, Smolík J, Rimnác M (2011) Atmospheric aerosols in suburb of Prague: the dynamics of particle size distributions. Atmos Res 101:539–552
52. Aitken RJ, Creeley KS, Tran CL (2004) Nanoparticles: an occupational hygiene review. Research Report 274 Institute of Occupational Medicine for the Health and Safety Executive 2004, p 113
53. Dawson NG (2008) Sweating the small stuff: environmental risk and nanotechnology. Bioscience 58:690
54. Roco M (2005) International Perspective on Government Nanotechnology Funding in 2005. J Nanopart Res 7:707–712
55. Peralta-Videa JR, Zhao L, Lopez-Moreno ML, de la Rosa G, Hong J, Gardea-Torresdey JL (2011) Nanomaterials and the environment: a review for the biennium 2008–2010. J Hazard Mater 186:1–15
56. Helland A, Wick P, Koehler A, Schmid K, Som C (2007) Reviewing the environmental and human health knowledge base of carbon nanotubes. Environ Health Perspect 115:1125–1131
57. Brouwer D (2010) Exposure to manufactured nanoparticles in different workplaces. Toxicology 269:120–127
58. Fujitani Y, Kobayashi T, Arashidani K, Kunugita N, Suemura K (2008) Measurement of the physical properties of aerosols in a fullerene factory for inhalation exposure assessment. J Occup Environ Hyg 5:380–389
59. Donaldson K, Tran CL (2004) An introduction to the short-term toxicology of respirable industrial fibres. Mutat Res/Fundam Mol Mech Mutagen 553:5–9
60. Xia T, Li N, Nel AE (2009) Potential health impact of nanoparticles. Annu Rev Public Health 30:137–150
61. Choi J-Y, Ramachandran G, Kandlikar M (2009) The impact of toxicity testing costs on nanomaterial regulation. Environ Sci Technol 43:3030–3034
62. Schmid G, Baumle M, Geerkens M, Helm I, Osemann C, Sawitowski T (1999) Current and future applications of nanoclusters. Chem Eng Soc Rev 28:179–185
63. BSI (2007) Nanotechnologies – PD 6699-2:2007. Part 2: guide to safe handling and disposal of manufactured nanomaterials. British Standard Institute, p 32. , ISBN 978 0 580 60832 2
64. Han JH, Lee EJ, Lee JH, So KP, Lee YH, Bae GN, Lee S-B, Ji JH, Cho MH, Yu IJ (2008) Monitoring multi-walled nanotube exposure in carbon nanotube research facility. Inhal Toxicol 20:741–749
65. Yeganeh B, Kull CM, Hull MS, Marr LC (2008) Characterisation of airborne particles during production of carbonaceous nanomaterials. Environ Sci Technol 42:4600–4606

66. Muller NC, Nowack B (2008) Exposure modeling of engineered nanoparticles in the environment. Environ Sci Technol 42:4447–4453
67. Handy RD, Henry TB, Scown TM, Johnston BD, Tyler CR (2008) Manufactured nanoparticles: their uptake and effects on fish-amechanistic analysis. Ecotoxicology 17: 396–409
68. Handy RD, von der Kammer F, Lead JR, Hassellöv M, Owen R, Crane M (2008) The ecotoxicology and chemistry of manufactured nanoparticles. Ecotoxicology 17:287–314
69. Valant J, Drobne D, Sepcic K, Jemec A, Kogej K, Kostanjsek R (2009) Hazardous potential of manufactured nanoparticles identified by in vivo assay. J Hazard Mater 171:160–165
70. Bystrzejewska-Piotrowska G, Golimowski J, Urban PL (2009) Nanoparticles: their potential toxicity, waste and environmental management. Waste Manag 29:2587–2595
71. AQEG (1999) Source apportionment of airborne particulate matter in the United Kingdom. Report of the Airborne Particles Expert Group, p 158. http://www.environment.detr.gov.uk/airq/
72. Baldauf R, Thoma E, Hays M, Shores R, Kinsey J, Gullett B, Kimbrough SJB (2008) Traffic and meteorological impacts on near-road air quality: summary of methods and trends from the Raleigh near-road study. J Air Waste Manag Assoc 58:865–878
73. Chan YC, Simpson RW, Mctainsh GH, Vowles PD, Cohen DD, Bailey GM (1999) Source apportionment of PM2.5 and PM10 aerosols in Brisbane (Australia) by receptor modelling. Atmos Environ 33:3251–3268
74. Fraser MP, Yue ZW, Buzcu B (2003) Source apportionment of fine particulate matter in Houston, TX, using organic molecular markers. Atmos Environ 37:2117–2123
75. Harrison RM, ApSimon H, Clarke AG, Dewent RG, Fisher B, Hickman J, Mark D, Murrells T, McAughey J, Pooley F, Richards R, Stedman J, Vawda Y, Williams M, Coster S, Mayland R, Prosser H, Hall I, McMohan N (1999) Source apportionment of airborne particulate matter in the United Kingdom. Technical Report, Department of Environment, Transport & the Regions, the Welsh Office, the Scottish Office & the Department of the Environment (North Ireland)
76. Schauer JJ, Hildermann LM, Mazurek MA, Cass GR, Simoneit BRT (1996) Source apportionment of airborne particulate matter using organic compounds as tracers. Atmos Environ 30:3837–3855
77. Wåhlin P, Palmgren F, Van Dingenen R (2001) Experimental studies of ultrafine particles in streets and the relatioship of traffic. Atmos Environ 35:S63–S69
78. Johansson C, Norman M, Gidhagen L (2007) Spatial & temporal variations of PM10 and particle number concentrations in urban air. Environ Monit Assess 127:477–487
79. Kumar P, Fennell P, Langley D, Britter R (2008) Pseudo-simultaneous measurements for the vertical variation of coarse, fine and ultra fine particles in an urban street canyon. Atmos Environ 42:4304–4319
80. Kumar P, Fennell P, Symonds J, Britter R (2008) Treatment of losses of ultrafine aerosol particles in long sampling tubes during ambient measurements. Atmos Environ 42:8819–8826
81. Kumar P, Robins A, Britter R (2008) Fast response measurements for the dispersion of nanoparticles in a vehicle wake and a street canyon. Atmos Environ 43:6110–6118
82. Kumar P, Fennell P, Britter R (2008) Effect of wind direction and speed on the dispersion of nucleation and accumulation mode particles in an urban street canyon. Sci Total Environ 402:82–94
83. Mejla JF, Morawska L (2009) An investigation of nucleation events in a coastal urban environment in the Southern Hemisphere. Atmos Chem Phys 9:7877–7888
84. Shi JP, Evans DE, Khan AA, Harrison RM (2001) Sources and concentration of nanoparticles (<10 nm diameter) in the urban atmosphere. Atmos Environ 35:1193–1202
85. Wehner B, Wiedensohler A (2003) Long term measurements of submicrometer urban aerosols: statistical analysis for correlations with meteorological conditions and trace gases. Atmos Chem Phys 3:867–879

86. Li Y, Suriyawong A, Daukoru M, Zhuang Y, Biswas P (2009) Measurement and capture of fine and ultrafine particles from a pilot-scale pulverized coal combustor with an electrostatic precipitator. J Air Waste Manag Assoc 59:553–559
87. Hu S, Fruin S, Kozawa K, Mara S, Winer AM, Paulson SE (2009) Aircraft emission impacts in a neighborhood adjacent to a General Aviation Airport in Southern California. Environ Sci Technol 43:8039–8045
88. Mazaheri M, Johnson GR, Morawska L (2008) Particle and gaseous emissions from commercial aircraft at each stage of the landing and takeoff cycle. Environ Sci Technol 43:441–446
89. Saxe H, Larsen T (2004) Air pollution from ships in three Danish ports. Atmos Environ 38:4057–4067
90. Hansen D, Blahout B, Benner D, Popp W (2008) Environmental sampling of particulate matter and fungal spores during demolition of a building on a hospital area. J Hosp Infect 70:259–264
91. Kumar P, Mulheron M, Som C (2012) Release of ultrafine particles from three simulated building processes. J Nanopart Res 14, 771, doi: 10.1007/s11051-012-0771-2
92. Buonanno G, Morawska L, Stabile L, Viola A (2010) Exposure to particle number, surface area and PM concentrations in pizzerias. Atmos Environ 44:3963–3969
93. Reche C, Querol X, Alastuey A, Viana M, Pey J, Moreno T, Rodríguez S, González Y, Fernández-Camacho R, de la Rosa J, Dall'Osto M, Prévôt ASH, Hueglin C, Harrison RM, Quincey P (2011) New considerations for PM, Black Carbon and particle number concentration for air quality monitoring across different European cities. Atmos Chem Phys 11:6207–6227
94. Wåhlin P (2009) Measured reduction of kerbside ultrafine particle number concentrations in Copenhagen. Atmos Environ 43:3645–3647
95. Harrison RM, Jones AM (2005) Multisite study of particle number concentrations in urban air. Environ Sci Technol 29:6063–6070
96. Dall'Osto M, Thorpe A, Beddows DCS, Harrison RM, Barlow JF, Dunbar T, Williams PI, Coe H (2011) Remarkable dynamics of nanoparticles in the urban atmosphere. Atmos Chem Phys 11:6623–6637
97. Atkinson RW, Fuller GW, Anderson HR, Harrison RM, Armstrong B (2010) Urban ambient particle metrics and health: a time-series analysis. Epidemiology 21:501–511
98. Harrison RM, Giorio C, Beddows DC, Dall'Osto M (2010) Size distribution of air-borne particles controls outcomes of epidemiological studies. Sci Total Environ 409:289–293
99. Morawska L, Johnson G, Ristovski ZD, Agranovski V (1999) Relation between particle mass and number for submicrometer airborne particles. Atmos Environ 33:1983–1990
100. Nel A, Xia T, Madler L, Li N (2006) Toxic potential of materials at the nanolevel. Science 311:622–627
101. Int Panis L, de Geus B, Vandenbulcke G, Willems H, Degraeuwe B, Bleux N, Mishra V, Thomas I, Meeusen R (2010) Exposure to particulate matter in traffic: a comparison of cyclists and car passengers. Atmos Environ 44:2263–2270
102. ICRP (1994) ICRP Publication 66: human respiratory tract model for radiological protection. A report of a task group of the International Commission on Radiological Protection, pp 1–482
103. EC (2008) Directive on Ambient Air Quality and Cleaner Air for Europe (Directive 2008/50/EC). http://eur-lex.europa.eu/LexUriServ/LexUriServ.do?uri=OJ:L:2008:152:0001:0044:EN:PDF. Accessed on 20 March 2010
104. Li XL, Wang JS, Tu XD, Liu W, Huang L (2007) Vertical variations of particle number concentration and size distribution in a street canyon in Shanghai, China. Sci Total Environ 378:306–316
105. Longley ID, Gallagher MW, Dorsey JR, Flynn M, Allan JD, Alfarra D, Inglish D (2003) A case study of aerosol (4.6 nm < Dp < 10 μm) number and mass size distribution measurements in a busy street canyon in Manchester, UK. Atmos Environ 37:1563–1571

106. Agus EL, Young DT, Lingard JJN, Smalley RJ, Tate EJ, Goodman PS, Tomlin AS (2007) Factors influencing particle number concentrations, size distributions and modal parameters at a roof-level and roadside site in Leicester, UK. Sci Total Environ 386:65–82
107. Shi PJ, Khan AA, Harrison RM (1999) Measurements of ultra fine particle concentration and size distribution in the urban atmosphere. Sci Total Environ 235:51–64
108. Can A, Rademaker M, Van Renterghem T, Mishra V, Van Poppel M, Touhafi A, Theunis J, De Baets B, Botteldooren D (2011) Correlation analysis of noise and ultrafine particle counts in a street canyon. Sci Total Environ 409:564–572
109. Weber S (2009) Spatio-temporal covariation of urban particle number concentration and ambient noise. Atmos Environ 43:5518–5525
110. Voigtlander J, Tuch T, Birmili W, Wiedensohler A (2006) Correlation between traffic density and particle size distribution in a street canyon and the dependence on wind direction. Atmos Chem Phys 6:4275–4286
111. Puustinen A, Hämeri K, Pekkanen J, Kulmala M, de Hartog J, Meliefste K, ten Brink H, Kos G, Katsouyanni K, Karakatsani A, Kotronarou A, Kavouras I, Meddings C, Thomas S, Harrison R, Ayres JG, van der Zee S, Hoek G (2007) Spatial variation of particle number and mass over four European cities. Atmos Environ 41:6622–6636
112. Boogaard H, Montagne DR, Brandenburg AP, Meliefste K, Hoek G (2010) Comparison of short-term exposure to particle number, PM10 and soot concentrations on three (sub) urban locations. Sci Total Environ 408:4403–4411
113. Paatero P, Aalto P, Picciotto S, Bellander T, Castano G, Cattani G, Cyrys J, Koster M (2005) Estimating time series of aerosol particle number concentrations in the five HEAPSS cities on the basis of measured air pollution and meteorological variables. Atmos Environ 39:2261–2273
114. Roth E, Kehrli D, Bonnot K, Trouve G (2008) Size distribution of fine and ultrafine particles in the city of Strasbourg: correlation between number of particles and concentrations of NOx and SO2 gases and some soluble ions concentration determination. J Environ Manag 86:282–290
115. Ondráček J, Schwarz J, Zdímal V, Andelová L, Vodicka P, Bízek V, Tsai CJ, Chen SC, Smolík J (2011) Contribution of the road traffic to air pollution in the Prague city (busy speedway and suburban crossroads). Atmos Environ 45:5090–5100
116. Gomiscek B, Hauck H, Stopper S, Preining O (2004) Spatial and temporal variations of PM1, PM2.5, PM10 and particle number concentration during the AUPHEP–project. Atmos Environ 38:3917–3934
117. Morawska L, Thomas S, Hofmann W, Ristovski Z, Jamriskaa M, Rettenmoser T, Kagerer S (2004) Exploratory cross-sectional investigations on ambient submicrometer particles in Salzburg, Austria. Atmos Environ 38:3529–3533
118. Bukowiecki N, Dommen J, Prevot ASH, Weingartner E, Baltensperger U (2003) Fine and ultrafine particles in the Zurich (Switzerland) area measured with a mobile laboratory: an assessment of the seasonal and regional variation throughout a year. Atmos Chem Phys 3:1477–1494
119. Rodríguez S, Van Dingenen R, Putaud J-P, Dell'Acqua A, Pey J, Querol X, Alastuey A, Chenery S, Ho K-F, Harrison RM, Tardivo R, Scarnato B, Gianelle V (2007) A study on the relationship between mass concentrations, chemistry and number size distribution of urban fine aerosols in Milan, Barcelona and London. Atmos Chem Phys 7:605–639
120. Gómez-Moreno FJ, Pujadas M, Plaza J, Rodríguez-Maroto JJ, Martínez-Lozano P, Artíñano B (2011) Influence of seasonal factors on the atmospheric particle number concentration and size distribution in Madrid. Atmos Environ 45:3169–3180
121. Ruuskanen J, Tuch T, Ten Brink H, Peters A, Khlystov A, Mirme A, Kos GPA, Brunekreef B, Buzorius G, Vallius M, Kreyling WG, Pekkanen J (2001) Concentrations of ultrafine, fine and PM2.5 particles in three European cities. Atmos Environ 35:3729–3738

Index

A
Accumulated population exposure index (APEI), 266
Accumulation mode, 312
Acidification, 32
ACTRIS, 303, 308, 315
Administrative deficiencies, 21
Aerodyne aerosol mass spectrometer (AMS), 169
Aerosol optical depth (AOD), 111, 269
Aerosols, 101, 219, 297
  number concentration, 297
  number size distribution, 297, 339
  secondary, 219, 260
Aethalometer model, 128
Agriculture, 213
Air exchange rate (AER), 330
AirMonTech, 289
Air quality, 3, 13, 31, 196
  assessment, 20
  directive, 3
  limit value, 3
  modelling, 261
  ozone, 56
  regulation, 277
Aitken mode, 110, 299, 312, 343
Aldrin, 77
Alternative fuels, 31
Ammonia (NH3), 38, 172, 213, 262
  impact of regulation and climate change, 141
  inventories, 141
  models, 141
  spatial distribution, 141, 149
  temporal distribution, 141
  trends, 141
Ammonium nitrate, 210
Animal houses, ammonia, 143
Anthropogenic contribution, 240, 348
Anthropogenic factors, 5, 18
Arctic size distributions, 310
Asbestos, 182, 283, 348
Asia, air quality standards/monitoring, 282
Athens, 222
  Greece, 22
Atmospheric transport, long-range, 75

B
Back-trajectories, 195
  modelling, 201
Balkans, 219
Banana event, 345
Battery-electric vehicles, 47
Benzene, 284
Benzo[a]pyrene, 79, 82, 93, 284
Biodiesel, 46
Bioethanol, 46
Biofuel, 46
  nanoparticles, 340
  wood burning facilities, 213
Biomass fires, 102
Biomass-to-liquid procedures (BTL), 46
Black carbon (BC), 82, 108, 111, 123, 173, 287, 290, 333
Black Sea, 219

B

Brake wear/abrasion, 168, 211
  emissions, 182, 211, 348
Britain, 308

C

Calcium, 209
CALINE4, 265
Carbon monoxide (CO), 37, 114, 281, 283, 285
Catalytic converter, 34, 37
CCN, 297, 299, 308, 315
Chemical composition, 101, 195, 240
Chemical transport models (CTMs), 195, 201, 240
Chlordane, 77
Cloud droplet number concentration (CDNC), 299
CMAR, 347
Cold condensation, 83
Combustion, 26, 35, 209, 224, 243, 247, 299, 343
  control, 36
Convention on Long-Range Transboundary Air Pollution (CLRTAP), 33, 58, 155
Criticality/noncriticality, 4

D

DDT, 77
Decay, 327
Deposition rate, 327
Dieldrin, 77
Diesel heavy-duty trucks, 45
Diesel oxidation catalyst, 38
Diesel particulate filter (DPF), 39
Differential mobility particle sizer (DMPS), 302
Dioxins, 77, 212
Dispersion models, 200
Domestic heating, 26, 212
Dry deposition, 87
Dust, airborne, 175, 195, 226
  Sahara, 208, 227

E

Eastern Mediterranean, 219
Eco-driving, 47
Elemental carbon (EC), 173, 333
Emissions, 5, 12
  control, technology, 31, 36, 44
  future evolution, 48
  sources, 219

Endrin, 77
Engine type, approval, 41
Environmental Protection Agency (US-EPA), 82, 179, 280, 288
Europe, 75, 101
  air quality, 283
  northwestern, 240
EUSAAR, 303, 317
Eutrophication, 32, 58, 155, 278
Exhaust emissions, 165, 168
  on-road, 40
  particles, 172
Exhaust gas recirculation (EGR), 37
Exposure, 259
  modelling, 263
Exposure–response doses, 339

F

Fertiliser, 247
  ammonia, 142, 144
Financial deficiencies, 21
Finokalia, Crete, 222, 231, 311
Fireplaces/stoves, 126
Fires, 102
Flame retardants, 77
Fluoranthene, 82
Forest fires, 104, 228
Fuels, alternative, 45
  quality, 45
Future developments, 277

G

Gasoline direct injection (GDI) engine, 37
Gas–particle exchange, 84
Gas-to-particle-phase partitioning, 75
Global distillation, 83
Gothenburg Protocol, 155
GPS/GIS, 290, 323
Grazing animals, ammonia, 147
Greece, 219
Green Public Procurement (GPP), 25
GUAN, 303

H

Health effects, 259
  nanoparticles, 339
Heavy metals, 168, 176, 200, 212, 283
Hemispheric background, 55
Heptachlor, 77
Hexachlorobenzene (HCB), 76, 79

Index 369

Hexachlorocyclohexanes (HCHs), 76, 78
Hot-spots, 195
Hydrotreated vegetable oil (HVO), 46

I
Individual exposure model (IEM), 265
Indoor air, 321
  quality, modelling, 324
Indoor/outdoor relationships, 328
Industry, 18, 212
  measures, 26
Infiltration, 321
  factor, 330
Integrated Pollution Prevention and Control (IPPC), 141, 156
Internal combustion engines (ICEs), 32, 47
Iodine oxides, 345
Ireland, 308

J
Japan, air quality standards, 283

K
Košice (industrial region), Slovakia, 22
Kraków, Poland, 22

L
Latitudinal fractionation, 83
Lean NO_x trap (LNT), 39
Lenschow approach, 195, 198
Levoglucosan, 107
Lisbon, Portugal, 23, 268
London, United Kingdom, 23
Long-range transport, 75, 101
Long-term trends, 55

M
MACC, 6
Macrotracer approach (MTA), 129
Mann–Kendall test, 64, 67
Manufacturing processes, ammonia, 148
Manure storages, ammonia, 143
Marine aerosol, 199, 225
Mass, aerosols, 301
Mass closure/tracer-based approaches, 199
Mediterranean Basin, 219
Mercury, 284

Microelectromechanical systems (MEMS), 289
Milan, Italy, 23
Mineral dust (MD), 209, 242
Mineral fertiliser, ammonia, 144
Mirex, 77
Modal split, 20
Modelling, 259
MODIS, 108
Monosaccharide anhydrides (MAs), 107
Motor exhaust, 165
Mountain stations, 309
Multicomponent analysis, 289

N
Nanoparticles, 298, 340
  airborne, 343
  emerging (ENPs), 341
  engineered, 339
  regulatory measures, 356
  respiratory deposition doses, 355
  sources, 344
National Emission Ceiling Directive (NEC), 27, 141, 155
Natural contribution, 240
Natural sources, 5, 10, 17
NEC Directive. *See* National Emission Ceiling Directive (NEC)
Networks, measurements, 303
New European Driving Cycle (NEDC), 42
New powertrain technologies, 47
Nitrate, 333
  secondary aerosol, 210, 230
Nitric oxide (NO), 32, 36
NO_2, 3, 13, 167, 287
Nonexhaust emissions, 175
NO_x, 13, 31, 33, 165
  on-road, 40
  reduction, 41
  trap, 39
Nucleation, 110, 172, 262, 299, 305, 343
Number size distributions, 304

O
Off-road machinery, 25
Old vehicles, 223
On-board diagnostic (OBD) systems, 42
Operational street pollution model (OSPM), 264
Organic carbon (OC), 87, 123, 233, 242, 334

Organochlorines, 75
Organohalogen compounds, 76, 78, 123
Outdoor, 321
Oxalate, biomass burning smokes, 107
Ozone, 55
　photochemical formation, 55

P
Paris, France, 23
Particle number (PN), 41
Particle number concentrations (PNCs), 173, 299, 323, 339
Particle number distributions (PNDs), 304, 343
Particles, deposition, 343
　penetration, 321
　resuspension, 343
　salt, 343
　size, 321
　size distributions, 344
　ultrafine, 321, 339
Particulate matter, 36, 101, 123, 219, 259
　exposure modelling, 266
Pavement abrasion, 179
Penetration, 321
　efficiency, 326
　factor, 330
Pentachlorobenzene, 77
Perfluorooctanesulfonic acid (PFOS), 77
Periodic technical inspection (PTI), 42
Persistent organic pollutants (POPs), 75, 82
　atmospheric deposition, 86
　historical trends, 88
Personal protective equipment, 347
$PM_{2.5}$, 108, 240
PM_{10}, 3, 13, 108, 195, 240
PMF, 195
Pollution control, regional, 55
Polybromodiphenyl ethers (PBDEs), 76
Polychlorobiphenyls (PCBs), 76, 80
Polychlorodibenzofurans, 7
Polychlorodibenzo-p-dioxins, 77
Polycyclic aromatic hydrocarbons (PAHs), 75, 81, 123, 166, 283, 334
Population, 11
　exposure, 16, 286
Portable emission measurement systems (PEMS), 42
Porto, 267
Potassium, biomass burning smokes, 107, 113, 129
Public procurement, 25

R
Residential wood burning, 123
Respiratory deposition doses, PNCs, 355
Resuspension, 165
Road dust, 219
　resuspension, 176, 227
Roadside environments, PNCs, 353
Road transport/traffic, 31, 165
Road vehicles, nanoparticles, 348
　NO_x emissions, 35
RUPIOH, 330

S
Sahara dust, 18, 87, 179, 208, 210, 219, 226, 248, 253
Scanning mobility particle sizer (SMPS), 302
Sea salt, 111, 209, 243, 344
Sea spray emissions, 199, 207, 247, 343
Secondary aerosols, 210, 230
Secondary inorganic aerosol (SIA), 242
Secondary organic aerosol (SOA), 125, 166, 207, 233, 269
Secondary organic carbon (SOC), 233
Selective catalytic reduction (SCR), 38
Selective trapping, 83
Sewage sludge, ammonia, 148
Ships, 25
　emissions, 219, 229
Site types, 285
Smoke, aerosol, organic/elemental carbon, 108
　LRT, 104
　particles, 109
Sodium, 209
Sofia, Bulgaria, 24
Soil erosion, 199
Source apportionment, 202, 240
　modelling, 21
Spain, 240
Spatial increment, 197
Statistical receptor models, 199
Stibnite, 183
Stuttgart, Germany, 24
Sulfates, 210, 224, 229, 230, 242, 255, 289, 333
Sulfur dioxide, 172, 196, 212, 229, 262
Sulfuric acid, 172, 213, 246, 344
　nucleation, 344

T
TAN (total ammonia N), 142
Technology trends, 277

Terpenes, trees, 345
Thessaloniki, 222
Three-way catalytic converter, 37
Tire wear, 348
Toxaphene, 77
Traffic, 210
  ammonia, 148
  analysis, 20, 210
  counterproductive community measures, 25
  management, 48
  measures, 24
  primary/secondary emissions, 165
Transboundary pollution, 5, 10
Trans-European Transport Networks, 25
Transport, 31
Twomey effect, 291

U
UK, air pollution, 279
Ultrafine particles, 340
  aerosol (UFP), 214, 331
Urban air, 195
  quality, monitoring, 277
Urban background, PNCs, 354
Urban increment, 12, 245

Urea, 38
USA, legislative monitoring requirements, 280

V
Vehicles, approval, 41
  emissions, 31
VOCs, 56, 124, 169, 173, 285

W
Water-soluble organic carbon (WSOC), 108
Wear, 165
  brake, 168, 182, 211, 348
  emissions, 181
  tires, 348
Wet deposition, 86
Wildfire, 101, 102
  aerosol, chemical composition, 111
  LRT, 103
Winds, 309, 343
  erosion, 176, 207, 248
  speed, 7, 9, 142, 205, 248, 345
Wood combustion, fine particulate matter, 125
  residential, health effects, 116, 123
Wood fuel, 125

Printed by Printforce, the Netherlands